Why IPTV?

Written by people in the know, the **Telecoms Explained Series** for Telecoms Professionals will

- Demystify the jargon of wireless and communication technologies

- Provide insight into new and emerging technologies

- Explore associated business and management applications

- Enable you to get ahead of the game in this fast-moving industry

Written in a concise and easy-to-follow format, titles in the series include the following:

Convergence: User Expectations, Communications Enablers and Business Opportunities
Saxtoft
ISBN: 978-0-470-72708-9

Triple Play: Building the Converged Network for IP, VoIP and IPTV
Hens & Caballero
ISBN: 978-0-470-75367-5

Why IPTV?
Interactivity, Technologies and Services

Johan Hjelm
Ericsson

A John Wiley and Sons, Ltd, Publication

Registered office
John Wiley & Sons Ltd, The Atrium, Southern Gate, Chichester, West Sussex,
PO19 8SQ, United Kingdom

For details of our global editorial offices, for customer services and for information about
how to apply for permission to reuse the copyright material in this book please see our
website at www.wiley.com.

Library of Congress Cataloging-in-Publication Data

Hjelm, Johan.
  Why IPTV? : interactivity, technologies, and services / Johan Hjelm.
    p. cm.
  Includes bibliographical references and index.
  ISBN 978-0-470-99805-2 (pbk.)
  1. Internet television. I. Title.
  TK5105.887.H58 2008
  384.550285'4678—dc22

                                                    2008021444

A catalogue record for this book is available from the British Library.

ISBN 978-0-470-99805-2 (PB)

Set in 10/12pt Optima by Integra Software Services Pvt. Ltd. Pondicherry, India
Printed in Singapore by Fabulous Printers Pte Ltd

Contents

Acknowledgments

This book is dedicated to three people: Loretta Aniana of the European Commission, who put me on the right track; Örjan Sahlin, the most persistent IPTV guru you will ever find; and Murakami-san, who said "but why don't you write a book about IPTV?"

My friends in Ericsson who have (unwittingly) helped me write this book (in no particular order – Oda-san, Matsumura-kun, Andreas, Martin, Jan, Bo, Bobbo, Justus, Robert, Ayo, Micke, Helena, Guylaine, Sergey, Mattias, Theo, Charis, Thomas, and Thomas); and in Sony (Nobori-san, Kobori-san, Takeyari-san, Igarashi-san). And in particular, Andreas, Martin and Mikael who provided comments on the draft.

And a special thanks to my wonderful wife Mikako. I will show you how to make an interactive cooking show now.

Preface

It used to be very simple. When I was born, switching on the television set meant that it started receiving the channel – because there was only one channel. Life soon became twice as complicated, because there were now two TV channels. Until my 20s, there was still a monopoly over the airwaves, with only two available channels. Then, satellite technology made commercial television possible, and all of a sudden there were tens of channels.

In other countries, like Finland and the United States, there have always been more television channels, however, when the videocassette recorder (VCR) made video available on demand, the impact was worldwide (until then, you had to have a very special interest to buy the expensive video discs). The DVD attempted to simplify life even further. And then, video moved to the Internet – although not to television; not yet.

If each change doubles the complexity, Internet Protocol Television (IPTV) is many, many times more complicated than watching black and white television with only one channel! What has also happened is that television is moving out of its ebony tower of isolation and is becoming connected to other services. The big driver for this, as for so many other changes, is the Internet. IPTV is not a new idea. Streaming video over the Internet has been around since the mid-1990s, and traditional linear television was added a few years later. However, there has not been a significant number of users for the technology, partly because of the bandwidth needed, and partly because it did not add any value to the television experience over cable or terrestrial broadcast.

If you are working with IPTV, especially if you want to implement a system based on it, I hope this book can help. There are many books about the high level of the technology behind IPTV, and there are books about how to program the interactivity with set-top boxes, but this book takes a different angle – it tries to help those who have experience from developing web applications to understand IPTV.

Today IPTV has millions of users worldwide, but the speed at which it is being deployed means that this is the first stage of the adoption curve. The ease of converting the existing video feeds to IPTV means that IPTV can be used as a compliment to existing distribution technologies, including mobile video. This is the final driver behind the revolution, the openness of IPTV that allows developers to modify and add to the IPTV streams.

IPTV is not YouTube. While many components are identical, the idea behind YouTube – a database of thousands of uploaded video clips – is the same as a traditional video-on-demand (VoD) system, with shorter clips and different navigation. It is in the navigation that the new possibilities created by the web and Web 2.0 will really be leveraged.

It must mean something when Bill Gates speaks out on the future of IPTV, but he is as likely to bear the brunt of the IPTV revolution as anyone. Microsoft, one of today's market leaders, has captured the wave of the 1990s, providing a great VoD system, but not a good system for broadcast television over IP. And the future of IPTV is a combination of these two, leveraged by the interactivity, which is made possible by more advanced Electronic Program Guides (EPGs) – based on the various XML formats that have emerged. But wait. If IPTV was the hot topic of the 1990s, why has it not taken off until now?

The world today is not the same as in the 1990s. Then, the bright ideas behind IPTV were either practically impossible or not viable. Broadband was not broad enough even for Web TV, there was no workable quality of service, and interaction was not feasible because the systems were not powerful enough – and the standards were not there. Today, peer-to-peer traffic is vying with YouTube and other centralized systems for upload and download of video clips about being the biggest contributor to Internet traffic, much to the detriment of users who would like to watch other things – or sometimes, even causing problems for people who wish to send e-mails. Obviously, quality-of-service techniques have finally found a place where they can be useful.

Two other trends contribute to making IPTV a technology whose time has come: HDTV and digital terrestrial broadcasts. In many countries, the analog networks for television broadcast are being switched off, and digital terrestrial broadcasts are taking their place. This has become a driver for IPTV both because the content has to be encoded for digital transport by default; and because – paradoxically – the user experience has become worse. Introducing an additional few seconds of channel switching for broadcast has removed one of the major advantages of analog over digital. Add to this the fact that as cable-TV providers have increasingly gone digital, the barriers from streaming video over a digital cable to adding IP networking to it have been significantly decreased.

If you only provide the same thing as a cable-TV provider, however, you are engaged in a race to the bottom: the only way to present users with more value than the incumbent service provider is by lowering the price. Instead of doing that, the providers of Internet services (many of whom are desperate for new revenues) could engage in a race to the top – by creating something different from standard technologies. However, having standards is important, because they make it possible to plug in things and plug them out again. That goes for software as well as for hardware, and it helps create a simpler, more efficient platform for the services.

Chapter 1: Interactive, Personal, IPTV: From TV over Internet and Web TV to Interactive Video Media

Video can be experienced from the sofa, or from a chair by the desk: laid back and/or leaning forward. How the programs are distributed is secondary. IPTV can offer totally new things in terms of user experience, which is why it is so exciting. Not that the video itself changes – while there are different ways of telling stories with moving pictures than those we are used to today, the social conventions of video have become so ingrained that programmers will change it at their peril.

Introduction to IPTV

Interactive TV is not new – it has been around at least since the end of the 1990s – but it is still a rather stiff and artificial experience. Interactivity, where the users can change things happening in the story as the program progresses,

Why IPTV? Interactivity, Technologies and Services   Johan Hjelm
© 2008 Johan Hjelm

works when you rely on the participants. As in a computer game, the actions of the user can change what happens on the screen. And games technology is probably one key in creating this new extension of the medium.

That said, there are plenty of experiments with different ways of storytelling, for instance nonlinear videos (think of it as curved loops of stories turning back on each other), which create a different experience, but the existing, linear, format is likely to dominate IPTV programming for a long time to come. However, if the "TV" part is resistant to change, the "IP" part will make it. When broadcasts were analog, there were always pioneers trying out ways to interact with the audience through chat and web pages, and although the formats were interesting, they were never successes.

Interactive programs have not been a success in most of the world. In general (apart from the UK), there has not been a widespread deployment of interactive video applications, although there is one exception: programs where the viewers can vote.

Users tend either to interact at any time (e.g., when they are using the service to get additional information – during sports events for statistics – and are interested in getting information all the time, not just when a player scores), or once the linear program has ended ("half-time factual and learning" viewers). The main reason to interact is to get a more convenient and enhanced experience, and to engage in the program to be entertained in a richer way. Usage peaks after the TV program is broadcast, even if it is made available on video on demand (VoD). The most efficient trigger for interaction is the call for interaction from the presenter – in other words, when the viewers are asked to interact, they will interact, if they know how.

Interactive TV is not the web, however. On a website, there are hyperlinks, which make the site into a big ball of interconnected pages. There is no single "right" way to go through it. A television show is different – it has a linear story. The storyline may be fixed in time (which is usual, since that is how people experience the world); but it can also be fixed in space, and in relation to other stories. Although spatial stories are more complicated to tell, these are where the next generation of user experiences are likely to happen.

The most successful interactive service is betting. Even if you regard it as user-provided content, the function of betting is to intensify the user experience, while at the same time it creates an additional revenue source for the broadcaster (however, note that betting is forbidden in many countries). There is one thing that can be gained from the betting experience: if the content and the interactivity work together, instead of being disconnected, they enhance each other. This also makes the case for live interactive TV, which is also cheaper to produce than TV programs built out of chunks of video by an automatic system on the fly.

In interactive TV, the content creator works more like an advertising company than a traditional broadcaster. It produces content for which it sells the rights; if the buyer is a broadcaster, the broadcaster gets the rights to show the program a number of times, under certain conditions. Usually, the content provider produces the content when commissioned by the broadcaster, not on speculation.

Viewing has become increasingly decoupled from the original transmission, and users do not want to be slaves to an arbitrary schedule which says that "Children's programs are broadcast at 6pm, no matter what". They want to be able to decide. However, when the nature of the program is an event, they are perfectly willing to follow it live. Sports events are one example.

The Value Chain

The value chain (see Figure 1-1), the organization of the industry working in IPTV, is not very different from that of traditional television, and today – since there are so few IPTV systems actually deployed – not very different from its big brother, digital cable. In the US, these two are positioning themselves as competitors, but in reality, digital cable is just one way of carrying IPTV.

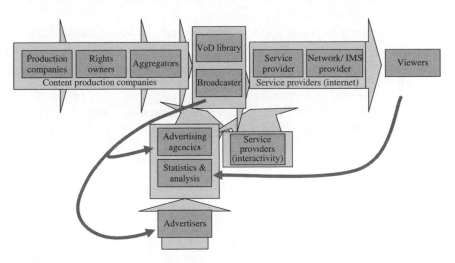

Figure 1-1. Value chain for IPTV.

The value chain looks different depending on who draws it. It depends on what you want to show, and who you are. As always, there may be national variations as well – different countries have different regulations, for example, how much advertising may be included in editorial content. Such regulations, as well as regulations on what data can be used and which audience can be targeted (in some countries, advertising towards children is forbidden), may put constraints on the system.

One constraint that has to be taken into account is privacy. Laws about which information can be given out to whom are nowadays strict in almost all countries around the world, except the US. The strictest laws when it comes to individual permission are those in Europe. These laws are based on an EU directive, and one of the provisions is that the express permission of the user has to be obtained before any data is used, and data may only be used for the purpose for which it is collected. So an advertiser either has to very painstakingly ask everyone to whom he wants to provide information whether this is allowed, or the service provider has to gather the information with the explicit purpose of providing it to advertisers.

To get user consensus, it is probably sufficient if the subscription agreement contains a provision that the service provider can use the data; there is no need to ask for information every time. Periodic checkups may be required, but the laws vary in different countries – the European directive is a minimum stipulation.

In the ecosystem of the earth the majority of life is driven by energy coming from the sun. In the ecosystem of IPTV, all the actors are driven by energy coming from the end-user. The end-user pays in three ways: a subscription fee; a connection fee; and with his attention when he is provided with advertising. Interactivity adds a fourth way, which the broadcaster currently shares with a number of service providers.

The IPTV value chain is likely to be the same as the traditional television value chain at first. It will start diverging, and in a few years the picture may look completely different. Table 1-1 indicates what it looks like today. The roles do not necessarily happen in all companies. Many of them are the same, but different parts work in different parts of the chain, and in a variety of ways. To confuse it a bit, these roles often overlap. A production company is frequently the rights owner of its productions; a broadcaster can be a production company.

The value chain ends with the viewer, since it is from the viewer that all the revenues come in the end. Users want to have the same services that they are getting today, but better and cheaper. Television is, despite the rise of the Internet, the most viewed medium. Attractive as it may seem to add the web to television, things are not that simple. Over 10 years ago, Web TV (later purchased by Microsoft) tried to make the television the information terminal of the home, by providing a web browser. Tempting as that may seem, it is not a way forward: the television is a lean-back device; the web requires the user to lean forward, to be active. Marshall McLuhan, the last great media philosopher of the twentieth century, characterized television as a "hot" medium, which engaged the user and forced them to focus on the content provided; as opposed to the "cool" medium of radio, which fostered detachment. The PC is a "lean-forward" device, where we have to act to interact, press keys or move the cursor to make things happen. Games are the same. The television is a "lean-back" machine, where the user is not engaged – other than when the television shows become social objects, and you have to watch *Hannah Montana* to be part of the gang of girls at school.

Role	Function	Example
Production company	Creates the program (and the advertisements) which are going to be shown	
Rights owner	Owns the rights to the production, may lease them to production companies and broadcasters	Endemol
Aggregator	Aggregates content, perhaps according to a type of event, and resells the aggregation	Formula One
Advertising agency	Purchases advertising time for the advertiser, manages the production and insertion of the advertisements	Havas
Statistics & Analysis	Tracks usage, according to demographics or individual preferences	Nielsen
Advertiser	Purchases time in programs to leverage the captive attention of the audience with commercial messages	Unilever
Service provider (interactivity)	Provides voting services and aggregation, e.g., SMS aggregation.	Netsize
Broadcaster	Produces and sends out the program, manages the advertising time	BBC
VoD library	Provides old content either for a fee or free	iTunes
Service provider	Provides the technical resources for the broadcaster and the VoD library	Akmai
Network/IMS provider	Provides the network and the identity management and other services (today assumed to be the same actor)	BT
Viewer	Consuming television	

Table 1-1. The roles in the value chain.

Business Models and the Value Chain

A value chain reflects a chain of business models. The viewer pays a license fee to the broadcaster (directly in some countries, indirectly in others, not at all in some). However, to get the content from the broadcaster, there has to be an Internet provider, who provides the connectivity; and a service provider, who provides the servers from which the content is delivered. In some contexts, the viewer pays with his attention, not with money, to view the program. Advertisers pay for access to the audience that is watching the show.

If the broadcaster is providing interactive TV today, they are probably using a service provider for the service. This is an aggregator of SMS messages or premium phone calls; the aggregation can be done at a national level, but if it is to be profitable, it has to be done for many countries and operators. The user pays for this, too, but over the telephone bill. If one of the middlemen could be disintermediated, it would mean more income for the broadcaster and the other parts of the chain.

Video on demand is popular also for traditional television shows. Many public broadcasters are putting their programs online (some charging for it), and people do use them: every month, 7 million program instances are watched from the Dutch public broadcaster; and on YouTube 70 million videos are watched every day (although those are mostly short). If users can delay their television viewing to a more suitable time, they will do so – 50 % of users in the UK with Sky set-top boxes already do. And enabling this in IPTV is easy.

Most IPTV services – especially VoD services – are not free. They are based on the user paying a monthly subscription. In some countries, there are free-to-air channels, which are financed by license fees or taxes on television sets or by similar means; they have to be shown to anyone who has a television set. Often, this means cable systems must carry them; and while the rules are not clear when it comes to IPTV, it is not unlikely that IPTV providers must also carry the free-to-air channels in countries where they exist. This is, of course, a constraint on the business model – on the other hand, the user has to have a network connection, and that has to come from a network provider. In this book, an IP Multimedia Subsystem (IMS) provider is also included, but for practical reasons that is likely to be the network provider. Even though the IMS standard talks about the possibility of roaming and interoperability (and we do too in this book), there is no way to do it today. Anyway, the IMS operator has to be able to interact with the network infrastructure to provide the service in an acceptable way, as we will see later.

At the beginning (or the end, depending from where you see it), there is a different group of companies: those who work with content. In the television industry, broadcasters outsource the production of television series and programs to independent production companies. Their role is to coordinate the programming, sell advertising, and act as an interface towards the IPTV service providers and network operators – they have established themselves in the role as a middleman. The media industry is based on maintaining copyrights, and while there may be other ways to measure and meter content usage, digital rights management (in the widest sense) has emerged as the favorite method of the industry. However, the methods that are applied today, tightly coupled to devices and charging, may be diminishing user interest.

Content production and IPTV

The content is normally created by specialized companies, or units within the large companies. In the old days, only a large broadcaster such as the BBC could afford a unit to produce a drama series, and these were sold to other broadcasters around the world. Smaller companies started taking on the production role, however, they did not become really interesting until they started taking a different role: not just as a producer of a TV series on order, but as a creator of a concept and a packager. Sometimes, the broadcasters take on the role as content aggregator and content producer, in addition to being the service provider. Other times, the service provider is the network operator. There is no standard in the industry for how the roles are distributed – this will depend on the local economy and regulations. But the roles exist in most, if not all, IPTV systems. If a company is a bespoke producer, it is unlikely to be very much impacted by IPTV; if it takes a bigger role, it can use IPTV to its advantage.

The pioneer here was the Dutch company Endemol, which made its name with the Big Brother television show – which by combining television shows, live Internet webcasts, chat (in the early editions), and viewer interaction (to vote out participants) was also a pioneering multimedia experience. The interaction really changed the way that television was produced. The producers had no idea what would happen in next week's show. They did not have any control over it either. They gave that up to the users, in exchange for their curiousness – did the person you voted for get booted out? What happened next? And who would you vote for next week? Strictly speaking, the Internet was just an additional peep show – the big money in the show was made from the advertising, and the voting. Attempts to sell the naughtiest bits on DVDs and as private material did not turn out to be big successes.

Endemol is a content rights owner in this picture, a packager of content and deliverer of it to content aggregators. Aggregators can be broadcasters, but also, for example, specialized sports channels, or companies that create golf news for television channels, by combining coverage from several live events. The rights owner can be very powerful in the television industry, since copyright gives them a very strong tool to ward off anyone who might use their content without permission. Broadcasting without permission means a lawsuit.

The content industry is large – there are specialized trade fairs in both Europe and America where content owners can offer their content to aggregators and other distributors, as well as make deals among themselves. At these companies, interactive TV is usually very sparsely represented. The number of companies that work with interactive TV and productions directly for IPTV are low compared to the number of companies that work with Internet content – even if the techniques of production are largely the same, as we will see later.

It used to be that the broadcaster was the company that owned the studios, the equipment and all the resources required to produce television programs. This was when production equipment was expensive; nowadays, a professional television camera does not cost much more than 10 times a good amateur camera, and often it is hard to tell the difference in the result – the skill of the filmmaker is becoming more important than the technology. Creating a television studio is not a matter of expensive investment in recording equipment, it is more a matter of creating a workable space for the recording.

In traditional television, the broadcaster used to be the operator of the radio network (and still is in many countries). In other places, the radio network is run by a specialized operator, and the broadcasters pay a fee for the broadcasting, just like the broadcasters who go directly to satellite. In IPTV, the radio network is replaced (from that point of view) with the Internet. The model that is emerging is more similar to cable-TV, however, where the user pays a subscription fee to get the service, and the operator pays a fee to the broadcaster to get the content.

In cable-TV, the user fees finance the purchase of a number of channels, packaged by broadcasters, some of which have a very high number of subscribers, others of which are more specialized and have fewer subscribers. The low-subscriber channels are often packaged with the popular ones as part of the conditions from the content owner (who may wish to promote a channel that users do not yet know about, or be able to claim to advertisers that the channel has a certain number of subscribers even though they did not choose it).

In the same way as users have stopped buying CDs, in preference of buying individual songs, television viewing in the US has gone towards individual program viewing – a trend enhanced by personal video recorders such as the TiVo. Some cable and satellite network operators offer this as part of the service subscription, but this is an expensive proposition – you need high economies of scale to be able to make enough to offset the costs. Smaller operators do not have that option, and for them providing a VoD service in the network is cheaper and also provides an opportunity to sell advertising, however, this brings a heavy penalty in terms of network traffic, which requires an expensive network and very tight control over it.

While the network operator is rarely the same as the broadcaster, the IPTV service provider often is. This is because they need the tight control over the network, but it is not two roles which marry easily. The network provider wants to provide a network that is optimized for the transmission of IPTV, but the IPTV service provider wants a network that always gives the absolutely best quality of service. This duality is likely to become disruptive in a few years' time, as network providers start devolving their IPTV services (if they have not started out by subcontracting the IPTV service).

If all the service provides is cable-TV over a different cable, then it is not adding any value to the user; and the only way to give the user value and make him switch to the IPTV service is to lower the price. For most IPTV operators, this means lower than zero, since the broadcast is provided free to air. They have to add other values, most often VoD libraries. Very seldom do they try to add interactivity, despite its excellent track record in the UK as an additional source of income. Sometimes they are constrained by legal frameworks, it is true; but often, it is simply because they have not thought about the possibility.

The value chain of IPTV is not different from the value chain of television, especially interactive television, but the technology used to deliver it is different – it is what this book is about. The difference is most marked for two roles: the user, and the network provider. Another party that will see a significant advantage is the advertiser, since the interactive advertising models, which have emerged in the digital cable and interactive television industries, will get a significant push by the IPTV technology.

The consumer electronics manufacturers are working towards IPTV. They have created the basic standard, leveraging Web 2.0 and IMS. The transport of the television stream is standardized, the interaction mechanisms about to be. This book describes one set of interaction mechanisms, based on the IP Multimedia Subsystem (IMS) standard. There are others, but they are not as flexible, or as good – though the threshold for developers may be lower. Consumer electronics manufacturers are slowly getting into the act. Television sets are already delivered with Ethernet connections for IP connectivity (at least in Japan), and built-in web browsers. Full IPTV clients are not far behind. This is because Internet usage is increasing, and one way it is used is watching video.

At the same time, there are many more devices today that can be used for viewing television than there ever has been, from handheld mobile phones to 100-inch plasma television sets, not to speak of televisions, PlayStations and Xboxes. There will be more in the future, and those suitable for IPTV viewing will have greater capabilities and possibilities, which can be leveraged by IPTV viewers. And the set-top box, which has created so many constraints for the television industry by trying to conserve the cable-TV model, is far from dead. Rather, it is set to get a new life – as the interaction box.

Advertising is the primary means of monetizing television today. However, as more television channels have become available, the "mass media" effect – to place advertisements in front of millions of people at the same time – has diminished. IPTV technology can be used to help advertisers find **the** right audience for their advertisements – by adding statistics.

Advertising in IPTV

Traditional advertising came about when early television networks time-sliced the programming time previously dedicated to one single sponsor (e.g., the US Steel Hour, the Palmolive Soap Theater – the latter the origin of the term "soap opera"). As big a revolution as that was, it was driven by the ability to splice content from different sources together to form one single continuous show, although interrupted by advertising. Advertising interruptions have now become so familiar that we hardly react to them anymore, and according to some researchers mentally tune out during the advertising breaks, if we do not even take this opportunity to change channels. In time-shifting systems, such as the TiVo and other set-top boxes with local memory, users can decide when they want to see something, instead of having to wait for the time that the broadcast planner has decided (which in turn

(Continued)

is based on measurements allowing them to pinpoint the desired demography of viewers, in terms of age, type and group).

Advertisers on traditional television have a problem: the audiences are leaving. Despite the program producer's best efforts, more and more viewers are turning to pre-recorded programming (either which they have recorded themselves, or which has been recorded for them). When you buy television programs or films from a site on the Internet, they can either be downloaded to your computer – or the recorder in the set-top box – or they can be streamed to your television directly, with the storage at the service provider's site. It becomes very easy to fast-forward past the advertisements.

Advertisers need to measure the result of their advertising, and traditionally this was done by questionnaires and different kind of sampling techniques, including putting boxes in people's homes to measure what they were watching. Since it simply would not be economic to cover all households, this meant putting boxes within a statistically significant sample of viewers. As target groups become narrower, it becomes impossible to do this using traditional broadcast.

However, it becomes easy to do using IMS-based IPTV, because the broadcaster can collect the statistics in real time (even to the extent of telling the advertisers how many people are watching their advertisements just now). The issue becomes more one of aggregation and anonymization, since it is not at all certain that the users want all and sundry shoe and soap companies to find out what they are watching, or even what they have watched in the past.

Selling video films on the Internet – or from the portal of an IPTV service provider – has its own challenges, as we will discuss a little in this book. But it is broadcasters who have the biggest problems. The audiences are disappearing from viewing the advertisements, so why should advertisers pay for the attention of the audience, when they are not reaching them?

There are three ways out of this dilemma for broadcasters, the first two are: they can make better programs, so nobody wants to leave before they know what happened; and they can charge more for advertising – making it interesting enough for viewers to watch. That is happening to some extent.

With today's technologies, even in the most sophisticated cable-TV networks, you are stuck with guessing who your audience may be. There are attempts to measure who watches what, but all methods are based on sampling and statistical analysis. There is no way of either telling who has watched an advertisement – or who that person is. Demographic information is only available on a very general level. And even if you can profile the household, that information does not say anything about who in the household watches the show. A few years ago, it would have been a foregone conclusion that if the cartoons were watched at 5pm, that was the children, and the economy news at 7pm was the father. But nowadays the father is equally likely to watch the cartoons. And the one watching the economy news may be the grandmother. There is no way of telling with the existing technology.

To provide an IPTV service, which adds more value than traditional broadcasts and video rentals, the next-generation IPTV systems have to both make registration of user data easy and practical, and give users control over it. And luckily, IMS comes to the rescue again. There are alternative ways of handling user identification and data, but only the Liberty Alliance protocols allow for federation of data across different service providers. Liberty Alliance and IETF Geopriv both give users some control over their personal data, but they do not have any idea about the data structure – which is one of the features of the personalization system in IMS.

So this book is about the third way to create value for the viewers: turning them into participants. It also has another consequence: since it is far cheaper to make live programs which adapt to feedback (just watch any shopping channel) than to create compositions out of recorded videos, this will mean a resurgence of the live format. This will have consequences for the technology of television, as well.

There is another constraint on the business model: very little professionally produced content is available for free. The rights owner has set the price for his content so that it fits the existing market, but the world is changing. While it may have been possible to buy productions at the prices demanded when there were just a few television channels and everyone had to watch, the industry has changed. Now, there are thousands of television channels, with much fewer viewers, and the advertisers are discovering that they are being sidelined by technology.

Changing the role of Digital Rights Management (DRM)

Up until now, the broadcaster has had the monopoly of bringing the user what he sees. Not so with mashups. Deciding who owns the interaction ability – and hence the ability to make money from it – is likely to be a major struggle in the media industry, which has already moved to use legislation to gain control over subtitles, crows and other content additional to the television show. In some countries (like Japan) legislation actively forbids any overlays over the content sent out in the television channel. But fighting users who want to add value is a losing proposition, as almost 15 years of web experience should tell us by now. Enabling the possibility to create added value to the television show, rather than constraining it, will mean driving new business, rather than locking in old.

The broadcaster can commission content, and hence get the rights to it, but the rights to the content can be sold to others as well. These include aggregators, who take many different types of content and sell them on (e.g., creating golf shows for sports channels). The content industry is a large industry with well-established actors, and since the existing content is seen as the key to making IPTV take off, it is not likely that it will go away. This is different to what happened

> **(Continued)**
>
> *on the web, where the existing content owners were sidelined by individuals and small companies developing new content. While this is likely to happen once IPTV becomes widespread enough for dedicated channels to take off, television is dominated by popular series and movies, and will likely remain so for many years to come. We will look more at copyright and DRM in Chapter 6.*

The content aggregator usually sells the content to a broadcaster, which acts as an agent for the content; it sells the content rights on to the broadcasters who are interested in showing it to their viewers – and think they can monetize it by selling advertisements in it. Advertising comes from advertising agencies (it may be produced by the same companies who produce the content) and is inserted into the programs by the broadcaster. How this is done, and how the process can be partly automated, we will look at in Chapter 5.

The distribution technology is the same as most users today use to receive an Internet service, which is why it is attractive to network providers: they can get more users for the networks they have already built, and they can get existing users to pay more for the new services. It is also attractive to broadcasters: they can get an additional customer in addition to cable-TV providers, which means that they can sell their programs one more time. It is attractive for users as well, since having one additional service provider will create price pressure on the Internet service, as well as the IPTV programs. However, IPTV is not just a new distribution technology.

If the same service is offered to all users, regardless of delivery network, network and service providers are indeed caught in a bind: there is no way they can get more for the service, since there is no reason for users to pay more, and the only way they can compete is through price. If operators want to be able to create something new, they have to do two things: they have to offer the same (or better) service as their competition, primarily cable-TV networks but also broadcast; and they have to provide something new and attractive.

If the content is so attractive that people want to pay for it, they will be charged for it. Movies and other content where the audience is highly immersed, thanks to the plot and production values, are costly to make, and every returned cent counts. Pay per view has had some success with certain types of programs. Interactivity today is not a large part of the revenue stream for most operators – but it is a source of worry for users, who have to pay every time they want to interact with the television show. Charging a flat fee has increased usage in any other medium; it is very likely to be true for interactive television. And with IPTV, there are mechanisms to do it. They come as part of the parcel when you use a standard called IMS, the IP Multimedia Subsystem.

Interactivity in Reality: The British Red Button

The only country where interactive television has become widespread is the United Kingdom, where the set-top boxes deployed by BSkyB, the broadcasting company of Rupert Murdoch, have been based on the otherwise less than successful WAP standard. The success carries an important lesson for the future of interactive television, both in terms of what content is most appreciated, and how users want to interact with it.

The interaction model of WAP, originally developed for mobile phones to interact with information services in a web-like way, was based on Apple's HyperCard, and instead of pages, the user interacted with a deck of cards, which were interlinked by a scripting language. Overlaying the deck on top of the television signal enables the user to interact with the service. Since WAP was designed for the early mobile networks, the transmissions are extremely compressed and latencies become low even over a dedicated telephone line.

However, in the BSkyB service, as well as in many other services (commercial or experimental), it is not the services that the broadcasters expect will become popular. Interacting with the television programs themselves is less popular than interacting with the dedicated sites which content providers can create. While actual interactions with programs is becoming possible in real time when the user is connected over the Internet – television becoming almost indistinguishable from games, the only difference being the interaction model – there is one type of content which is likely to suffer and flourish at the same time in the new IPTV systems, and that is advertising.

Four levels of interactivity

There is a lot of confusion about what consists interactivity; with some even counting channel switching as an interactive activity. Looking at user behavior in combination with technology and content, however, four levels of interactivity become easily evident.

Level one is where the user interacts with the meta-information about the content, such as the program guide. This includes video on demand, setting personal video recorders, and selecting content in an Electronic Program Guide (EPG).

The next level is where the user accesses external information, which is not necessarily related to the program. This includes Teletext or on-device portals. The user can get news and other information, but the interactivity is limited to pointing and clicking, perhaps with pages pre-adapted according to user preferences. Examples include Bloomberg, a finance and economics show, which displays stock prices and charts; users can call up new charts to see market fluctuations. Voting is another feature, although you cannot vote on the stock price or trade stocks.

The third interactivity mode is where the user can influence the program by voting. This includes programs such as Big Brother and American Idol, and can also include chatting and other interactions with other users through the mediation of the television and the communications device. This is very popular: 27 % of

(Continued)

all young European owners of mobile phones had voted or participated in game shows on television at some time. In the UK, the red button on the remote control connects users to the interactive services, which is very popular during the same type of events: 58 % of the audience used the service during the 2004 Olympics, according to the BBC, of which more than 60 % watched it in an interactive way for more than 15 minutes. Of all the viewers who had access to digital television, more than 40 % have participated at least once in an interactive television game. The Olympics (a special event, if there ever was one) aside, there have been a number of successful interactive television shows in Europe. In different events, such as Formula One and football leagues, the user can choose which camera angle to view the action from. There are automatic systems which track the ball on the pitch and select the best camera for viewing, as well.

The fourth level is where the actual story changes depending on how the user interacts. This includes both explicit interaction, where the user makes a choice of how the program should proceed; and implicit interaction, where previous user actions are taken into account to change the program. This type of interaction approaches games, and while the games industry is much larger than the IPTV industry at present, there are lots of things interactive television developers can learn from games – since games include many interactive television features. Still, only 81.3 million euro were spent on interactive television games during 2005, with UK users contributing 42 %. SMS can be used for more than voting, however. In Italy, MCS Tutte le Matine, a popular television show, is available in sign language. Users can interact with the subject of discussion, and they can get SMS reminders about when their requested subjects will be discussed.

Interactivity has also been extended to advertising. There are service examples in the UK where interactive TV is used for advertising campaigns with some regularity and success, for instance Adidas. Increasingly, VoD services are placing personalized advertisements into the programs, and making it possible for the user to select an extended version of the advertisement ("tromboning" in IPTV-speak).

Traditional television has tended to see itself as the focus of attention, its schedule binding the user's time, but personal video recorders have turned that around: the user now sees the television signal as a distribution system for the video he wants to watch after recording it. This contributes to the fragmentation of the television audience – but also the opportunity to target the viewers with advertising, since a fragmented audience is a number of specific audiences. This is also the key to success for broadcasters. Offering the right audiences (not always the same as the biggest) will mean revenues from advertising – and, if the audience interest is captured, interaction.

The most convenient way of access to content will always win, in particular in the convenience-driven television medium. Users could not care less if the video they rented was delivered on a disc or via the cable, as long as they can enjoy it whenever they like (or at least within the terms they have paid for). An exciting game can be sent via radio or cable, and the cable can be analog or digital; the radio waves can come from a satellite or a tower. This does not matter for the user experience (except in setting up the service, waiting for the cable-TV technician to show up, and so on). But not all users are the same. They want to have their own choices. These choices have to be presented in a comprehensive and easy-to-understand way. The least intrusive user experience will be the most attractive. Individual interaction is hard for television programmers to handle – it was not until the interaction turned from the individual to groups, in the shape of voting, that interactivity started taking off. Users could make programs different by voting, for example, in the *Eurovision Song Contest* and *Big Brother*.

That voting is handled through the telecom operators, and users vote through their phones or mobiles. A service provider aggregates the votes, and presents them to the program. The service provider also aggregates the revenues which the television station receives – typically a percentage of the (premium) cost for the message or call. The votes are presented on the air, and the results will affect the program, which is what interactivity means (that you can select different variants of something, but not change it, is not real interactivity). There is a way to charge users for interaction; and you know for sure that those who voted watched the show – since it was sent live, and any voting done after the show would be meaningless. This means that the format of the show has to be adapted to the technology, as with so many other shows which follow a similar model: the viewers have to be given a reason and an occasion to vote.

Interaction works best (according to the EU LIVE project) in documentaries and news. In fiction, people want interaction to be as unobtrusive as possible, not to disturb the viewing and the immersion in the plot and storyline. For the producer, having a number of short video clips, which are mixed together based on user interactions, can be problematic. The shorter the clip, the greater the loss of the meaning the user perceives; and the bigger the opportunity for remixing and sampling.

In the minds of the first designers of interactive television systems, the user configures the system precisely to his needs, and sits back and watches it happens. But most users are not programmers, and they may not know their own needs. They also like some serendipity, surprises in what is being presented to them. This became evident as early as the end of the 1980s in the MIT individualized newspaper experiment, Fishwrap, where the users quickly got bored with their own selections of news, and started asking for an editor to come up with some surprises. Nobody wants to be completely alone, even though we may want to be individuals; belonging to a group, and knowing it, is one of the strongest motivators of human actions. And, even though the next generation of IPTV can be made both personal and interactive, there is also a way to capture the groups the user belongs to, and use that in creating the shows. There are many ways to do this, but unless you build them into the system from the start, they require significant effort to implement – and will feel clunky and pasted on. But if you use the technology, which is the base for the IPTV system described in this book, IMS, you get a solution that is part of the parcel.

How IPTV Services Work

IPTV is different from traditional interactive television, because the backchannel is built in. Today, the best backchannel for interactive television is the mobile phone (or traditional telephone), and that has its own problems, among others, in that the interaction has to pass through a service provider, and that it is rather expensive for viewers to interact with programs, so they tend to refrain other than in special circumstances – and the content producers tend to think about the interaction in terms only of interaction with special events.

Figure 1-2 shows how an IPTV system works at one level: how the request for a television signal gets from the television set to the server delivering the content as a data stream – the streaming server. In the telecom industry, the "signaling plane" is often separated from the "media plane", a convention we will follow in this book, since it makes it easier to talk about the next generation of IPTV. The interactivity we will discuss in this book is part of the signaling plane, at least if you use the IMS-based solution. Other types of interactivity, such as a user providing content, become part of the media plane; and channel switching (which is not really interactivity, since it does not change the program) also takes place on the media plane.

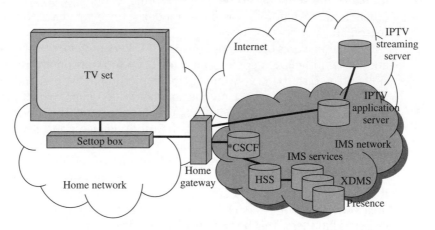

Figure 1-2. How the IPTV system handles communications with the media server.

The killer application: video to television?

If you ask people in the telecom industry, the killer application is video telephony. Despite the fact that it has been promoted as the killer application since the 1950s (but is yet to take off in a big way), and several of the interactive television trials which were conducted during the 1990s showed that people want to watch television, not take video calls on it, technology pundits continue to promote it as the application that will sell almost any new network technology.

This also extends to IMS, one of the cornerstones of the technologies discussed in this book. While the European Telecommunications Standardization Institute (ETSI; despite the name a global standardization organization) is now working on a standard for IPTV using IMS, it has already standardized four other "services" – which means profiles of IMS that can be installed within an existing client. These are "Multimedia Telephony", a new name for what was previously known as video telephony; Instant (or "Immediate") Messaging; Push to Talk over Cellular, an unwieldy name for a service that lets users send short voice clips to each other; and Presence – which makes it easy to keep track of what your friends are doing.

In particular presence makes it possible to create a completely different service offering. While video telephony certainly has a role, it is not making users take telephone calls on their television sets. It is enabling them to call the television show. Interactivity at such a highly personal level is somewhat out of the scope of this book (since only one user at the time can interact with the host).

However, the same technology that enables video telephony makes presence and messaging possible – and those are the basis for the interactivity described in this book. So we will look deeper into how they work.

The layered view of the network helps in understanding and modeling – but it is a model, not the network itself (even if the distinction is dubious when talking about software). If you show the media delivery instead of the signaling layer, you get a different picture – the same components of the IPTV system, but with different relations between them. In IMS, two other layers are used to describe the system: the application layer and the network layer. Applications use signaling to set up media over the network; if you show only the network layer, a different set of components comes into play. For example, there has to be a DNS server, which allocates the IP addresses used in the network; we will not go into how that is done, since that is pretty much standard today. Nor will we look at how IPTV works over ADSL or fiber to the home.

We will look into the home network, since it is important to understand how the different components work together. And since there is a fight in the IPTV industry, despite hardly being standardized yet, over who should own the user interface, and who should create the middleware – and where it should be. There are two extremes in this view:

- Either the middleware, the software that works with the media stream to create the IPTV services, is in the set-top box, and there is hardly any software in the network – only servers getting signals to deliver data.

- Or the software is in the network servers, and the set-top box is just a dumb box (if it exists at all), which forwards the input from the user's remote control.

The reality will be somewhere in between, but there is also potential for IPTV systems to look very different and yet have the same basic functions.

And since the "managed" network of the IMS is an overlay on top of the "regular" Internet, in a part of the Internet where the service provider owns the routers and other systems, a very large part of the system design is already given. The IPTV system also has to work on home networks that may look completely different, depending on the users who built them – both from the perspective of networking and the perspective of signaling.

From a system designer's perspective, the layering (see Figure 1-3) makes things easier. The important thing becomes the interfaces: if the clients and servers conform to them, the software developer can use his time to make things that run more efficiently.

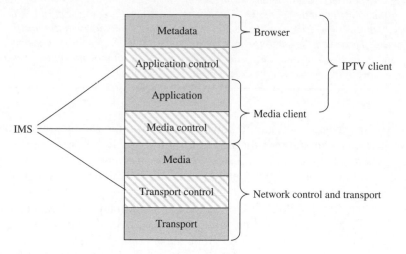

Figure 1-3. A layered view of the IPTV system.

An IPTV system with interactivity will consist of the following parts as shown in Table 1-2.

Most of the components of the IPTV system are already in place. Video servers, for example, are not new – development started in the 1970s. The networks are the same as those used to deliver Internet service today, and there is neither need nor possibility to replace them. This does not leave much space for developers. The client, the application server and servers, which provide services to the applications, are all that is left to make innovations – but this is more than enough, as hardly any of the existing components that enable new types of services are in place today.

In reality, the user wants to see a television program. He clicks on the "start" button on the remote control. When he does that, the IPTV system registers with the IMS Core, to verify his subscription and make sure any profile information is applied. This registration may go through a separate gateway, or it may go directly (as will all the following requests within the session this sets up). As part of the confirmation of the session, the user gets the media stream – either as a stream directly from the streaming server; or as

What it is	How it works	Why it is needed
TV set/Media renderer	Receives the IPTV signal and renders it on the screen	Makes it possible to see the video content
Set-top box/Interaction device	Captures user input and sends it to a central interaction server	Makes sure the user's interactions get to the IPTV service provider, and can be used to change the content of the show
Home network	Connects the different types of equipment in the home together	Makes it possible for different media stores, renderers and interaction devices to interact with each other, and services on the global network (Internet)
Home gateway	Manages addressing in the home, registration with the service provider, and filtering of content (the last two functions can also be performed by the set-top box)	As a firewall and address management system, and to ensure that the user's actions are authorized
IMS proxy	Captures the request for the video service and makes sure it gets to the right receivers, including the QoS system	Interconnects the network and signaling planes of the system, and makes sure the service requests get to the right nodes. Also connects to the profile management system
QoS system	Instructs routers in the network how their queuing mechanisms should be set up	Without QoS, video can be delayed and result in degraded user experience
IMS identity management	Makes sure the user is who he claims he is, and connects the use of the identity to the relevant subscriptions (and hence charging)	Without identity management, anyone could use anyone else's services; the charging systems would have to work offline and with special tokens to keep track of who should pay for what (as it is now)
IMS presence and profile management	Keeps track of what the user does and has done; makes sure this is registered in the system	Makes it possible to personalize content, and to know what other users are watching (if they allow the user to see it)

Table 1-2. The components of an interactive IPTV system.

What it is	How it works	Why it is needed
IPTV streaming service control	Manages the video stream, including switching to a different video stream when the user selection demands. Note that this is not the same as channel switching	Makes sure that the program is started when requested, and eventually creates programs from different video sources automatically on the fly
IPTV streaming service delivery	Handles the streaming of the content over the network. Interacts with the QoS management	Makes sure content gets where it is supposed to go
Advertising insertion	At selected points in the media stream, pastes in video sequences which contain commercial messages (although this could be a generic mechanism)	Puts advertising in the right place in the program
Interactivity server	Captures the interactivity requests (from the user's IPTV session), collates them (if required), and sends to the appropriate server(s), such as charging, profile management and streaming service control	Interactions which come from more than one user need to be collated and coordinated, otherwise they will not result in anything

Table 1-2. (Continued).

a multicast address, where his IPTV set can join an existing multicast group. The streaming starts to the television, and the user can watch it – since it is encoded in a standard format. At the same time, the television program the user is watching is registered in the presence server, and the user's nominated friends are informed that he is watching.

When the user wants to interact, for example, to comment on something stupid the quizmaster just said in a quiz show, he presses the red button. This makes a menu appear on his screen; the menu can either be fetched from the IPTV Application Server (AS; "Application Server" means something special in IMS) when he makes the request, or he can get it from the IPTV AS separately. How it is displayed depends on which standards the IPTV set implements; but if it uses the latest standards from the consumer electronics industry, it can display the content on top of the television program.

The user selects the interaction from the menu, and this triggers a message to the IPTV AS, which includes it in the system used to either select the video clips (e.g., advertisements) to be displayed next, or to change the script on the teleprompter in front of the quizmaster. At the same time, the profile of the user is updated, so the user's preferences can be taken into account (even though it may be hard to draw conclusions from the user's interactions in a quiz show). The selection of the video clips can be individual (if the user is watching a "unicast" data stream); or it can be done on a group basis. If it is done for a group, the profile of the group is compared with the description of the video clips in the metadata.

What the user is watching can also, after being anonymized, be sent back to the advertiser, or others who are interested in finding out how to monetize the programs.

What is Next for IPTV Users?

It takes a daring (or desperate) producer to invest in a completely new format for television programs – or government funding. In Europe, the latter is the case. The development comes under the scope of EU research projects, funded under the 7th Framework Program.

The EU aims wide in its research frameworks. The intention is to involve a wide selection of countries and organizations. The goal of only strengthening the European industry and creating more jobs through research was widened greatly in the 6th Framework, and this widening is even more pronounced in the 7th. Hence, many more organizations and countries are eligible for research project funding – and will be able to participate on an equal footing with companies and universities from the EU proper.

What this means is an unrivalled funding opportunity for companies and universities in Europe. And for companies and universities outside Europe, it is an unrivalled way of getting involved in very interesting, very directed research projects. However, the process of deciding on a research framework in the EU is nothing if not complicated, and an interesting reflection on the processes behind the EU.

The EU 7th Framework Program

Since 1984, the EU has allocated money for research programs in its budget. These programs, like the rest of the budget administered by the European Commission, are an attempt to create more research-driven industries in Europe, and hence more jobs and more growth. They are framework programs, which means that they have several subprograms, which are intended to cover different aspects in the i2010 plan – to make Europe the most advanced knowledge economy in the world by 2010. All the framework programs are heavily laden with political agendas, covering everything from computer support for the elderly to ethical aspects of research. The 7th Framework is simply the seventh in number.

The process for framework program approval is as follows. The initial proposal comes from the European Commission. This is commented on by the European Atomic Commission (since some of the financing goes to European atomic research); and the Committee of Regions. The comments are then read

(Continued)

by the European Parliament and the Council of Europe, who may suggest proposals for amendments (and then there is some iteration). If all the amendments are accepted (something that does not happen), the program is approved. As it was, the European Parliament did not approve the proposal, and sent it back to the Council with changes, which were discussed, and read a second time by the Council. In the current process to approve the 7th Framework, the European Parliament has taken a much more active role than before.

When the European Parliament (the final example of approval for the EU budget) approved the funding for the 7th Framework, the European Commission (who of course had been preparing for a long time) set its wheels in motion. The process for the 7th Framework is essentially the same as for the 6th, but with some tweaks and additions. In addition, navigating the process requires a great deal of experience in itself.

The framework programs are a budget, as well as a strategic direction. And the budget is big: 3 % of the EU budget should go to research. Two-thirds should come from industry, and one-third from public sources. And only one-third of that is the EU budget for the 7th Framework. This still makes it the second biggest part of the EC budget, after agriculture. The framework will not end until 2013, and there will be a budget revision in 2011, but if the previous framework programs are anything to go by, the lion's share of the budget will be allocated early, during 2007 and 2008. By far the largest part of the budget goes to the information and communication technologies.

The EU-funded research projects have some unique properties, which are not found in other research projects – or even project organizations. The first property is that the consortium signs a contract with the European Commission to undertake a certain piece of work – including disseminating it to a wider audience. These days, this means that the resulting software can be made available as open source, and that reports are expected to be public. The contract is binding to the partners in the consortium, who have to be about half from universities and from many European countries – and can come from outside the EU, as well.

The proposals for projects are evaluated by a group of experts, who look at the technical excellence (note that the EC and other parts of the EU do not have any say in the results). The evaluation is driven by the call for proposals, and the scientific and technical excellence of the proposed results – while a proposal may be politically correct, it will not pass the evaluation if the technical excellence is too low. A proposal which falls through in one call might have a chance in a later call – if it is appropriately modified.

"Research projects" do not mean men in white lab coats chasing white mice with cyclotron beams in imaginary labyrinths. Or at least, not only. It means projects that are aimed at "advancing beyond the state of the art" – adding to what is known in an area. What this is, and the method for it, will depend very much on the area. The EU funding frameworks are intended to drive applied research, which means that projects are expected to lead to practical applications of theoretical and long-term research; however, product development as such is not funded. Drawing the line between these is not simple – especially in the area of future interactive television, where a new format can be deployed overnight if it turns out to be successful. And part of the research funded by the EU is about new television formats.

Shape-Shifting Television: New Media for a New Millennium

The New Media for a New Millennium (or NM2 for short) was an EU-funded project in the 6th Framework which ran until 2006, and which tried to create a new type of user experience – with IPTV in mind. Participants came from several European telecom companies, and the inspiration was in equal parts computer games and interactive films, a genre which has a small but thriving subculture in art cinema.

The common idea behind the "shape-shifted media", which NM2 created, was that productions would be based on pre-produced content, produced by professionals and to professional standards. However, the users would determine the narrative by voting and sending messages to the show; this was possible because the project used set-top boxes which provided for this type of interactivity. The interactions could determine more than just what would happen in the next scene – also the length of the show, the location, narrative perspective, camera angles – as long as it was available within the pre-recorded content (which is a big constraint on systems based on stitching video clips together). At the same time, the producers were careful to make sure that the changing story did not disrupt the viewing pleasure. Of the eight productions, one was actually broadcast (in the cable network in Helsinki, Finland). The others were made available on the web.

For anyone familiar with the history of hypertext (before and after the web), the storytelling model seems hauntingly familiar. A website, as most designers have realized by now, is a mesh, not a tree, with several possible starting points. Any interactive television production today will have to compete with the ubiquity of the web, and while it is not much harder to create a branching script than one built for linear viewing, it requires a different type of thinking – just as the first websites could claim to change the way people thought about media, when they had to leave the linear way of writing behind.

According to the NM2 project, in a reconfigurable media experience, the storyline is not determined in advance. The production team has a database of footage (either new footage or archive material) that can be edited in numerous ways, with or without preconceived scripts or plots. Depending on the input of the users the story is shaped and configured. The tools used should make it possible to create flexible narrative structures, called narrative arcs by the NM2 project. Narrative arcs consist of a number of video shots based on a particular structure. The tools should be able to model and structure narration automatically. If the system can automatically define the narration based on information gathered from the user and from the production team, this saves a lot of time.

In an interactive drama it is also very important to keep track of the exact timeline and how much time is left until the end of the program (and for the advertising). The typical slot for the broadcast is 28 minutes, and this should be indicated by the tools. In addition repetition of clips should be avoided, which can be implemented by the use of rules.

Figure 1-4 illustrates the various tools developed by the NM2 project.

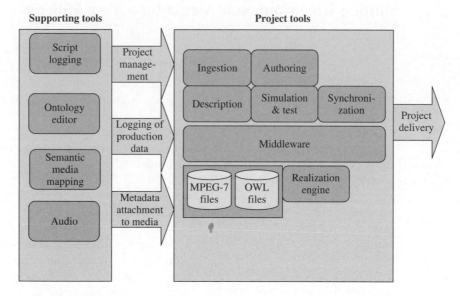

Figure 1-4. NM2 tools.

- The Script Logging Tool is a standalone application, which enables metadata relating to media and narrative objects to be captured at an early stage in the production process, for example, when writing a script, shooting scenes or searching for archive material.

- The Ingestion Tool is the means by which metadata from a variety of sources can be imported into the NM2 tools in order to define media and narrative objects, concepts within ontologies, and other information.

- The Description Tool enables media objects within a project to be created, edited, modified and deleted. It provides hierarchical management of objects and multiple options to review and append metadata to one or more objects. Both MPEG-7 and ontology-based metadata can be expressed in the tool, which also provides a framework for automatic content analysis.

- The Authoring Tool enables the creation of interactive narratives by means of a unique interface consisting of a hierarchical "canvas" on which narrative objects can be positioned and interconnected. Media objects from the Description Tool can be added to the canvas, and specific rules and heuristics can be entered to define the logic of the narrative.

- Simulation and Test provides functions which enable an interactive narrative to be checked and reviewed from within the NM2 tool application environment, prior to deployment on a delivery system (such as an IPTV service). This includes the ability to simulate multiple user inputs and review a synchronized media output, which is representative of the intended user experience.

- The Middleware Framework provides common functionality, which can be accessed by all NM2 tools encapsulated within the Application Environment. In addition, it exposes Application Program Interfaces (APIs) which can be independently used by production-specific delivery systems.

The most important functions provided by the middleware are persistent storage of metadata for both media and narrative objects, including the project's narrative structure itself, and access to the Realization Engine in order to execute a narrative in accordance with external inputs. The Realization Engine within the Middleware Framework is the execution engine for interactive narratives. It combines a narrative structure defined by the Authoring Tool, media object metadata from the data stores, and external input from a delivery system to progressively generate a multilayered media playlist for input to a media composition engine either on the client or server side.

The NM2 project identified six trends, which would make "shape-shifting media" take off:

- A massive uptake of digital networks.

- The convergence of PC and television in devices that facilitate personal media experiences.

- Games becoming truly interactive media productions.

- Mobile phones allowing for media consumption anytime and anywhere.

- The rise of Web 2.0.

- Young generations guiding "us" into an interactive future.

While this sounds like a wish list of media executives from the early 2000s, they do not automatically apply. First, the younger generation is probably the driving consumer, but that makes demographic assumptions which are not sustained in a family context, and that is where television is typically viewed. Mobile phones have become media consumption devices – in Japan and Korea – but they are not used for interactive media, they are used for playing music and watching traditional broadcasts. And the massive uptake of digital networks has already happened, with the emergence of the Internet and the television industry going digital.

The three trends that do matter are as follows:

1. The convergence of the PC and television – but not in the simple way that you put them both in the same device; just like the web combined features from newspapers with features from computer media, the resulting system is both, and neither.

2. The second trend, the emergence of Web 2.0, probably holds the key to the predictions of NM2 becoming reality. Web 2.0 makes new types of user interactions – using new types of devices – possible.

3. The third trend, that games should become truly interactive media productions, has already happened to some degree. It is in massive multiplayer games such as Second Life that the real change in the way users perceive storytelling has the opportunity to take off. Machinima is already an established genre.

NM2 also rightly saw games consoles as a crucial emerging technology. While designed to be viewed on a PC or using an interactive set-top box, their productions clearly pointed the way towards an IPTV world where games consoles are also set-top boxes (something that is happening with Microsoft's Xbox, and also Sony's PlayStation).

The productions of NM2 used different engines for the system, but the idea was the same: a characterization of the media clips in metadata made it possible to take the user's interactions (triggered by traditional interactive graphics) and select the clip which branched the story in the selected direction. Here is one weakness of interactive media: the broadcast makes it difficult for an individual user to select a favorite direction. The broadcast, although stitched together from different clips, has to appear the same to all users. Hence, it is likely that shows such as those produced by NM2 will work better as video on demand.

The work of the project did not only include the productions, which are interesting enough (and hard to show in this medium, so anyone interested will have to look up their website), they were also able to draw a set of conclusions from existing media and business models, which are worth quoting:

1. Interactive audiovisual formats are not provided as a standalone service, but are added to television programs or offer new ways of exploring and using broadcasters' audiovisual archives.

2. Most of the business models in the cases analyzed depend on strengthening a particular brand and generating audiences and buyers for other, related services. They are not designed to be profitable in themselves.

3. Business models for interactive content are still developing. Although the case studies are successful in terms of numbers and popularity, proper and

fully developed business models are in most cases still lacking. Exceptions were those where existing payment systems could be leveraged; or the brand and model of the broadcaster could be used (e.g., for selling subscriptions).

4. Interactive, nonlinear audiovisual formats need to be highly modular in order to be able to refresh content regularly, and thereby create customer loyalty.

5. Notwithstanding increasing possibilities to distribute content through decentralized and peer-to-peer (P2P) networks, the nonlinear audiovisual formats in the case studies are all centrally operated to prevent misuse, copyright infringements and guarantee a certain quality of service.

Common bottlenecks and dilemmas were:

1. Scarcity of attention: a scarcity in distribution channels has been replaced by scarcity in attention. Companies offering online digital content need to invest in cross-media promotion, search engine marketing and creating a strong brand in order to attract sufficiently large audiences.

2. Copyrights issues remain an obstacle to exploitation of content in new ways. It is time consuming to clear copyrights on archive material and for newly produced material it is difficult to agree on exploitation contracts between right owners on the one hand and distributors and packagers on the other.

3. Digital and online media offer potentially interesting ways of tracking and registering user behavior, thereby enabling new forms of targeted advertising. However, reliable, standardized audience measurement methods upon which all stakeholders agree are still lacking and interactive and targeted advertising are not yet used to their full potential.

4. Fear of piracy and format copying leads to centralized concepts of distribution and DRM. This might not always be the most efficient way of handling and distributing content.

5. Public broadcasters find themselves in a contradictory position. Offering access to publicly funded material in broadcasters' archives and offering their viewers access to programs and related services on digital platforms can be considered as a key part of public broadcasters' remit. On the other hand, offering (free) access to public broadcasters' audiovisual archives might and public broadcasters' expansion on the Internet and in digital domains might be subject to accusations of unfair competition.

6. Public broadcasters generally want to uphold certain standards of quality and objectivity. This prevents them from fully embracing user-generated content as a means to expand and open up their offer to contributions of users.

7. In order to reach audiences public broadcasters need to cooperate with owners of distribution channels and portals. However, public broadcasters are limited in collaborating with commercial partners as they are usually not allowed to directly contribute to profit making of third parties. This makes it more complicated to enter public–private partnerships. More importantly, cooperation between public service broadcasters (PSBs) and commercial partners can be politically controversial and clash with PSBs' professional culture. The main issue to be solved seems to be how to maintain the public broadcasters' integrity and independence, when offering content in the context of commercial services. Also revenue sharing and customer ownership have to be negotiated between PSBs and commercial partners such as network operators.

All the work of the project is, as usual with EU projects, documented on its website.

Project LIVE: Interactive Sports Events

The nonlinear storytelling with which the NM2 project experimented can only work for content that does not follow a timeline. However, there are many types of content being broadcast on television that are forced into a certain frame by the nature of the event they are presenting, for example, sports. Watching a marathon race backwards can perhaps be amusing, but it is viewing an event from start to finish which brings value – especially in sports where the timeline does not force the action in the same way as a race, such as football or ice hockey.

To understand what it would mean for a broadcaster to show a story from several different timelines at the same time, the LIVE project (with participants mostly from European research institutes) designed a system that was able to adapt the storytelling around a sports event to user feedback, and also include other stories which might unfold alongside the main story. See Figure 1-5.

The difference from other interactive television systems (such as that of NM2, or the ability to select camera angles in Formula One) was that in addition to the multiple live audiovisual streams, there was an ability to create transition points, where the consumer could be invited to switch to another subchannel. Zapping between different viewpoints became equivalent to navigation through the event, getting multiple points of view.

There are three keys to making the LIVE system work:

- Annotations of the sports event (which have to be created in real time, since there cannot be any delays in a live broadcast).

- The use of the annotations to select content from the different streams, based on the user's interactions.

- The "staging" of the system – the creation of the transition points and multichannel coordination, which makes it possible to handle the shifting between different channels.

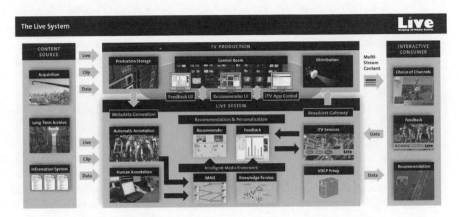

Figure 1-5. The LIVE technical system. Reproduced by Permission of © 2008 Live Consortium, http://www.ist-live.org.

To create the broadcasts, the project designed a console that could be used in combination with a broadcaster's production system. Using the console would not be too different from the way that current production systems are used.

During the broadcasts, four different types of feeds are automatically analyzed and processed into the system:

- live event multi-stream video feeds;

- related event database services;

- relevant archived material (clips); and

- consumer feedback data.

Part of the analysis of the live event feeds is the metadata extraction and human annotation components – where a producer or director can take the material and put annotations on it, for example, describing the event. But automated analysis is also an important component – if the system can identify one runner, then the relevant data for that runner (including previous races) can be attached to that runner. Metadata in the system was handled through an "Intelligent Media Framework" (IMF). Its role in the production process was to accept and handle partial information about particular media items (derived through the application of the automatic and manual annotation), to add semantic information to the items and to infer and attach contextual knowledge to the items probably related to the staged event. It also provided knowledge services that offered controlled vocabularies related to the current context of a stream, to guarantee the unambiguousness of the terms used.

The IMF also had an important role in the generation of the user experience of the multichannel viewing. It included a messaging system that enabled the real-time aspect of the staging process, by receiving triggers from internal and external metadata generators, which mainly include the metadata generation system (automatic annotators and the human annotator), as well as accessible external information systems (e.g. providing event and timing information of sport events). The IMF was responsible for aligning all these triggers to the already available knowledge of the event and to propagate the resulting messages by using the Action Message Queue to the other components of the LIVE production support system.

If you are interested in marathons, half-marathons may also be of interest to you. That was the basis for the recommender system built into the LIVE system, and which could determine from the user's choices and the metadata what media streams and events may be interesting next; in reality, the selection was primarily done from annotated audiovisual material from the television archives. The goal of the Content Recommender System in the production of the multi-channel television program was to provide automatic selection of suitable content from the pool of available live or archive content. The content selection procedure primarily focused on the selection of semantically annotated audiovisual materials from the television archives according to the preferences of the target audience. The receiver of the recommendations was not the end-users, however, but the director of the broadcast. By receiving content recommendations for each channel in the form of a list of audiovisual segments, he could review the material and decide if it was suitable to be included. The directing process meant that the director was able to instantly include recommended audiovisual segments. The result of this process is a television program composed of several channels. The resulting television streams were sent to the television viewers. The user could be guided through the event by a number of interactive television applications, which also managed the user feedback.

Multiple streams are hard enough to understand for an average user, but understanding when it is a good idea to switch between them is more difficult. When there are multiple channels, the channels, their content, and interrelations need to be defined simultaneously. This may mean different channels have different topics (e.g., you can choose the home or away team angle in a football game; or watch it from the point of view of the referee). In the LIVE system, this was referred to as the staging process, and it was used to assign a profile to each channel, which the recommender system could use to find a suitable set of recommended content. The actual composition of the channels, and the points where a user should be able to transition between them, was not automatically created. This role, often taken by the producer in traditional productions, was given to a "video conductor", who mostly resembles a video jockey, a disc jockey working with video. Whether that can be automated, or whether it is a skill which requires human intervention, remains to be seen.

As with all EU projects, there is much more material available on the project website.

Me on TV: Five Minutes of Fame for Everyone with a Mobile Phone

Andy Warhol famously quipped that "in the future, everyone will have 15 minutes of fame". If fame is the same as being on television, he has already been proven right, and in spades. What he did not foresee was that you could make yourself famous, by putting your face on television.

So the experience is there. The question is: How do you go about designing in the user into an interactive television show? Dutch content producer Endemol (of *Big Brother* fame) made sure the viewer could participate in the *Big Brother* finals – by using his mobile phone to call into the television show, and then be seen on the television screen (superimposed on a green or blue-screen, a surface where a picture can be projected). This was "productified" by telecommunications company Ericsson, enabling "citizen journalism", where everyone with a mobile phone can be a reporter (or a paparazzi) – not just for photos, but also for video. See Figure 1-6.

Figure 1-6. The Ericsson Me on TV system. Reproduced by Permission of © 2008 Ericsson.

Getting an additional video stream from a 3G mobile phone is not high-tech today, and video telephony was built into the standards from the start. The problem is how to manage many users calling in at the same time, and how to display the phone call on the screen. There also has to be an application on the screen, which helps the user to create the content. For example, tagging of content by the user is also possible, even while recording, so the editor can know that something interesting is going on.

Using a mobile phone during a video call, the picture size becomes 176 xy 120 pixels, and is suitable to be used for a "Picture in Picture" display (the more advanced H.264 standard, essentially a mobile version of MPEG-4, is not yet available in phones). This means displaying the user on a green or blue prop surface in the studio (or splicing the picture directly into the video stream). The video stream from the mobile video gateway is no different from any other video stream. The user calls into a gateway, which forwards the call to a server, where it is tagged and managed; a management tool interfaces to that, and makes it possible for the producer to ensure that the right video stream gets on the screen at the right time.

When using "Me on TV" the user experience is no different from other programs – the only difference is that part of the content comes from other users. However, it comes through the mediation of the producer, who is still in charge of the user experience. It can be viewed on any television set, but to interact with the content, there needs to be some additional support in the home terminal.

Chapter 2: IPTV Standards and Solutions

IPTV changes the user experience, but if one user gets the program guide when pressing the red button, and another gets a betting site, then it is like different car makers being able to decide for themselves whether the brakes should be a pedal on the right or the left, or even a lever. Actually, it is worse: some car makers could decide to drive on the left, some on the right; and some road builders could decide that only cars of certain makes should drive on their roads.

That is not an open market, and open markets make the industry grow faster – it did not take the Internet to prove that. The Internet is the prime example of an open marketplace, and to work together, open systems need standards – whether *de jure*, agreed in a formal standards body and made into law, or *de facto*, agreed among the major industry players, does not matter (even if some claim to care – but then often follow their own *de facto* standards anyway).

Why IPTV? Interactivity, Technologies and Services   Johan Hjelm
© 2008 Johan Hjelm

Standardization of IPTV

If all broadcasters used their own encoding of the video, would-be viewers would have to buy different television sets when they wanted to see a different television channel. Now, the differentiation is on a continent level, the Europeans use PAL, Americans NTSC and the French SECAM – for analog television. It has improved a little with digital terrestrial television, but it is still not true that broadcast television is standardized. Telephony, the Internet and a host of other services are completely standard. Without standardization, the services could not exist, and there would not be a media industry.

Standards have another great advantage: you can disconnect one piece, and if there is a standardized interface, hook in another – probably saving money in the process. Unfortunately, in the case of IPTV, there are too many standards. The basic technology is the same in all of them, but since there are several ways of putting that technology together, the resulting standards are different. In an ideal world, there would only be one standard, and it would be easy to build systems that worked together based on what is specified. In the real world, there are several competing standards, and building working systems is difficult.

The IPTV industry is trying to standardize itself. Old, traditional forums such as the ITU-T are mixing with new forums such as the Open IPTV Forum. They all have different scopes and slightly different takes on things, but together they at least form a finite set of standards – so any service provider, broadcaster and operator looking for IPTV services can select the appropriate ones. Some are so entrenched that they are more or less default (and not mentioned in Table 2-1 below), since they have been around for a few years and everyone else is using them – XML and MPEG are the best examples of such foundation technologies. SIP, HTTP and RTSP are three other examples of foundation technologies, which the Internet Engineering Task Force (IETF) has standardized, and which have hardly changed since they came about more than 10 years ago.

Standardization is important in the telecommunications industry, and even more so in the Internet – although the IETF, the main standards body for the Internet, has a sometimes just claim that the telecom industry overspecifies and underimplements. That has changed over the years, and ETSI now arranges "plugfests" to make sure that things are interoperable by allowing vendors to test their products together. The specifications are written in terms of "reference points", the points where a node connects to some other node over a network. These would be called "interfaces" if they just referred to a single node; as it is, they essentially describe what happens when two interfaces meet, and what they should do together. It is arguable what the precise distinction between the two is; still, reference points are used in the telecom industry. The "nodes" were, when the telecom industry started making specifications, physical entities such as switches and telephones, but software

Name	Focus	Type of organization	Primary industry
Open IPTV Forum	End-to-end IPTV service, including interaction and quality of service	Industry consortium	Telecom
ITU-T	IPTV Focus Group	Formal standards organization	Telecom
ETSI TISPAN	IPTV based on IMS, and referencing relevant standards for the transport layer	Formal standards organization	Telecom
ATIS	IPTV for cable-TV providers	Membership organization	Cable-TV
SCTE (Society of Telecommunications Engineers)	Technologies related to digital cable television	Industry association (standardizing through ANSI)	Cable-TV
DVB Forum	IPTV and interactive television, primarily for broadcasters	Industry consortium	Broadcasting

Table 2-1. The standards organizations driving IPTV.

has changed all that, and the nodes in a telecom network nowadays are functional entities, which do not have to be put in a specific hardware, and can be co-deployed with each other in various configurations. This is a different way of thinking for telecom companies; but it is even more different for consumer electronics companies, who find themselves in a world where the television sets and stereos they are selling do not have to have the same configuration when they are used as when they were sold.

However, just like in the consumer electronics industry, there are too many options, and to work together to provide services, an implementer of a system needs help in selecting which options are useful, and which should not be used. The mandatory set of services provide the basic functions – just like a car, which can be customized with a souped-up engine, aluminum rims and leather upholstery, it still only needs four wheels, an engine, steering wheel, controls, a body, and some other things to work. So there is a need to de-select the options. That is a role some standards bodies have taken onto themselves, in addition to certifying the devices that conform to the profile of the specifications they have created. The most famous among those is probably the Digital Living Network Alliance (DLNA), which takes the UPnP standard (and some others), selects the options and how they should be used, and has a certification program to demonstrate that the device works seamlessly in the home with other devices.

The Open IPTV Forum set out to restrict the options in the end-to-end chain. So far, it has published a set of specifications which describe the architecture at a high level. It does not completely map to the main standard for IPTV with IMS, which comes from the Telecommunications and Interent converged Services and Protocols for Advanced Networking (TISPAN) technical committee of ETSI, but comparing the two makes it possible to see what is required to build an interactive IPTV system.

In addition, there are a number of different standardization bodies – everything from industry consortia to *de jure* standards organizations – that work on different aspects of IPTV standardization. Some of them leverage the work of others – and everyone uses the standards from the IETF, which has standardized the Internet.

It is possible to do things differently, and still build along a set of *de facto* standards. The most famous example of a working IPTV solution that follows standards in this way is probably Microsoft, which has built a solution which follows standards – except for some crucial parts, such as the identity and profile management, which are proprietary. There is a big risk in creating a *de facto* standard in this way, since it assumes that there will always be a good relationship between the vendor and the IPTV company. But if something goes wrong, say the technology or even worse the business relation changes, and the relationship breaks down, not having well-defined interfaces can be as messy as any high-profile divorce. Clearly defining the responsibilities between components is an easy way to make sure that the system works, whatever happens to a vendor – and also, as most telecom operators know, a supreme way of being able to choose according to requirements. Within the standard, there is plenty of room for innovation, in terms of creating a faster system, or being more robust, or with additional functions. But if an operator decides to take a chance on a small company, and that company goes belly-up, it is still possible to put in a server from some other company – and it will still work.

Standards are especially important when talking about such a complex system as the Internet. And as we have seen – and will see – the IPTV system is large and complex, and requires many standards for the user's vote to be securely recorded when pressing the red button on the remote control. Hence, some of the standardization bodies, which determine the future of the industry, have become involved. The main standards body for IPTV has emerged to be ETSI, through the branch that develops standards for the "next generation networks", which many telecom operators are deploying. The ETSI IPTV standard, the first coherent example of a standardization of IPTV, was published in March 2008 (as Technical Specifications 181 014, 182 016, 181 027, and 181 028). In this chapter, we will scratch the surface of what that standard means – and in the following chapters, delve a bit deeper on what the different parts are and do, and what is missing or could be done better in the standard.

The Open ITPV Forum Architecture

The Open IPTV Forum was set up by a group of companies in 2006 to create an architecture for the future IPTV services. Having published an architecture document in November 2007, it fell seriously behind its original timetable of publishing the full set of specifications in December 2007, probably due to the influx of new members who also wanted their say in how things should work. The published document, however, is different from most other standards in that it considers the entire chain of IPTV service delivery, from the end-user to the service provider, and the network in between (although the network is half-hearted, and anyone who wants to build an IPTV network should probably follow the ETSI standard).

There are five domains in the architecture view of the Open IPTV Forum:

- the consumer domain;

- the network provider domain;

- the platform provider domain;

- the IPTV service provider domain;

- the content provider domain.

Sometimes these will overlap, in particular, the content provider, IPTV service provider and the platform provider will occasionally be the same. However, it is also perfectly possible (although hardly foreseen in the standard) to imagine the consumer as a content provider – this is already true on the Internet, where YouTube and similar systems dominate.

In the view of the Open IPTV Forum, the consumer domain is where the IPTV service is consumed. It can be one single terminal, or a network (i.e., a home or office). The device may be a mobile phone, in which case the network of the network provider has to be a mobile network, with all the consequences that this brings. The normal assumption is that the network provider is a company such as a telecom company or a cable company, with a physical connection to the end-user's premises and the systems, which enable it to deliver IP data packets over that connection. Most of the functions in this domain are standardized in other places, of course – and the Open IPTV Forum seems to shirk away from the hard questions about how to manage quality of service, which ETSI has taken on board.

If the network is a "managed" network, in which the service provider can control the signaling to ensure that everyone gets the television signal at the same time, it is likely that there has to be a close relation between the network provider role and that of the platform provider. The platform provider in the IPTV system typically has the biggest role in providing the identity management and security – and also the charging; and in extreme cases, the hosting of the service for the IPTV service provider.

The IPTV service provider is where the actual IPTV service is provided – the data is streamed, and the control is managed (like interactivity). However, the IPTV service provider can also be a middleman between the content provider and the other actors. The content provider owns the content and manages licenses. As we saw in Chapter 1, this is a role that has multiple faces, combining the aggregator, the content producer and the rights holder. The Open IPTV Forum has attempted to standardize some functions for content providers – such as DRM – but its main work has been on the "middle" roles – the platform provider and the IPTV service provider, and how they should relate to the devices the consumer uses to watch and interact with the IPTV service.

The Open IPTV Forum architecture (Figure 2-1) is designed to accommodate both "managed" and "unmanaged" services, where the "managed" services are those taking place in a network controlled by IMS, and the "unmanaged" those taking place in the ordinary Internet. The quotation marks are the author's, since it is not true that the Internet is unmanaged – Internet service providers do a lot more to ensure the quality of the network than regular users may appreciate. Both the IMS and the Internet, however, connect to a residential network – the network in the user's home. To make it work, there has to be gateway functionality which enables the "OITF" – the Open

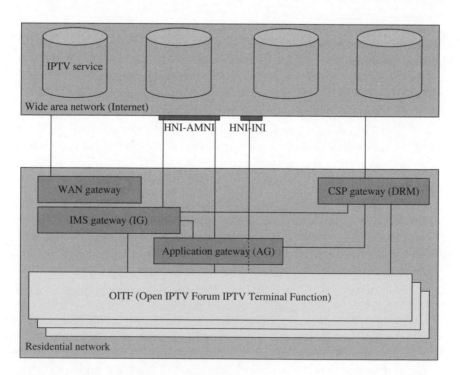

Figure 2-1. Open IPTV Forum architecture.

IPTV Forum Terminal Function that enables the user to consume the service – to interface with the IMS control system of the network service provider, and that manages both the network connectivity (by controlling the router in the home) and the identity and profile management that IMS also provides. This is the "IG", the IMS Gateway Function, which interfaces the terminals with IMS. The OITF does not have to be an IMS terminal; it can work through the gateway.

The gateway can have a number of other functions (although these are optional) in the "Application Gateway" (AG) function. These include functions to connect terminals that follow the DLNA standard, which was designed to be used for consumption of media in the home, leveraging the Universal Plug and Play (UPnP) standard. Other optional gateway functions enable the gateway to interact with the media streams, to manipulate the content guides, and to provide "remote user interface" functionality – essentially setting up a TCP connection through the home network, and then connecting the terminal to the gateway, and sending display primitives to it. This includes a declarative programming language, and since this is based on the CEA-2014-A standard, it is ECMAScript, perhaps better known as JavaScript. The idea from the Open IPTV Forum is that should be possible to download these functions into the gateway, although how this is supposed to be done is not defined. There are also functions to handle DRM, and the actual connectivity, in the WAN Gateway.

The OITF has functions to manage user profiles (using the OMA XDMS standard), manage sessions and streams (which can be done using the UPnP AV Control functions), and to receive streams and buffer them if required. It also contains the codecs (decoders) needed to display the streams, and functions to handle multicast, which is the way media is distributed when it is not on demand.

One of the trickier aspects is how to discover the service. When connecting to the network, it is not like the analog television system, where just scanning a part of the radio spectrum would give you a number of signals that could be mapped to channels; or even like the digital spectrum, where the multiplex that is broadcast provides the information about the channels, and a structure for how to receive them. On the Internet, there can potentially be an infinite number of channels, all broadcast in real time. Discovering what is available is a matter both of network availability and of service availability – that is, the service availability does not only depend on the network (in which case everything could theoretically be available always over the Internet) – however, the main issue is commercial. There has to be an agreement between the service provider and the content provider to provide the service, and the network provider has to have the right agreement with the service provider to be able to provide the service. Since this is already a well-established industry (that we talked about in Chapter 1) there is no need to go deeper into it here, except to note its consequence: discovering a service means discovering both how it is technically available, and what it contains – for this particular user.

In the Open IPTV Forum system, the terminal function first discovers what network interfaces there are – in the managed model, this is tantamount to discovering the IG. Having discovered this, it can use the IMS system to query the network provider about which services are available in the network. This can be done using the standard IMS protocols, and adding metadata such as the DVB-IP Service Provider(s) Discovery Record, which in addition to the network address includes the names of IPTV service provider(s) and related attributes (e.g. a logo image of the IPTV service provider, the means to retrieve IPTV Service Discovery information, etc.). This information will be used by the OITF to perform IPTV service provider selection.

Then the user – or the terminal, automatically, depending on the subscriptions it has registered – connects to the service provider. The service can be video on demand, personal video recorder, or a broadcast (linear) television service, which is the one that is most difficult, and hence most interesting to look at how it works. The service then provides the metadata about what is available.

There are two assumptions in this: that the user is authenticated and authorized (i.e. that the user's identity is verified); and that the user identity has the rights to access the service. If this is true, the IMS system can help identify which services are available using a Public Service Identifier, which is a URI where the service (as well as the discovery service) can be located. This can be provisioned, i.e., provided from the network, by the network provider.

The authentication system is based on the identity management, and this is one of the major differences between the "managed" and "unmanaged" models. In the managed model, the authentication can be done using IMS, and the identity management by using the IMS management system; from this follows the profile management, and from the profile management the functions of the recommender system which provides users with lists of their potential favorites. One area where IMS shines is the security functions; as they are transparent once a user has been authorized, however, they will only be discussed briefly in this book. IMS provides a number of key management systems, and there is a single sign-on function (the 3GPP Generic Bootstrapping Architecture) which can be used to enable the user to sign on once, and then let the network handle the authentication and authorization once it is done the first time.

The consumer domain will always need to be connected to a service provider to receive the IPTV service; this means there has to be a service provider, and a network provider to connect the two. In the managed model, the network is managed end to end, which means that the network operator guarantees that there will be no degradation of the signal between the service provider and the consumer. This requires a number of restrictions on the service to be delivered. On the other hand, in the unmanaged case, the service provider has no way of knowing – and no guarantees – that the user will receive the service as he had designed it. There may be packet losses, the signal may be degraded to a lower quality due to network constraints – all of these may happen – and there is no way of control.

The assumption is that in a managed network, the same actor will deliver the service, the network and the content. This is not how things work today, but the specifications of how to get the authentication information from one managed network operator to another are still kind of sketchy. Since this is based on IMS, it should work transparently for a service provider when delivering service to a customer in a different network – in an ideal world. However, today the assumption has to be that service provider and consumer are in the same IMS network. This can easily be solved by bridging two IMS networks using an application server which sits in both networks – but how to do that is out of the scope of the Open ITPV Forum specifications (and this book).

In addition to the linear television delivery over multicast, the OITF architecture also covers "content on demand", probably better known as video on demand. Here, a user selects an item from a catalog and it is delivered to the user, that is, there is a one-on-one relationship between the user and the service provider. This is how movie catalogs are provided over the network, and by creating his own videos (through recording the program, either by scheduling it or pressing the record button), the user gets a personal video recorder (PVR) in the network (known as an nPVR). Using a PVR is also known as "timeshifting", and this is changing the viewing culture for television in many countries. From a technical perspective, having a PVR, which is accessible from the network – either provided as a service by the service provider, or by the user himself, accessing his home PVR through the network – is no different from VoD.

What is different, also in the Open ITPV Forum specifications, is the management of "linear television", or scheduled broadcasts, or a number of other pseudonyms for traditional television. Delivering television services over the Internet is not particularly different from any other streaming, except for one major detail: everyone has to receive the streams at the same time. This is crucial in the case where there is a result that rides on the reception, such as betting or voting on a television show (horse racing, or even stock trading, would not work if some people received the information about who won the race a couple of minutes before the rest). So there has to be some real-time guarantees for interactive television services to work – and this is where IMS comes in.

In the Open IPTV Forum standards, the idea is to provide the linear television through multicast, a technique where the stream is sent to a number of receivers at the same time. And with the managed network functionality, there can be guarantees for the simultaneous delivery of the stream. It is possible to zap between channels, just like in traditional television, since the terminal will join the multicast group associated with the channel and start receiving the stream when the user presses the button – there is no need to re-authenticate the user. However, a user is allowed to join a multicast group only if there is enough bandwidth with the right service priority to handle the requested stream within the access network. Otherwise the service can result in a bad user experience and poor picture quality. And during channel zapping, interaction or handshake between the network entities related to bandwidth, service priority or admission control have to be optimized. This saves precious time and contributes to a faster channel zapping speed, which is important – since a delay in switching channels will result in user irritation, and hence a less pleasant user experience. Even in terrestrial digital broadcast, this can be too slow and result in the user zapping by the channel he really wanted.

The ETSI IPTV Standard

The ETSI and Open IPTV Forum standards complement each other; where OITF defines how the consumer electronics equipment should interact with the service, and does not delve too deeply into the network aspects, that is exactly what TISPAN defines, leaving the "user equipment" a mere footnote at the end of the service delivery chain.

Published in March 2008, the IPTV standard, TISPAN, is more oriented towards the network, and less towards the consumer electronics side of the network than the Open IPTV Forum architecture – this is as to be expected, considering that the consumer electronics companies have a much higher weight in the Open IPTV Forum. It focuses on describing how the network functions will relate to IPTV services.

The standard, however, tries to paint a complete picture of the system – in a slightly different way from the Open IPTV Forum. By focusing on the service control and media delivery – the "service layer", and the network control – the transport functions – it also reaches a second goal: making it clear how to integrate the standards-based systems into the Next Generation Networks (NGNs), which are also being standardized by TISPAN. This is a major project, which focuses on providing a complete architecture for fixed networks, including everything from the physical cabling to the control system.

This means that the system covers a number of areas – including authentication and authorization (two key issues), user profile management, content protection, exchange of capabilities information between the server and the client, resource management, management of policies in the network, and charging. However, it is not enough to separate the system into applications, transport and services. The TISPAN committee separates the service layer further into those functions that are used to provision the system, that is the functions used to decide which services an individual user can receive (including such functions as parental control rules, which determine what a user is not allowed to watch); and those which are used to show the actual service, and to select it. These then connect to the NGN using the standard interfaces to provide the policy control of the network, the resource reservation, and other functions associated with the transport. This is demonstrated in the ETSI TISPAN reference architecture, which is shown in Figure 2-2.

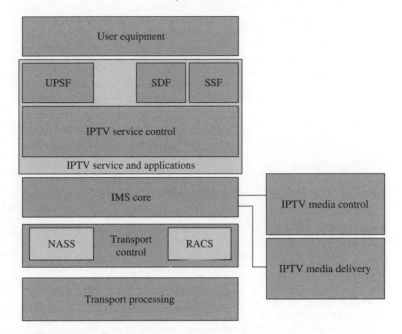

Figure 2-2. How IPTV is included in TISPAN's NGN architecture.

Three types of services are supported by the TISPAN IPTV architecture – plus electronic service guides (ESGs) to help the user navigate. They are Content on Demand (CoD), Broadcast, and networked Personal Video Recorder (nPVR). CoD is perhaps better known as video on demand (VoD), but the generalization means any content can be handled this way, as long as it can be distributed using the mechanisms that TISPAN provides. The important thing is the streaming; there is no direct method of providing downloaded content in this architecture. Timeshifting of content – recording something now for viewing later – is included, however, through the use of the nPVR function. It can also be pushed directly from the service provider to the user's nPVR; in this case, the standard prescribes that the expiration time should be longer than if the user himself had recorded the program. Access to timeshifted content is done through the same mechanisms as the CoD system. However, since CoD supports "trick play" – starting, stopping, rewinding and fast-forwarding – the use of nPVR in combination with Broadcast makes it possible at least to rewind a real-time broadcast, for example to see what the host just said (it is, even for such a complex system as IPTV, difficult to fast-forward in a real-time broadcast).

The Broadcast services handle the live streaming, and work in a very similar way to CoD, except of course for being live and real time. This part of the system is – like the functions providing CoD – split into a service control part and a media delivery function. The control function is the part which talks to the core IMS system, to manage the network, get identity information, and so on. The media goes directly from the media delivery function to the user equipment, in keeping with the layering of IMS to separate signaling (the control of an application) from media (the actual delivery of that application). The same is true for the nPVR functions.

Since the main focus is on those functions that are touched by the NGN, there is hardly anything in the specification about the content provider functions. The user equipment is only briefly mentioned; it is supposed to handle the control and media signals for the IPTV services, and display the Electronic Service Guide (ESG) information. The ESG, in the definition of TISPAN, is an Electronic Program Guide (EPG) plus interactivity data; the EPG contains the navigation information for the user about the program.

EPGs are built from metadata and formatting information. The metadata in IPTV can come from many sources: users, service providers (network or IPTV), or the content provider – or a number of other parties. Metadata is usually difficult for humans to read (unless they are skilled at programming computers); it is intended to be read by a machine, and interpreted in a way that humans can easily access, for example, through formatting the information. This can include filtering and reformatting the metadata based on preferences in the user's profile. User interaction with the EPG is usually limited to forwarding to the next page and selecting a link in a table (e.g., by clicking on the symbol or word which embeds the link); this navigation is heavily inspired by the web model today. There are a number of possible metadata standards that can be used in building EPGs, none of them is better than any other and they are all likely to be supported by the architecture, since they have the same aims and are built on the same base technology (XML). There are several ways in which an EPG can be sent to the user equipment – it can be pushed by the service, so it is always fresh; or it can be pulled by the service, so the fetching is under the control of the user; or a combination of the two.

TISPAN also differentiates between an EPG and the ESG, that is, the user interacts with the service through the ESG. This model may work, but it is certainly not how systems are implemented today; it does, however, leverage the separation of the content provider (which provides the EPG information) and the service provider (which provides the interactivity that makes the EPG an ESG).

Before the user can use the services, he has to have a session with the control function of the IPTV service provider, and to set up this session (through the terminal he is using) he has to discover the service. IMS provides functions to handle this, as long as there is a well-known address registered to identify the service. In the TISPAN view, the CoD and nPVR services are provided using unicast, and the Broadcast services are provided through multicast – which means that there is a dedicated stream of packets for each nPVR and CoD user, but that users of Broadcast can use the same service, which is sent to a number of points where they can access it at the same time. A consequence of this (since this is a network-oriented standard) is that there has to be resource reservation for nPVR and CoD, which means heavy interaction with the elements in IMS that take care of the network; on the other hand, for multicast, the network provider only has to make sure that there is one channel constantly available. If the network is bandwidth-constrained, this has two consequences: first, there are fewer multicast channels per user, and because the service is pre-decided, then more resources can be allocated to them. Since unicast is unpredictable, at least to some degree (who knows when the user may want to watch television?), the multicast channels have to be given individual resources. To steer users towards the multicast channels makes sense for the network provider, and if there is a close relation with the service provider, is likely to make sure that the broadcast channels are more appetizing than the VoD offering – for example, by offering them in high definition, and with interactivity. VoD is more likely to be standard definition, since it is more costly not only for the network provider, but also for the other viewers sharing the same network.

When the user sets up a session to the service provider's control system, the quality of service is set up at the same time. If the user changes to a channel with a different quality of service (e.g., from standard definition to high definition), this means a new resource allocation has to be made.

When the user equipment (on behalf of the user – because sessions are tied to user identities in IMS) connects to the service provider, there are two sets of functions with which it interacts. The NGN applications include those available to the user for communication, for example, presence and IMS communications services such as multimedia telephony. They also include management – and are all part of the IMS standardized set of applications.

The new applications that TISPAN defines for IPTV are collected under the name "IPTV applications", and are those server-side functions required for IPTV, facing the customer and the rest of the NGN system. They provide the provisioning of the service to the user equipment, the selection of the IPTV service, and authorization of the user to the IPTV service. In the set of applications are also included those which interact on behalf of IPTV with the NGN, for example, content preparation, media management, content licensing, subscriber management, and user profile management.

Without user profiles there can be no personal television. Therefore, user profile management is one of the most important functions of the TISPAN standard – and the profiles can leverage the IMS standards, which provide a very good toolbox for managing user profiles. Actually, there are several different categories of profile information distinguished in the IPTV standard from TISPAN:

- IMS profiles, which include all the information required to establish the IMS session, and access IPTV services in application servers within the IPTV system.

- Communications services profiles, which are used to handle the IMS communications services, for example, how a VoIP call interfaces with a traditional telephone network. They also include presence, which is the most important service from the perspective of this book.

- IPTV user profiles, which encompass all the information required for the IPTV service, including the user preferences in terms of language and other personal preferences; the settings for the broadcast channel(s); and which broadcast service package a user subscribes to (a broadcast service package is a set of broadcast channels with the same authorization and charging policy, which means they can be accessed in the same way at the same time).

- CoD profile, which includes parental control rules.

- PVR profile, which includes settings for whether the PVR should be local or networked; and whether it has a storage limit, and what that is.

- User equipment settings profile, which includes the information related to the different capabilities of the user equipment (e.g., which codecs it has installed, and size of screen). Since, in IMS, it is quite possible for a user to be logged in from several terminals at the same time, there is a unique user equipment identifier associated with each terminal in this profile.

This is not an exhaustive list; there is still a need to download the capabilities of the terminal when it is to be used, because changes to the information are not covered.

IPTV service action data is a profile that contains information related to actions that the user may have taken while accessing services, for example, pausing a CoD session. The information in this profile includes the list of CoDs that the user has ordered, and their status; which broadcast television services the user has paused or bookmarked; nPVR recording information and status; and other recording information.

This profile is managed by the User Profile Service Function (UPSF) in the TISPAN standard; or by dedicated databases. In practice, this means they are likely to be managed by the XDMS system. How the data is accessed is transparent to the applications, however.

One way of providing data about the user – and in particular, about the user right now – is to use presence. In the TISPAN standard, this is based on the IMS standard for presence. The mechanisms will be described later, but they can be used both by individual users (to see what their friends are doing), and by service providers (to get a picture of what their users are doing). Presence information in the IPTV system includes not only the basic information about the user (what the presence system was originally designed for), but also information about the user's devices, and which services are being used – and how. Literally all services related to IPTV can use presence; it is possible to have presence-enabled EPGs (which are formatted according to what other users are doing), for example. Targeted advertising and other profile-dependent information can also be provided, once presence is available – since it can be used not only to provide a static profile, but also dynamic information of the user's current status.

The IPTV presence attributes – what is described in the service – includes information about the user, such as their ability and willingness to be reached for communication, connection status, location information, channel currently accessed, or acceptable communication means. That information may be used by another user to instruct his set-top box with a single click to switch to the identified channel, or to instruct his set-top box to keep following identified channel changes. Important in presence information is that the user is in control of who receives it – not just anyone can get everyone's presence information, the user can set filters which determine who gets what information.

Providing only profiles and their management will make for poor service, however. There have to be functions which deliver the media, and which control that delivery. In the TISPAN standard, there are two functions to handle this: the IPTV Service Control Function (SCF), and the Media Delivery Function (MDF).

The SCF is responsible for the interaction with other functions in the NGN to make sure that the user is accessing services that he has a right to. This means this function is responsible for the service authorization, which in IMS is done during the setup of the session (and when the session is modified). That includes checking user profiles, but also checking the user's credit and subscriptions in the charging system. It selects the relevant MDF for the user (e.g., depending on whether the user has the right to access VoD or not, he may be allocated to a different media delivery mechanism). The SCF is a SIP Application Server, which means it can access information from the IMS system through the Sh or ISC interfaces; these are interfaces that enable servers to get, for example, profile information and authorization information from the IMS core.

The SCF is also involved (as one of two alternative methods) in service discovery, when the user wants to find the actual IPTV service during the setup of the session. There is also a specialized Service Discovery Function (SDF), which is used together with the Service Selection Function (SSF) to provide the user with service discovery, based on the user's profile information (especially what the subscription entitles him to do).

The actual media delivery, however, happens through the MDF. It is controlled by a different function, the Media Control Function (MCF), but in practice these two would probably be deployed as one system, in a streaming server. Storage is also managed here, as well as the processing of media. Feedback also comes to the IPTV service control. It may also be the location for content protection – and is likely to be the primary interface for content providers. Figure 2-3 shows TISPAN reference architecture for IPTV.

While content providers are not the focus of TISPAN, it would be a poor service that lacked content. So there has to be at least interfaces for content providers to get their information into the system. The content providers are expected to provide not only the actual content, but also metadata about

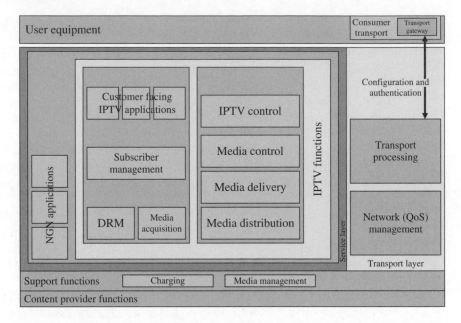

Figure 2-3. The relation between IMS and IPTV.

it, and information about which usage rights pertain to it. In combination with this, TISPAN has interfaces for the content provider's DRM system, but does not go into details of how this should work, other than that the content provider can provide keys for the content encryption functions included in the system.

Content management can roughly be broken down into three parts:

- Content acquisition, which is the process of getting content from the content provider and importing it into a suitable format and location in the service provider's system.

- Content validation and adaptation, which is expected to follow on, and makes sure the content is in the right format to be accessed – something which might be especially applicable when talking about content to be consumed on a different device from which it was intended (e.g., mobiles). However, this process is probably more applicable to metadata, since there are a huge number of different metadata formats, and the metadata is likely to be needed in providing the service, especially when automating the service management.

- Content distribution, which provides the lifecycle information for the actual file, that is, when it should be removed, what the licensing terms are, and so on.

The content management does not have to be done in real time, it is likely (especially when talking about VoD) that it will take place out of band from the actual process of providing content to the end-users. However, for user-provided content, the process may well be reversed, and in that case, the system will have to take into account users interacting directly with the system by providing content (as in the Ericsson Me on TV system mentioned Chapter 1).

For a content provider, it is quite important to know if the user has received the service in the way that he intended. This is covered in the management functions, where service fulfillment, service assurance and service billing are handled. The billing is actually a separate function from the IMS system, interfacing it through the charging system; the service fulfillment and service assurance interact with the SCF, and the SCF with the IMS Core, to provide the service in the way that the service provider has specified.

The IMS Core system is the core of the TISPAN IPTV system (pun intended). It is called the "IMS Core" because it is the system at the center of the IMS control mechanisms; it includes the parts that interact with the network, and also the parts that manage the user identities, including the authentication and authorization of the user's subscription. It interacts with charging and profile management; it really is the core of the system (although the name comes from the analogy with the "core network", which sits between the "access network" and the services). It is here that the flow control is facilitated and managed and that the security is managed; and it is through the IMS Core that the network provider ensures that the users get the appropriate network quality of service, and access to other resources in the system. Since TISPAN is a network-centric standard, this is reflected in how the relations to services are described; TISPAN describes the interfaces that relate to the network in great detail, but these would not be the focus for a server developer; however, the APIs are briefly described at best.

Setup of services in the TISPAN IPTV system is also managed through IMS. The service control function contacts IMS (when it, in turn, has been contacted by the user equipment) and sets up a session to the media control function; this session (in particular in the case of VoD) also involves the reservation of resources in the network.

The transport control functions in the TISPAN IPTV standard include the IMS functions used to manage the network (Resource and Admission Control Subsystem (RACS) and Network Attachment Subsystem (NASS)), and includes policy control for the services, as well as the actual resource reservation (through these functions). The transport processing handles the IP network itself.

Applying Standards to IPTV: An Implementation

Specifying how things ought to work together does not mean that they actually do. If you want to build a working IMS-based IPTV system, you have to set it up. And that is exactly what the German institute Frauenhofer Fokus has done. They have taken the idea of an IMS-based IPTV system, combined it with the IMS communications services (chat, multimedia telephony, presence and push-to-talk) into an architecture, which they then put together in a test bed (the "Open IMS Playground") to demonstrate (Figure 2-4).

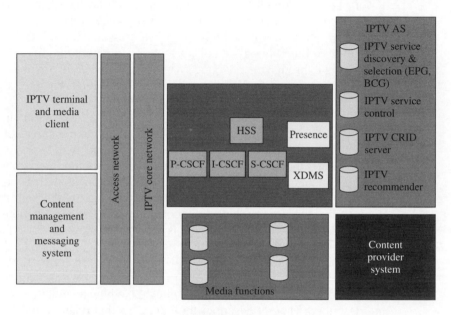

Figure 2-4. The layout of the Fokus test bed.

Working reliably and giving all the advantages that this book is built around, it is not a commercial system – although it is not free Open Source software, either. It was built using pre-standards, since the standards document was published in 2008, and the Frauenhofer test bed has been a going concern since 2006.

One of the major drivers was to test whether "triple play", when communications services and IPTV are provided in the same network, could be fitted into a working system. The main use cases for the system included the ability to redirect calls to the user's mailbox when doing something which does not benefit from interruption – for example, watching a movie. It also included displaying caller notifications on the user screen while watching television when a call came in, or an emergency message. Perhaps less realistic was the case where the user could make an outgoing call from his television set; although technically possible, video telephony has been one of the biggest duds of the telecom industry so far. More of interest to real users was multimedia messaging, where users could send messages to each other based on their presence, and invite each other to view the television programs. Frauenhofer Fokus developed a service called "see what I see", where the user gets a link to the content another user is viewing in a message; and – when including charging into the system – the ability to give away content as a gift can be added.

Discovery of services in today's IPTV system is typically based on the EPG, where the user essentially navigates the received program guide, and selects the service. There is no adaptation to the viewing device – something which is actually rather easy in IMS. Context information such as the capabilities of the user equipment, the location, current network usage and availability can be used to adapt the services in ways that make the consumption more enjoyable. Another side of this coin is the provisioning of services, where the information to be provisioned is triggered over the back channel in the IP network.

Services, as envisaged by Fokus, were run over multicast or unicast (meaning VoD), and the access was secured through the use of the TISPAN NASS and RACS to manage the access network. The user agent in the Fokus system was intended to run in a set-top box, but could potentially run on a home gateway, a home theater system, or even the user's mobile device (all viable IMS clients). The access network must support both multicast and unicast.

The user data – profiles and other user information – was stored in the HSS of the IMS system, but while the HSS data was quite sufficient for communications services, it was not sufficient when providing IPTV. Service and content providers need to get information from the user profile system for charging and personalization purposes. In the IMS standards, it is possible to have multiple public identities tied to one single private identity, for example, where the user is not necessarily a single person but a whole family. These public identities are user configurable to support service personalization. This, of course, impacts not only the identity management, and through this the authentication and authorization, but also services that depend on it, such as charging. The charging in a triple-play IPTV system needs to be done both online (when the user requests a service), and offline (tallying the usage of the user, and sending a bill afterwards).

The session management is not only the core of the IMS standard, it is also the core of the Frauenhofer Fokus implementation. Sessions are translated into threads in the experimental client, enabling network resource allocation, trick play, and simultaneous use of IMS communications services.

The service delivery, on the other hand, was based on an already implemented system (and standardized component), the Media Resource Function (MRF). This is intended to be used primarily for video conferences and other services where floor control is required between several simultaneous video streams. Floor control, is where a user can be given control for some time, for example, play a video (i.e. speak during a video conference) uninterrupted, and then hand back the control to the chair. This is similar to the way channel switching in multicast works, so the step is not very big, although that is not how the standard has ended up. The Fokus implementation, however, allowed Real-time Transport Protocol (RTP) streaming and RTSP-based control of the IPTV channel through a SIP interface.

However, using IMS brings more advantages. Since IMS is based on sessions, and IMS-based IPTV is as well, it is possible to move these sessions to other devices (because the session is anchored in the user, not the device). The user can shift the session from his mobile phone to the television set, or back – the picture moves seamlessly between devices. Ericsson, and others, have demonstrated this a number of times. But the way to do it is limited, as we will look at in the next chapter.

Chapter 3: The Next-Generation Consumer Electronics and Interactive, Personal, IPTV

There are very few homes, at least in the industrialized world, without a television set. Yet these same homes have a number of other audiovisual (AV) devices, and they usually have an Internet connection – at least, that is the premise of this book. These AV devices are connected together, and the services coming into the home can be consumed on a number of them, not just the one attached to the physical connection (as it used to be with cable-TV, and also terrestrial broadcast – the media business-speak for sending television over radio waves). However, there is a lot more to Internet television than the bearer medium – just as television is a lot more than sending pictures over radio waves.

The bigger picture: fitting the home into IPTV

The consumer electronics industry has developed tremendously in the past 20 years – but it has been an incremental improvement. A television set is still a

---

(Continued)

screen with which to receive a broadcast signal, even if it is HD, plasma and digital.

Consumer electronics have competition from the PC (or the Mac). While a home stereo or a television are specialized for what they do, the PC is generic – it can be a stereo, a DVD player and a games console. There is a tradeoff – it becomes good at so many things that it cannot be best at all of them – and the consumer electronics industry has stood up to the competition well so far.

The big threat to the consumer electronics industry is not the PC moving into the home, it is the network connection of their devices. A television set with network access can be used for web browsing (but not very well), it can also be used to interact with television programs. The consumer electronics industry has tried to meet the challenge by creating a set of standards for how a television – or a stereo, or any other device – communicates in the home. The industry has also defined how the devices work over the Internet, and over managed networks.

Apart from Japan, there have so far been very few networked television sets and video recorders – and even in Japan, they follow their own domestic standards. The consumer electronics industry is remarkably slow in implementing its own standards. This leaves the field open for new players, and there is a ready-made niche for them: the set-top box.

In cable-TV systems, set-top boxes were needed to render the digital television signal into the analog television set. Now that television signals are going digital all over the world, the set-top box manufacturers are seeking new niches. Those that have a position to defend are trying to introduce Java as an interaction technology. Those who want a position as a system supplier prefer to include manageability of the set-top box. But those who will be the winners are making small and cheap boxes, which connect to the Internet, overlay interactions and other features (including advertising), and turn the television into an interactive experience, using the very standards the consumer electronics companies have developed, but not deployed.

When services come into the home, instead of into specific devices, this changes the way we can interact with them. The technologies that make this possible are the focus of this chapter, but it should really be the way in which behavior changes when the telephone is no longer the only thing that you can make phone calls with, and the television is not the only thing through which we can watch television. If challenged to think about a second device in the home for these services, most of us would probably answer "PC", but it could equally well be a number of other devices – taking phone calls on your television is possible today, and watching television on the mobile phone a reality in many countries.

IPTV in the home is possible in two very different ways: connected to the IPTV service, and locally. Using the metaphor of the DVD is probably useful: you can watch the DVD at home, and you can have your own library of DVDs; you can also rent DVDs from a store, and bring them home. To start with, we will look at how the different services in the home are connected, and how you can watch IPTV locally (your DVD library). Next, we will take a look at what happens when you open the local media services to IPTV from the Internet or a managed network, which you need to be able to watch events from the outside (bringing home DVDs – and returning them).

When a user switches on the television set to watch IPTV, this does not only mean that the radio receiver is activated and starts receiving at the pre-selected frequency, and then sends the electric signals which result from this to the electron ray gun, which creates three different beams that are directed to the phosphorized glass screen through the mask that creates the pixels. Nor does it mean that the electric signals are sent to a decoder, which interprets the bits and creates instructions about which pixels to switch on and off on an LCD screen. It also means getting an IP address, setting up a session to the IPTV service, and start receiving the bits from the multicast channel selected. This does not necessarily imply that there is more delay in starting IPTV (although it may sound like there are a lot of additional operations that have to take place). Any digital system has to decode the received stream and turn it into a signal that can be displayed, and this is typically what takes time in a digital television system, IPTV included. The setup of the session is a very small part of the delay – although anyone who builds an IPTV system has to plan the location and relative position of servers and services so that the potential latencies are minimized.

When an IPTV set is switched on, there has to be a session. If the Internet was like the airwaves, there could be a broadcast signal which anyone could receive, but since IPTV is different, you have to have an address, which tells you where to get the television signal, irrespective of whether the channels are free-to-air channels or not. There does not necessarily have to be a sub-scription, but there has to be an authorization to use the system, since there is no public Internet service in the same way as there is a public service in radio and television in many countries (where legislation says that the public service channels have to be made available to anyone who wishes to receive them – since there is a license fee or tax, although this may not be associated with the consumption of the service itself).

So first, you have to get the IP address – that is part of the connection setup. Then, you have to get the service, and this is where the IMS registration comes in. IPTV is about more than replacing the wireless broadcast with different types of cables. IPTV is about making the user experience of television as different as when the first color television sets came out.

Home Connectivity: Ethernet, WiFi and Beyond

Home networking has two main challenges: connecting the equipment inside the home; and making it work in combination. Much of the attention has been directed at the first part, just like when PCs were being networked, but the real prize is in making equipment work together. Here, though, the battle may already be won thanks to two standards initiatives: UPnP and DLNA.

To the IPTV service, it does not matter what you use in the home network. However, if you want to leverage the opportunities that IPTV presents, and not simply use it as a cable replacement, it has to be connected to a network, not just a single connection (otherwise, you cannot shift sessions between devices – taking the television service with you to a different screen, for example). So the IPTV system has to be part of a generic network, just like any other connected appliance. The IP network is connected to the Internet, in order to get services into the network, but it also has to be managed so that the service does not degrade because somebody in the next building (or room) starts making a huge file transfer. Networks are a scarce resource, but the scarcity is not one of bandwidth in most cases, it is a scarcity of time. Latency, not bandwidth (unless your connection is slow), is the big problem for IPTV.

Any cabling will not do for IPTV

There is no need to go into the techniques of home networking – there are other books that do this. There are, however, two ways of connecting equipment inside a home: through cabling, or through wireless. Cabling nowadays means category 5e or 6 cable, and the network run on it may be 100 Mbps Ethernet or even 1 Gbps Ethernet. The equipment is within the price range of most home owners. The main work is setting up the cabling, which can be as big a job as installing electricity. For that reason, many prefer an option that is less costly in terms of grunt work: installing a wireless local area network (LAN). Wireless means radio, and the radio network involved here is 802.11. There are several variants of this, some of which are incompatible, but without going too deeply into the details, wireless has one big problem – because there is no physical separation between the radio waves, all signals are distributed to all users at the same time. While this is true for a traditional broadcast system as well, the number of channels is much fewer, and they have two constraints: the user may have paid for the content, and it may be personalized. An additional problem is that if I run a wireless LAN, it may interfere with the wireless LAN that you are running – and even if we share the same wireless network, my receiving television signals may drown out your attempts at getting web pages. Not to speak of the collisions that may happen if we are both trying to receive television over a wireless LAN.

While cabling a home can be a major headache, and wireless carries its own problems, the standard networks do well even for high-definition television. The cabling carries the digital signal over a cable, nowadays to a hub, which distributes it to other connected devices, which pick up the packets of the signal only when it is addressed to them. The speed can be up to 100 Mbps, which is fast enough for HDTV. But while it is fast enough, and definitely the cheapest option, it has a problem: if you watch more than one channel on more than one receiver, you risk blocking each other. And there is a second problem: the connection to the television station may not be as fast as the local network. As a matter of fact, it may be a magnitude slower.

The local network in the home is just that: local. It does not extend outside the house. A broadband connection is required to connect the home network to the television station – or its streaming server, as the case will be over the Internet. The Internet will be connected to the home through a router, which is a computer that transmits the packets over different connections, depending on which addresses they have. The router is connected to the wide-area connection (because the electrical signal of the home system can only go a certain distance), which usually is connected through ADSL, coaxial cable or optical fiber. There are many other books about how this works, but here, we only need to remember that optical fiber is up to 100 times faster than the other techniques.

End-users do not have to worry about the wide-area connection, since this is the concern of the network provider. The end-user only has to worry how to connect to the modem (in the case of cable or ADSL), or the fiber connector, and then the different components work together and the signal magically comes in.

However, a few things happen before the IPTV media stream can start flowing into the television sets in the home. First, there has to be a way to address the IPTV devices. While the addressing was not a problem when there was a physical connection between the set-top box and the head end, it does become a problem in the Internet model. When multiple devices are connected to the same network, there has to be a way to separate them. An additional complication – and one that ranks high on the list of candidates for headaches for IPTV providers – is that due to the way the Internet is built and organized, there are too few addresses for all set-top boxes, refrigerators, microwave ovens, and other equipment that potentially can connect to the Internet. At the heart of the problem sits the NAT – the Network Address Translator – and the way it handles data.

The way that the address problem is usually solved is through having one single address for the entire home towards the Internet, and then address the individual devices in the home with a different address space – something like having one street address for deliveries from the outside, but individual room numbers on the inside. There is an additional twist with the type of addresses used on the Internet: there are no fixed addresses on the inside, instead they are handed out on a first-come, first-served basis.

Any device connected using this technology will have to "re-apply" for the address at specified intervals, and automatically does so when it is shut off and restarted. For the devices, it is no problem (although the technology does create a few problems, in particular since the inside network has to be mapped to the outside). This can be used to increase the security of the home network, however, by not only keeping track of the different addresses in the network, but also what is being sent to and from them, unauthorized transmissions can be blocked – a rough approximation of how a firewall works.

While the address translation and added security can be handled, this means additional processor cycles and memory in the devices, which means extra cost to handle the simultaneous address management, firewall function and IPTV streaming. And for the addressing to work, the appropriate ports have to be opened – which has to be done from the inside of the home network. Making an HTTP request and receiving a progressive download, setting up a SIP session and receiving a Real-time Transport Protocol (RTP) stream within it, or sending a control command to establish the session using RTSP, and then receiving the media using RTP works – but in different ways, and with different implications for the endpoints. These implications will come later in this chapter; the different ways the protocols work will come in a later chapter, because the home network is only one of the endpoints, and there are intermediaries in the network that can be involved as well.

The connectivity issues are equally valid irrespective if the user has a set-top box or connects a DLNA device to the network. The receiver can connect anywhere in the home – there is no physical limitation. The only difference is that with a set-top box, there can only be one connection and one television set. With the DLNA, there can be a large number of them.

Making Home Devices Work Together: UPnP and DLNA

DLNA is often described as a standardization organization. It is, however, something different: it creates a profile of existing standards and makes sure that the devices that implement them work together.

Consumer devices are normally designed to work with little, if any, active configuration by the user. "If it does not work when you take it out of the box, it is broken" is a line which describes the customer perception of how things are supposed to be. Connecting two consumer electronics boxes together is meant to be as painless as sticking in the plug, and switching on the electricity.

In the Internet model, of course, there are multiple other boxes to take into account. Not only does there have to be a way to allocate addresses, there has to be ways of deciding what types of media are delivered where – and to whom. There has to be ways for devices to discover each other, and for them to pass messages to each other (that they want to set up a connection and start receiving data, for example). Instead of doing this considerable work from scratch, the DLNA took two existing standards sets and defined them even further than the original specifications (mainly by deciding which of the multiple possible options should be used).

When a standard is created, it is a compromise between different interests. In the environments of some of the actors – most often companies, but also universities and industry organizations – one way will work better for some than others. Implementations that already exist are different, as is the need for connections to existing equipment and services. Also, someone may have a patent, which potentially could block a way of doing things, and others may not want to license it and so create a workaround. Not everyone can agree on the way that things should be done, and while this may lead to long and drawn-out blockages in standards development, it can also be resolved by compromise. This means creating different options. Options can be simple, but have wide-ranging consequences; or they can be complex, but in practice make little difference – or anything in between. The only certain thing about options in a standard is that the more there are of them, the less likely the standard is to work – as a standard.

Creating a document that describes how things ought to work is only the first step in the standards process, even though it is frequently the last as well. In the days before the Internet, it was sufficient to do so, since the document also included a set of test cases, which a testing house would charge dearly to use to validate the implementations. When the Internet standards started to come about, one of the major differences was the emphasis of showing instead of telling, and interoperability between different implementations as a prerequisite of approval.

It is of course possible to implement all possible options in a standard, but this would be costly and it would require a mechanism to auto-select between the options, something which would make the implementation slower. It is easier to deselect a few options and streamline things – especially if you have a well-scoped purpose for the work, which is exactly how the DLNA has approached UPnP.

What is UPnP?

There are many different standards for connecting home devices, but they often start with the premise that devices should be connected using the same physical infrastructure (e.g., Firewire). The only one that is independent of the physical interconnection, and hence maps well to the Internet model, is the Universal Plug and Play (UPnP)Consortium. Originally this was created by Microsoft and Intel as a way to connect peripheral devices to PCs, however, in the process they solved many of the generic problems in connecting devices together in the home, and the standard has seen wide adoption outside its original target area.

UPnP is based on three premises: the use of the User Datagram Protocol (UDP) in the home network; that the devices are peers (all are clients and servers at the same time); and that some devices can be control points, which means that they order other devices to perform action on their behalf – for example, rendering a video from a content server. In each device, there can be a number of services. They are themselves resources, and can be discovered by the UPnP discovery mechanism.

The control point is one of the most useful features in UPnP, and at the same time one of the hardest to leverage when developing services – it is more than a remote control, but less than a master in a master–slave relationship. The control point controls the operations of one or more UPnP devices.

The interactions are independent between devices, so the control point can control different devices at the same time, but the logic that might make them act in concert has to reside in the control point. The control point is logically independent from the devices it controls, but in practice it can be built into them (the analogy is that you can use the remote control as well as the buttons on the television to control it).

UPnP is based on UDP and HTTP, which means that any device in a UPnP network must have an IP address. These can be fixed, but the assumption is that they would not be. UPnP uses the Dynamic Host Configuration Protocol (DHCP) to configure the IP addresses of the network. Each UPnP device must have a DHCP client, and the client automatically looks for a server when it connects to the network. It is then assigned an address from that server. Devices can also assign themselves addresses, but this works less well, since there is a potential of collisions. The standard HTTP uses TCP/IP, which means the transmission of packets is guaranteed; note that there is no such guarantee in UPnP (so discovery messages, e.g., are assumed to be sent several times). This also means that UPnP can add a control layer for quality of service (QoS), which means that the data packets can be prioritized differently. That in turn makes it possible to ensure there are no interruptions to streams, which might degrade the user experience.

All UPnP devices already have an address, the Universally Unique Identifier (UUID), which is a unique number (assigned by the vendor). This can also be used in addressing the devices, for example, when requesting a detailed description of it.

When it has an address, the device needs to discover what there is to network with, apart from the DHCP server, so the device multicasts a discovery message advertising itself in terms of its capabilities (the embedded services and other features it has). This is done to a standard multicast address, which control points listen to. It does this by advertising itself using the UPnP discovery protocol, which is based on the Simple Service Discovery Protocol (SSDP), with additions from HTTP and the General Event Notification Architecture (GENA) Protocol, defined by UPnP (actually an Internet draft in the IETF, long since expired). The headers in the message are borrowed from HTTP (using cache-control for the duration, for example), but there are also additions.

In this way, the control points can create a database of the devices that are available in the network. Devices also have to send a "bye" message when they are about to leave the network (to enable the control point to shut down any services using it and remove it from the database), and there is a "keep alive" message, which is not a part of SSDP, but is implemented in the UPnP event protocol GENA, based on HTTP. To make sure that the network is not flooded with discovery messages, the number of hops that the advertisement can be sent is by default limited to four (which means that a control point can only know about devices four hops away from it); this is configurable, but as usual, the preconfigured value is the one that prevails. The message also contains information about when the advertisement expires, which means that the device description should be removed from the database; it can also be re-sent, with a new duration, to be refreshed. So if the device is abruptly disconnected, by going out of reach or by a power cut, there may be a period when it will look as if it was available, but is not. The length of the expiry time is a matter of fine-tuning on behalf of the network administrator; the minimum is 1,800 seconds, but that may be so short that if there is a large number of devices, the network will be flooded with discovery traffic. Longer times, such as 86,400 seconds (a day), might mean that the devices disappear even though the description remains in the database. Messages are also expected to be repeated, since UDP (which is used by SSDP) does not give any guarantees for the transmission. Figure 3-1 illustrates the mechanism of UPnP.

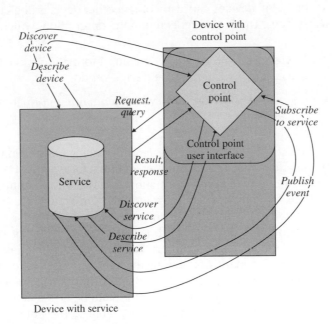

Figure 3-1. How UPnP works.

Conversely, the control point can also multicast a discovery message, in order to find any services that it can use. Since the original discovery message contained information about the type of the device, the control point limits its request for information to devices of a certain type (if you want a media store, it is useless sending a request to a router). This message is also sent to a standardized multicast address, but the address is different from the one used for advertising devices. All devices must listen to it, and respond if they fit the criteria. When the device is discovered, the control point sends a query to the device, requesting a description.

The original discovery message does not contain the complete device description (which is why it is necessary for the control point to request that information). It only contains the identifier, the type according to UPnP, and a URI that informs where to find more detailed information. The device types are defined by UPnP, and the type definitions contain an XML schema for what information is required; when a new type is added, a new schema is created. The UPnP architecture is defined to be extensible over time and with new device descriptions; it is possible for older devices to discover newer services, and as is usual in HTTP, what is not understood is ignored. When a device receives a request from a control point, it returns the description according to the template. The description can contain both vendor-specific information, and information defined by the UPnP working groups.

The UPnP devices can also contain other devices (logical devices), as well as services. This is also reflected in the description, which has two parts:

- a device description, which contains information about the actual physical device, and is written by the device vendor; and

- a service description part, which contains information about what the device can do.

Device descriptions include vendor-specific information such as the model name and number, serial number, manufacturer name, URIs to websites, etc. In addition, it contains information that is used to control the device, such as URIs for control, event messages and presentation.

A service is a functional unit, for example a media server, which does not modify the device itself but can be used by other devices. These are listed in the service description. For each service, the description includes a list of the commands or actions the service responds to, and parameters or arguments for each action. The service description includes a list of variables; these variables are used to model the state of the service at run time, and are described in terms of their data type, range and event characteristics.

The commands and their parameters are used to manage the discovered devices. When a control point wants something to be done by a device, it sends a control message to the URI it received from the device discovery; the control message is also an XML document using the Simple Object Access Protocol (SOAP), which is based on HTTP. It results in something happening in the device, or the service in the device. What that "something" is depends on the device (or service). In the device class Media Renderer, for example in a television set, it may be that a key is pressed to change the channel (which means go to a different URI in practice). The events in UPnP are described in XML messages, which are sent using the GENA protocol.

When a control point first discovers a device and wants to be able to control it, it sends a subscription message to the URI listed in the discovery message; this contains the names and values for the variables related to the event, and allows the subscribing control points to initialize its model of the state of the service. If there are multiple control points, they all have to be informed of any changes in the state, so they can update their model.

Finally, having performed the process of original and detailed discovery, and subscription to the device, the control point can present the status of the devices it controls. This can be done in the user interface attached to the control point (if information is available); it also depends on what presentation formats are available. Since the status information is also described in XML, it can be presented in all the same ways as an XML message.

When the discovery process is over (and the description may be lengthy, but the process is fast), the UPnP system can be used for media consumption. UPnP is a generic protocol, it is not limited to media consumption, and the UPnP standards include ways to control lights, air conditioners, and many other things. It can also be extended very easily. Its main use is undoubtedly for media rendering, thanks to the efforts of the DLNA, which has created a profile of UPnP and is currently certifying devices, so a user can be sure they work together. The DLNA is limiting itself to make media work in the home, which means it focuses on two UPnP device classes: the Digital Media Server (DMS) and the Digital Media Renderer (DMR). In version 1.5 of the DLNA standard, there is also a control point, the Digital Media Player (DMP).

When a user has selected a resource, the control points coordinate the Media Server and Media Renderer, for example through sending play, stop and pause commands. It is also possible to record, which is defined in the AVTransport – the control point directs the Media Server to send content to the Media Renderer, and then issues a Record action to the Media Renderer – which results in the content being saved, instead of displayed. How this is done is not specified, and AVTransport being optional, it also means the action is optional.

There can also be several examples of media streams, that is, they can be mixed together. In addition, a control point is capable of sending instructions about changing the brightness, contrast, volume and balance to a Media Renderer. To do this, the (optional) AVTransport service has to be implemented. And as services can be instantiated more than once, and given an ID of their own – the ExampleID is also bound to the connection, through the Connection Manager. It is also the Connection Manager that creates the ExampleID, through the PrepareForConnection command (this is also optional).

Media control very much depends on the protocols used for signaling and media transport – it is impractical to try to do a fast forward in HTTP, whereas it is possible in RTP, for example (since RTP has the notion of MPEG frames built in). However, RTP is not included in UPnP, and is optional in the DLNA.

One aspect of the separation of the media storage, media rendering and control is that the user interface is assumed to be in the control point. The Media Renderer, while it may be a big-screen television, does not necessarily have a user interface for the media control. That may be in a different device, such as the mobile phone.

Neither the Media Renderer nor the Media Server can initiate UPnP actions. Notifications can flow in the other direction, making it possible for the control point to become aware of the state of the device – for example, that the media is full and no more recording can be done. The control point is not involved in the actual transfer of media, and the flow control and other signaling goes directly from source to sink and back.

In most media scenarios, there is one device that sends the media, and another that receives it. The receiver is called the sink, the sender the source. The file transfer from the source (the Media Server) to the sink (the Media Renderer) is usually the response to a request, which may come from the same device or the control point. To make the control point work, it has to send a request to the source, which is sent on to the sink.

The Media Renderer can render content to a medium that the user can receive. This does not necessarily mean that the content is visually rendered; it can be rendered aurally, as well (and in principle through a couple of other ways as well – humans have five senses, and there are protocols defined both for how to send scents and touches over the Internet, although not yet taste). There are no specific rules for how the renderer should receive and handle content.

The Media Server in UPnP has access to a store for content, and it can be anything from a camcorder to a set-top box. This store can be somewhere on the network, but usually the assumption is that it is attached to the Media Server. There is no built-in constraint in UPnP about which content can be handled.

The Media Server in UPnP can be anything that can store content. Its real function is to enable the control point to display (in its user interface) a list of the content which can be rendered, for the user to browse, search, or otherwise access. There are two functions in the Media Server which are used for this: the Content Directory Service and the Connection Manager Service. The latter is used if there are multiple media streams coming from the same Media Server, since there has to be a way to manage the connections. It makes the control point several magnitudes more complex, since it is actually the control point that has to manage the setup and teardown of connections, instead of the Connection Manager itself. If there are multiple examples of a service in a Media Server, the control point can separate them using ExampleID. This makes UPnP's notion of service control rather similar to SIP – but it is optional in UPnP, since it only applies to the optional AVTransport and rendering control services.

The Media Server as defined by UPnP only stores data, it does not have any information about it. That is defined in another device class defined by UPnP: the ContentDirectory server. This provides the information about the individual content objects that makes it possible to browse and obtain detailed information about them. It also makes it possible to look up and manage the different media objects (a media object may be a file which contains music, a video, or something else that can be played). The ContentDirectory server provides a standard way of listing the object information.

The ContentDirectory lists characteristics of media objects using properties, which are expressed in XML (although they confusingly are not XML objects themselves); they represent aspects of the object such as its name, size, type, and so on. The properties can be associated together, and there can be independent properties (e.g., names) or dependent (e.g., size). The properties can be described using common AV properties defined by UPnP (which, in turn, are borrowed from other standards, such as the Dublin Core Metadata System).

The data about the media that the ContentDirectory can display is based on metadata. When used with IPTV, the metadata will probably come from an EPG, which most likely is defined in the TV-Anytime format, the standard that is used by almost all standards bodies. The DLNA, however, defines an EPG format, which is not the same as TV-Anytime, but it can be easily translated. The metadata also contains information about what transfer protocols and data formats a Media Server can support for the content item. The control point compares this to the capabilities of the Media Render, and ensures the content gets to it in the right format – or if there is no compatible format, ensures there is an error message (how that should look, however, is not defined in the specification). This is no negotiation, the control point does not actually try to find if there is a best match – it only looks at the list of capabilities, compares it with the list of file types, and then end of story. This process is shown in Figure 3-2.

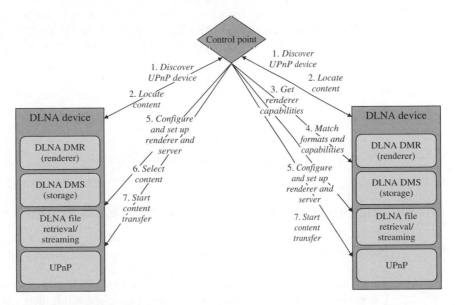

Figure 3-2. Watching content with DLNA.

> **HTTP in the home**
>
> *HTTP does not provide any real-time guarantees (which matters if you are trying to receive a media stream with minimal buffers, as will be explained in later chapters). There is no guarantee for when the packet comes, relative to the previous packets. This means that when using HTTP (and other similar protocols for content transfer), the buffer must either be full before rendering starts (e.g., by downloading the entire file), or there must be control mechanisms for the content transfer. Usually, the TCP mechanisms to manage packet transfer (which assure that packets get through in the right order, and if they lost, are re-sent and the transmission speed is slowed down) are quite sufficient. As long as more bandwidth than required is available, and the number of hops is sufficiently low, there is no special need for error correction mechanisms – which means that as long as the content transfer stays in the home network of the user, the content transfer will work without perceptible errors. However, if the network capacity is decreased, for example if the network is flooded with IPTV signals, HTTP runs into problems.*

UPnP provides very little help. The Media Renderer is supposed to implement a read-ahead storage buffer, which makes sure the data gets into the Media Renderer before it is supposed to be displayed. It depends on the AVTransport method, which is optional, so it is up to the individual manufacturers of UPnP devices to implement it. This is fine for downloads and for streams that are not time-sensitive (e.g., VoD streams), but it is not sufficient for television programs. Here, the DLNA provides for the use of RTP, which has all the functions needed, but it is optional and not everyone implements it.

There are no users in UPnP, only devices. Users do, however, exist; they initiate events using control points –there is just no way to distinguish Alice from Bob, as long as they have their hands on the control point. From an IPTV perspective, this causes a number of problems. In UPnP, the connection is abstracted away from the device; in IMS, the user is abstracted away from the device as well, and the subscription is tied to the user. A user can employ many different devices, but do so with the same subscription. The service provider might want to limit the number of simultaneous streams to the user – something that is both easy and possible with IMS, but hard to do in UPnP (there is no user, you have to limit the devices), and impossible in the model where a set-top box is equal to a physical connection.

Connecting the Home to the Outside: the Home Router

The home network described in this chapter contains servers, which enable you to store and navigate content; and renderers which make it possible to see and hear it. There are control structures to manage the servers, bringing much the same media experience to the user as today's disconnected devices. Once connected together with a network, then connecting that network to the outside world is a very small step.

The connection can be done over different physical media, which is not the focus of this book; it is sufficient that the physical connection has to be able to handle high data rates – bidirectionally, if you want to use the service for sharing your stored media with others. The network provides a very tight limitation on what services can be used, however. It is possible to transmit standard definition IPTV over an ADSL line, but not high definition, even with the highest compression possible. The network that connects the lines going out to the different houses (the core network; the individual lines to the homes are called the access network in telecom terminology) may have much higher bandwidth than the individual home connection – but it has to be shared by all users. This provides another constraint on the service delivery: it is only possible to handle "unicast" – where the client in the home connects to a server in the network on an individual basis – when there are few users. When there are many users, especially if the media streams they want are few, it becomes uneconomical and leads to a poor experience for all. The reason is that while the server can be designed to accommodate a very high number of streams, it is not likely that the network supports it. If the network is the Internet, all users will have to fight for the same bandwidth, and there is no prioritization. Everyone gets equally poor service. Building out the bandwidth solves part of the problem, but if all users are offered IPTV, even in the standard definition format, the cost of building out the bandwidth is more than most telecom companies (and cable companies) can bear.

If users are likely to want to watch the same programs, however, the case is different. Sending multiple examples of the same program is wasteful and does not bring any benefit to the user. Instead, multicasting makes it possible to address the program to the routers closest to the user, and distribute it over the individual connections from there. This can be done either by converting the multicast to unicast in the equipment that connects the individual household lines, or it can be done by making the user part of the multicast tree. The latter method is more efficient (since it reduces the number of nodes the media has to pass through), but it requires more from the home router.

Two factors come into play here. The first is that users have shown a high preference for consuming programs when they want, instead of when the television broadcaster thinks it is convenient. Known as timeshifting, this has been made out to be a problem for opponents of technologies that enable the user to record programs at home. In reality, moving the cache of content to the user's premises is probably the best way of saving bandwidth for all, since programs can be sent when it is most convenient for the network (e.g., if the core network also carries all the voice traffic of the operator, moving the IPTV traffic to night time is a clever way of making more use of a limited resource).

The other factor is that there are still many programs that users want to view as they happen. And broadcasters want them to do so too. These include, for example, real-time events such as election nights and football games – but also quiz games and reality television (where the surprise at the unexpected stupidity of the participants is part of the fun).

You may ask "why not simply use the radio spectrum to broadcast the television, it works well today". And indeed, the television antenna is not going to go away soon. But IPTV offers different possibilities (as we discussed previously), and it is those possibilities that makes IPTV worthwhile. A consequence is that bidirectional traffic has to be handled – it must be possible to send messages out of the home, as well as into it.

There are two basic ways of handling the connection from the IPTV client in the home to the server in the network, when IMS is involved. Either the client has its own IMS subscription; or it passes through a gateway, which manages one single IMS subscription on behalf of all the devices in the home. This means that the home network is decoupled from the IMS network in an effective way, which has two advantages: because the home network by definition consists of a number of devices which may come and go, it would not be feasible to have registration and management of them in one central database – that has to be done locally. And second, the security risk is limited – both for the user, who does not have the risk that a hacker discovers and hacks his IPTV set (it takes about three minutes for hackers to discover new equipment connected to the Internet); and for the operator, who does not have the risk that a large number of home computers become part of a "botnet" – a collection of computers infected with a program, which lets someone control them remotely, often for purposes of sending out spam or sharing files illegally.

Either way, the IPTV services are enabled, although it is a little more complicated in the gateway case. This also brings more possibilities, since the gateway can be used to let UPnP devices consume IMS-based IPTV services.

In principle, you can see most functions needed to provide services as services in the UPnP sense, which can be located in different machines (the work of the Open IPTV Forum is to define interfaces between them, so they do not have to be co-located in a set-top box or television set). However, some functions are dependent on others, and have to be put together. Other functions are dependent on physical properties – for example, the fiber or ADSL connection of the home to the outside world.

A firewall is pretty worthless if it is placed inside the network. It has to be placed at the edge of the network to be able to stop potential evildoers from getting in. So firewall functionality depends on being close to the external network connection. Conversely, media rendering is hard to do if you are trying to do it in the firewall – it has to be done close to the screen (or speaker). And QoS management, if it is implemented in the home network, has to be done in the router (since this is where all the network traffic in the home passes anyway).

The network, in this view, has two edges: the rendering point; and the network access. In between there can be a number of intermediary nodes or modules. Most of the functions required can be deployed anywhere, although some of them, such as the DHCP server needed to provide addresses to the devices in the home network, are normally built into home routers.

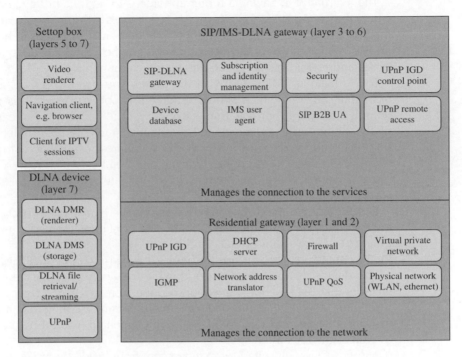

Figure 3-3. Router and gateway in the home.

One issue with the home router is that it probably acts as a Network Address Translation (NAT) gateway (see Figure 3-3). Since today's networks use version 4 of the Internet Protocol (IP), which determines the addressing on the Internet an all other networks that use the same underlying technology, the devices in the networks have to be given an IP address. The problem is that the number of IPv4 addresses is too small to give to all devices. There is a solution, in the IPv6 protocol (which has more addresses than there are molecules in the universe, and better built-in flow control than IPv4 as well), but that would require changing the software in all the routers in the network (both access and core), which is a pretty expensive proposition and out of reach for most telecom companies. It is easier to move the problem – and also slightly more secure.

Therefore most networks make use of network address translation: the router is given one single external IP address, and the network behind it uses DHCP to automatically get different addresses. These can be reused between networks, since traffic that does not go outside the network will look like it is coming from the single external address – all of it. The router has to use some tricks to keep track of who is supposed to get what traffic (mapping the internal addresses to ports in TCP and UDP, essentially creating a way of adding sub-addresses to the single IP address it has).

The alternative to having a router at home is to set up a bridged network, where the same address space is used for all devices in the entire network. This means having a very large number of addresses and difficulties in handling ad-hoc networks, where devices come and go (they would also require one of the addresses). This is expected to be typical of home networks.

If you have an IMS gateway in the home, you need to set up sessions from that gateway, which should control this port mapping. This sounds entirely backward, but it is true – if you want to redirect the streams out of the home again (going in and terminating there is no problem). The session setup information contains the IP address, and that will be the external IP address of the router plus the port number, which represents the internal device. There are attempts in the IETF to define how this should work.

UPnP comes to the rescue here yet again. There are several protocols to control routers in the home, but probably the most implemented is the UPnP Internet Gateway Device (IGD). And it has a further advantage: it is intended for the control point to be inside the home. The idea is to control all the aspects of the device that are required to manage the connection to the outside, whether that is a traditional telephone connection, an ADSL connection, or something else. UPnP defines how the interface can be configured, and how the DHCP server and dynamic DNS (if implemented) should be managed, as well as the LAN interface(s).

UPnP IGD is well suited for management from inside the home network, but there are occasions when there is a need for external configuration as well. For example, it is possible to provide keys for setting up Virtual Private Networks (VPNs) to the home router; or to provide a "blacklist" of websites and other parties from which you should avoid a connection. The protocols defined by the IETF to control routers from the outside are not suited for this; it is better to receive the configuration over a connection managed by IMS, and then have the IMS gateway forward it to the router. That way the chain of trust is unbroken.

The favorite protocol emerging in the industry is TR-069, defined by the ADSL Forum for configuration of ADSL modems, which is flexible enough to enable configuration of any device. It is based on SOAP, and since it is not the UPnP version of SOAP, the receiver has to have a web server to receive the POST message containing the SOAP document. However, there are a number of issues with any management, especially when it comes to who has the right to manage what, and there are solutions in the mobile industry that could well be applied here. There is an ongoing discussion among the standards bodies about how the solution should look, but it is unlikely to have emerged before this book is published.

A second aspect of management of the router is that it is transparent to the IPTV service. It can affect how services are performed, but it is essentially transparent. While security and performance (especially performance) are crucial, there is no need to go deeper into how they work here, because they do not affect the user experience – if they are done right.

The transparency also extends to another aspect of the home network's performance: the quality of service. It is very simple, really: it is a method to prioritize the packets that come into the router.

A router has interfaces where data packets come in, and where they are buffered, and then sent on through another interface depending on the addressing. What quality of service does is look inside the packets, and check what they contain; based on this, some packets are delayed in the buffer, to make others go ahead.

This sounds very simple, but it is not. First, you have to implement the mechanism that manages the queues in the buffers, which means looking inside the packets to understand what should be queued. This takes processing power. The buffers themselves take memory. There also have to be control algorithms to handle this. There is an entire branch of mathematics dedicated to this kind of thing. Finally, the algorithm has to allow for outside control, so the routers in the network can be coordinated (it is no help if one router prioritizes protocol X, but the next delays protocol X and prioritizes protocol Y). Since the number of routers is much higher in the telecom network than in the home network, this is where the most work has been done, and this is where IMS shines. However, it is useless unless the system can guarantee the quality of service end to end, and that means creating some way of making sure the home router is also able to provide quality of service.

One way is to assume that the home router is in an overprovisioned network with lots of bandwidth and low latencies. This is unlikely to be true – you may have fiber to the home and gigabit Ethernet connections everywhere, but if several people start watching HDTV on several screens in the home, the bandwidth will quickly run out, with resulting collisions (which in a UDP-based streaming results in packet loss, because there is no automatic resend; this means applying error correction, which is expensive in two ways: it costs processor resources, and it costs time, because the calculation of the replacement packet has to be factored in).

So, since all data packets in the home have to pass through the home router, it would be better to have it provide quality of service, and have a way to coordinate it with the external network. (Of course, if you have several subnets with different routers, then the situation is different, but that will hardly be typical of most home networks.)

UPnP has a solution for this as well. The architecture for it is very similar to the architecture in the industry, which has emerged through long and hard discussions and millions of hours of research. There may be better architectures, but the advantage of the QoS in UPnP and other places is that it works (the IMS architecture is very similar, but has different components – as we will see in later chapters). Figure 3-4 shows how these properties relate to each other.

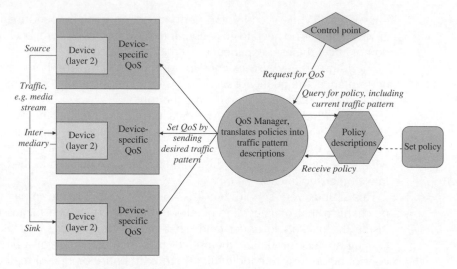

Figure 3-4. Quality of service in UPnP.

The control point, which as always in UPnP is the one that initializes the interaction, triggers the QoS manager to start managing. This is done based on the policies set by the user (through an unspecified interface – which could mean these policies are externally provisioned). The triggering, however, is not just a generic trigger, it is based on the specific communication. The trigger includes the address of the source and sink, and the parameters that define that particular traffic (the Traffic Specification, TSPEC). This is compared by the QoS manager with the policy set for the traffic. This is done by giving all the traffic types a number, which determines how important it is, and hence how it should be prioritized (there can be two different numbers, one for the traffic and one for the user's importance).

When the data packets pass through the network, they will now be prioritized according to the policy. In IPTV, the IPTV traffic will of course be prioritized – but there is a different type of traffic which also has to be given high priority, and that is the user feedback, which we will look at in the following chapters.

The Set-top Box Meets the Internet Model

Part of the reason for creating the set-top box was the need for a powerful processor and memory to decode the received signal into an analog format which could be viewed on a television set. While the television sets of yesterday were merely capable of displaying the received signal in a correct way, this is not true today. A television set can be expected to feature a processor capable of much more than the set-top boxes of a few generations back, and it is likely to feature a processor – and software – capable of this decoding itself. It is even unlikely to display content using a line which moves over the display surface, as flat-screen televisions are built on entirely different principles than cathode-ray televisions.

The set-top box as the sole connection point to the television system is an idea whose time has passed. However, the set-top box still has a life – even in highly digitized Japan, there are set-top boxes being sold for analog television sets, to make it possible to view the digital television signals. The replacement rate for television sets is around 10 to 15 years, so the flat-screen televisions bought today will need enhancements to work with IPTV. One way, and the most likely, to deliver that is a set-top box. Combining a digital television receiver with storage (acting as a PVR) and Internet connection, it is almost good enough to catch the user's interests, but the price is still a little too high to hit the sweet spot of the value proposition.

Cable-TV and Internet television have many things in common, but there are also major differences. Some of the connectivity issues are the same as with the connection through cable and satellite networks. Some of the major differences come from the separation of the network and the service. The solution for the satellite television industry was the set-top box. Receiving and rendering the television signal to the television set may be done in a separate box, but that box does not have to solve the same problems as the set-top box solved for satellite television.

Set-top box starting point: antenna termination

The first cable-TV networks connected the television set to an antenna. The connection became the central part of the connectivity. The set-top box was created, in a very different technological and business climate from that of today, to solve a set of well-defined problems – some of which no longer exist. The original set-top box model assumed that there was a physical connection between the receiver of the television or video signal (the set-top box) and the service provider. There was also a frequent assumption that there was a single connection between the set-top box and the viewing device (the television set) – although this is not necessarily true. The connection termination is the base for several of the main features of the set-top box model, for example, the delivery control and the subscriber management.

The termination of the connection in the set-top box meant that there was a one-to-one identification between the termination point and the delivery of the television signal. This also meant that this was a very efficient way of managing the subscription to the service: a one-to-one mapping between the termination point and the subscription point implied only the set-top box that received the signal was able to receive the service. The subscriber management in the set-top box model assumes that the subscriber has a set-top box, and that this set-top box can be activated and deactivated depending on the status of the user account (i.e., whether they have paid or not). This means that the service provider has to have a database with subscriptions, and that this database can be used to manage subscriptions.

In practice, this is done by distributing a card containing the decryption keys to users through an out-of-band method from the television signal. While out-of-band

(Continued)

distribution provides a high security, there are significant costs associated with physical distribution. Customer management also becomes more complicated, because the assumption that the content delivered to the user is completely out of the control of the service provider (be it from regulatory or legal reasons) means that there is no way to communicate with the user. Even more modern methods of doing so, such as using e-mail, are frequently not applied.

The set-top box manufacturers are moving away from this model, and upgrading the set-top box through their network (after all, they have to assume that there is a secure connection between the set-top box and the cable-TV head end – the part in the old cable-TV systems that converts the digitally encoded media from the satellite to analog cable).

The control of the decoding and encryption of the content means that the user must accept the offering associated with the decoding. In practice, this means that, while necessitated by agreements with the content provider, unprofitable channels can be delivered together with profitable channels, and the user can be enticed to pay for them as a package ("bundling"). Since the cost associated with creating a decryption system for the individual channel is lower, this means is preferred by vendors of set-top boxes.

In the Internet, there are two basic assumptions of how a service is provided: as a client-server system; or between peers. The client-server paradigm assumes that one node in the system provides services to other nodes, which consume it through dedicated receiver software. The peer model means that the nodes are interacting on an equal footing. It is important to note that these two models can comfortably mix, and are often used in hybrid settings. However, the functionality, which comprises a server (and a client), does not have to be executed in one node; it can be called from different nodes, and collated by one system, which acts as a front end to other systems or users. If the functionality available in the set-top box can be distributed in any node, it can be collected in one single node – the equivalent of today's set-top box. However, there is an important difference: the layering of the system implies that there cannot be a strong reliance on the hardware platform to perform functions that affect the service delivery. The receiver of the service can be any node that has the requisite interfaces – be it a television set, a PC, or a PlayStation 3 (or, for that matter, a refrigerator).

While the appliances in the home are already starting to be connected using the Internet model, thanks to the introduction of the DLNA and UPnP standards, these standards do not handle real-time television very well (although HTTP streaming is perfectly adequate for download of video, it does not do a good job when it comes to something which has to be handled in real time). The solution is not to use a set-top box. If you view the IPTV service as a silo, isolated from all the other services a user might want to access in the home – and impossible to blend with them – you have essentially disconnected from the future. And while the delivery of on-demand media is already a fact on the Internet, although more widespread in the form of short clips from services such as YouTube, it is not yet the dominant way of receiving movies. However, it does have the potential to become that, as we will see in later chapters.

While the old set-top box model assumes that there is a single, well-defined endpoint for all connections, the Internet model assumes that endpoints are ad hoc, can be distributed, and do not have to be well-defined. There does not have to be a connection in the set-top box sense. The difference starts with the Internet being a packet network, which means that the connections between nodes can be routed in many different ways. It continues with the addressing in the home network being dynamic, where the address does not have to be tied to a fixed device. This also means that the endpoint for a media stream can be shifted, ideally without perceived disruption, while the user is watching the stream. The separation of the network, session and presentation layers that are described in the OSI model make for a very different paradigm than the starting point for the new set-top box.

One major issue with the model that dominates the industry today (the model where updating a device requires replacing it) is that it is impossible to implement new software without replacing the physical box. This is not sustainable, because there are a number of profiles of MPEG-4, some of which include extensions, which would make it possible to create more interesting programming. The only consumer electronics devices in the home which can be updated remotely today are the games consoles and the ADSL modems – and some set-top boxes. The games consoles and set-top boxes often use a proprietary format, which is easy for the manufacturer, but can be very problematic, for example, if there is no relation between the maker of the device and the service provider (with games consoles, of course, the service provider and the manufacturer are usually the same company).

If the manufacturer of the box and the service provider are unrelated, there needs to be some way to define how the two boxes should interact. This can be done through the service provider providing a specification, which all device manufacturers have to follow; this only works if the manufacturer is dominating. The other way is that the device manufacturer provides the specification, and the service providers have to follow it. This is how it used to be in the IT industry, when first IBM and then Microsoft dominated. This is no longer true, however, and both these models are moot. The new model is standardization.

The layering of functionality is based on the seven-layer Open Systems Interconnect (OSI) model, and the five-layer model that has been derived from it to describe the systems of the IETF. These models have proved valuable not only in modeling systems, but also in designing them. The separation of functions according to the layers has enabled the optimization of each layer, and the use of well-defined interfaces has enabled easy communication between the optimized parts. Layering the system has enabled not just optimizations within each layer, it has also enabled the creation of standardized interfaces. These may reach through several layers – for example, the interfaces defined in IMS reach from the application layer to the network layer – but they still conform to the layered model.

That said, there are a few features that are required both by legal agreements (between service providers and content providers) and by regulatory requirements (mandated by different agencies). The owner of the rights to content has a right to determine how it should be used. This is somewhat counter-intuitive, and while it is frequently abused by the owners of content rights, it does have both legal and moral justifications. The mechanisms, which have been created to enforce the rights of the content owners, usually impose a burden on the end-user, however. Shifting the burden of enforcement this way does not help the case of the content rights owners, and it is somewhat doubtful if digital rights management actually has a business case – it is hard to calculate, but money spent on building enforcement techniques could have been better spent on educating users.

DRM and the set-top box

In the original cable-TV model, there was an assumed association between the viewing device and the sender. In practice, the endpoint should be user-selectable, because it is the subscription, and the user who is using it, that determines where the media should be viewed. While there is a perfectly valid argument that the content creator has a right to determine how media should be viewed, there is another perfectly valid argument that the user should have the right to determine how he views the media he has viewing rights to.

Digital rights management is a way of associating the rights, and control mechanisms to enforce those rights, with a media object such as a file or stream. The offshoot of applying the traditional cable-TV model to the Internet-model connected home is a requirement for a mechanism which enables the user to select the endpoint, but allows the service provider (together with the operator of the Internet service) to control the distribution. Such a mechanism exists, in the form of SIP and the additional security mechanisms provided in IMS.

In the OSI model, which constrains itself to the definition of an architecture for networking, the hardware layer has a well-defined role: it acts as the physical substrate for software. Nothing says that this software cannot be embedded or built in (indeed, it often is), but there is a vast advantage to having software that is loaded in a rewriteable memory: it can be changed remotely.

Software management does not only mean replacing the individual programs in a single device, it also means the ability to reconfigure systems, and create a merged version of some types of documents. The mechanisms to do this are starting to be put in place, but they are by no means there yet. In brief, remote software management is a sliding scale from replacing files on host devices (with the associated issues of access rights addressed) to synchronizing the views of the system between devices, including the way that user interfaces are presented. This does not imply that the view always looks the same, rather it can be modified based on who the sender is, and who is the receiver.

While the mechanisms exist, there are many issues with managing software, especially if you try to do it in equipment owned by someone else – not in the least liabilities which may hit the service provider, if something goes wrong in the home.

Another similar but different issue is that of user profiles. The main difference is that user profiles do not change the function of the device that receives them by changing its software. The main similarity is where to draw the line of responsibility – because there is, in the next generation of IPTV systems, a need for user profiles to be used both at the user and the service provider side. At the same time, it is possible that both do not want to give up parts of the control either. There is also a requirement to get some information from the user, such as statistics, which the advertisers will require. Without advertising there would be no service, since the cost to buy content and build infrastructure would be prohibitive.

User profiles, at their most basic, contain a unique identifier for the user; and some facts about that user which are relevant for the current service (e.g., if the user has paid or not). However, very few user profiles are this sparse. Most contain a plethora of facts about the user, some of which are relevant. These are often of interest to other service providers, and while there have been some attempts to federate profiles, the existing architectures do not apply easily to the federation of user profiles. Liberty Alliance, for example, which is the leading standardization body when it comes to federation of profiles, makes assumptions about ownership and responsibilities that suited the originators of the standard (banks and telecom companies), but which do not work in the IPTV area.

While federation is difficult, there is a standard distribution format in the IMS world, which easily lends itself to use in the world of IPTV. The *de facto* format of profile distribution is XML, and mechanisms which leverage this, in combination with user identity management, have been standardized in several places. The most comprehensive open standard system to manage user profiles to date is OMA XDMS. It creates a model for how to manage and distribute profiles securely to the relevant stakeholders using an XML-based format and the IMS infrastructure. But while the profiles contain all the information relevant for the IMS system, there is as yet no specific profile for IPTV.

What will make the cable industry (and the set-top box industry) change is the need for software updates that are managed remotely. This includes updates to the decoders in the boxes. And it requires a different security – and management – model than that which is in use today. Set-top boxes will, in this regard, become very similar to mobile telephones, which can be updated through the network, by the operator. With the security and management model in place for mobile equipment, the telecom operators will realize that this would be useful for other types of equipment that they control – such as ADSL modems, and set-top boxes.

For ADSL, there is already a system in place to handle these updates. It is based on HTTP, adding a semantic layer for security and software management, and was originally intended to update the software in ADSL modems. While it is very useful for this, it does not really fit into the IPTV world, where the needs are different, and there is a different model for security – and where the economics are different, as well. The model where the operator updates the modem may be suitable if the operator owns the set-top box and rents it to the user. However, it was more than 30 years ago since telecom operators stopped leasing telephones to users, and opened up their networks. The openness of the model changes the need for management, and while there are ways to do it, there is not yet any agreement in the industry.

The elemental streams of MPEG-4 could, as mentioned before, be stored or transported independently. Storage of MPEG-4 data is done in the .mp4 format, which we will not go into here; but the transport of MPEG data bears some mention, because it has effects both on the user experience and the interaction with data.

There are two basic ways to handle MPEG-4 encoded video. The first is to send the elemental streams and object descriptors to the client, and compose the stream there. This method has a lot of theoretical advantages, but they turn out to be just that – theoretical. There is no way to leverage them in reality. Two things come in the way: the limitation of the end-user device, and the limitations of the network.

The end-user devices are limited because they have to be inexpensive. If all end-user devices were personal computers (and today's television sets have far more processing power than the early personal computers had, so maybe we will get somewhere there) we would have a lot of processing capacity and memory to spare. Especially memory; some set-top boxes are scaled-down PCs, and what they save upon is the memory, which turns out to be expensive when incremented (you must have one processor, but there is a big difference between having 512 MB and adding another 512). As it is, the margins are already razor-thin, because there is so little distinction between devices that users select on price. This will continue to be true until users find the features required for interaction worth paying for. It is even more true when the service provider is paying for the device, such as the set-top box.

The addition of a client is an expensive business, which is what makes it interesting to reuse the DLNA for IPTV (although it currently lacks the inter-activity features). Adding an IPTV client, such as the OITF (or a standardized receiver for IPTV, in other words) is, however, not very expensive in terms of memory and processing power, and then you can offload the IMS logic to the home gateway. So what comes to the end-user device is a stream and control messages.

One reason to believe the set-top box is doomed, even in the Open IPTV Forum architecture, is that it is so easy to remove. Ericsson and Sony are working together in many fields besides mobile telephones – the most successful example. The "Connected Home" is another example where the two companies have been working together to leverage each other's strengths and to create new opportunities by exploiting synergies. In Berlin in October 2007, Sony, Ericsson and SonyEricsson demonstrated how the IPTV world can flow seamlessly into the connected digital home.

Connecting two "virtual homes" and a SonyEricsson P1 mobile phone using the Ericsson Research Home IMS Gateway and an experimental media server jointly developed by Ericsson and Sony, they showed IPTV services combined with presence, and being moved from one television screen to another. Users were also able to take photos with the mobile phone, which could be uploaded to the home and displayed on the television screen.

The demonstration leveraged the capabilities of the DLNA to easily access and display services, as well as the capabilities of IMS to handle user login, user profiles and presence. On the television screen, the service did not appear significantly different than any cable-TV service of today, the main distinction being behind the screens. In that demonstration, there was no set-top box. The televisions used an enhanced version of the DLNA, and a gateway translated the IMS signaling from the Internet.

The Browser in the Set-top Box

Surfing the Internet from the television set is sometimes touted as the killer application for Internet connectivity for television sets; but the first initiative, WebTV, which was one of the early dotcom companies and eventually bought by Microsoft, was never successful.

Why would you not surf the web on your television set? There are no technical hurdles – modern television sets, as well as set-top boxes, come with a built-in browser and Internet connection (at least in countries such as Japan) – it's just that surfing the web on the television set is a very painful experience. The navigational model of the television set is completely different from the navigational model of the PC, or indeed the mobile phone. The difference in the navigational models is also why web surfing on mobile phones has been so slow in taking off, plus of course mistakes in encoding, which make the mobile browser render pages in an unreadable way.

Rendering web pages on the television set was a hot topic at the same time as the mobile web started coming onto the radar screen of the telecom and web industries – while the technology is there, the navigational model is missing. For television, the user experience is completely different from expected – as well as having a different control device. In television sets, the control device does not even have as many keys as a mobile telephone, and there is a different expectation for how these buttons should be used. As we will look at later, the model for developing applications in IPTV is very different from the model for developing applications for a PC-based web browser, notwithstanding that the technologies used are similar. There are two major constraints: the linear flow, and the limited navigational device (the remote control).

The underlying technology, however, has been standardized by the consumer electronics industry. The CEA-2014-A standard from the Consumer Electronics Association is a profile of several web technologies to create a very specific environment for browsers, which should go in set-top boxes, television sets and the like. Unfortunately, the profile does something which is very bad for interoperability testers and developers: it not only constrains the existing standards by excluding elements, it also adds elements.

CEA-2014 contains a number of components used to render and control the services. The rendering is done using XHTML and CSS, the control through ECMAScript, and the networking through UPnP, XHTMLHTTPrequest, and a special socket interface between the client and the server. Notable is that the audiovisual profile, which you would expect to have a central importance in any such standard, is optional. So it becomes possible for consumer electronics companies to pick and choose from different components. There will not be one single IPTV device; instead there will be a number of devices that combine the profiles developed by different standards bodies, and make it possible to deliver an interesting IPTV service.

XML and Style Sheets – Format and Structure for Metadata

Since the first thing the user sees is the EPG, it is the way a service provider can influence the user's choices. And while users may use the EPG several times during a viewing session (instead of zapping through channels using the navigation arrows on the remote control), the time they use it is usually short. So grabbing the attention of the users and steering them towards a program they may want to see – or which the service provider wants them to see – is a matter of putting the most important things in front. In many ways, designing an EPG is no different from designing a newspaper.

The navigation model – how the users find the content they are going to watch – is also based on the same technology, even if the ways it is presented can be very different. The user interface design, while it is the part that makes the most difference in the eyes of the user, can be varied in almost infinitely many ways using the same underlying technologies – the same technologies which underpin the World Wide Web. XML, the eXtended Markup Language standardized by the W3C, enables the creation of documents, which can be rendered in a large variety of ways, and is not constrained to the list or tables that are the typical presentation of the EPG.

XML has established itself as **the** structured format for data, and it has found use in such diverse applications as database information interchange, user interface description and program code interchange. XML, in other words, is very versatile. But this versatility – that the structure can be adapted to many different types of data – is also the greatest weakness of XML. To display XML documents in older browsers, they have to be encoded in XHTML, the XML version of HTML. Newer versions of browsers support the Cascading Style Sheet (CSS) format, which is an encoding for how objects should be placed on the screen. If this is combined with Scalable Vector Graphics (SVG; a technique to create animations and display images) and ECMAScript (better known as JavaScript) you have Ajax in your television, and it is enabled to do everything Web 2.0.

It is not too hard to understand how a table can become a string of text entries – just take the columns and put commas between them. Crude as it sounds, this was actually the state of the art in the database industry for a long time (of course, this was when most terminals were VT 100s, and computers were minicomputers, as well). Based on this idea, and the difficulty of keeping track of which entry in a string of 100 comma-separated entries was the one you were looking for, XML was born (there was, admittedly, somewhat more to it).

XML is based on enclosing the data in elements, which in a way are similar to the column and row headers in a table. Since it is possible to take one table and fill it with different content, you can generate a description of the table, which decides what is allowed in the different fields of the table (e.g., that "age" can only be a positive integer larger than 0). This description is called a schema, and XML documents also have schemas which describe what is allowed and not, what the elements are named, and how they are handled (e.g., in one cell in a table, you can have another table – this is allowed in XML too, but it is expressed by enclosing elements in other elements).

From the XML schema, you can create an XML document. But what is more, you can make a program which reads the schema and, since it contains the rules for what is allowed in the different XML elements, can interpret that document.

One such interpretation may be to take the values and stick them back into a database. Other interpretations may require the use of other programs (such as assertion engines for ontologies). A third may be presenting it to the user. And here is where the problems start.

The mathematical operations performed on a database are unambiguous – "add one and multiply by four" can only mean one thing – but when you present something to the user, there are no rigorous definitions of the method. Does "H1" mean 36 point Arial centered, or does it mean Times Roman 24 point left-justified with a 0.5 line space to the next line? And does it mean the top heading in a hierarchy, or is it an independent expression of a typographical design, which does not necessarily have a semantic consequence? And what do you do when one service provider has one graphical profile, but another service provider wants to present content completely differently? Not to speak of how content should be presented on different terminals (the users would be very dissatisfied if the font size was the same when the document was presented on their television as their mobile phone).

Users confronted with XML normally see it through a piece of software which executes the commands embedded in the document. On the PC, and often on the mobile phone, this is a browser. Browsers for television do exist, but attempts to transfer the browsing navigation model from the PC to the television have been more or less disastrous. The browser-based model of navigation does not work well on the television, however, there are other models that do, and there is nothing that says that the receiver of the XML document has to be a browser. It can be any software.

HTML or XML?

In the PC and the mobile browser, the presentation mostly depends on the default settings of the browser manufacturer (since most users never get around to changing them). To the extent that an artistic expression is required, the browser does not automatically allow for it. Something more is required to give the content creator control over the user experience – or at least influence it.

In HTML, the presentation is embedded in the document. The fonts and font sizes are set by the browser manufacturer, but in XML, there are no presentation rules. While you could conceivably guess that H1 should be bigger than H2, what is the relation in the presentation between <actor> and <director>? Far from being automatic, this leads to guesswork from the users when showed documents, and hence confusion.

In many people's minds, this means it is better to stay with the conservative option: use HTML (or its XML version XHTML), which has a nice fixed set of elements that forces presentation into a single mold. People interesting in creating a better user experience have a different option, however: use a control language to decide what the presentation format for the different elements in the XML document should be (remember, they are all listed in the XML schema).

These presentation format descriptions are either collected in a separate document, or embedded in the XML document. They use a different format, called Cascading Style Sheets (CSS) (the "cascading", confusingly, is the reverse of "nested"). Here, the typography and layout of the different elements can be listed, and also how they should be presented in relation to each other. That makes it (conceivably) possible to have one single EPG for all service providers, and present that in different ways depending on the particular service provider (although there are other reasons this may be a bad idea, as we will describe later).

Creating one CSS document per service provider is not particularly difficult, but it does require graphics designers to think differently about how they handle content – no longer is it something that can be replaced by "lorem ipsum", but a different entity which cannot be squeezed to fit the pages. And if that is difficult, it is nothing compared to multi-device presentation.

When viewing the same EPG on a mobile telephone and a television set, they should not look the same. Apart from the font size issues, it is likely that the presentation on the mobile would have a different focus, for example presenting the entries in a sequence, instead of in a table. This means a different presentation – which can be accomplished by applying a different CSS document – but it may also mean a completely different content selection.

As luck would have it, the database roots of XML come to the rescue. While it is easy to select the appropriate CSS document to control the presentation automatically, it takes a bit more effort to select the appropriate XML elements – and their content – to present in a specific device. This requires both more work *and more thinking about how to go about it. And above all, it requires knowledge about the end-users.*

You can apply many different layouts to the same EPG, since the layout is decoupled from the content, but since XML is a machine-readable format, there are many different things that you can do with the content itself. One of these is simply sorting the content differently. While the software to do sorting only is rather simple (it can and has been implemented in many web servers), the algorithms for doing the sorting can be way more complex – as can the systems which implement them.

A personal presentation of an EPG depends – naturally – on the user's personal characteristics. This will mean knowing who the user is, and tracking the identity of the user; as well as what characteristics and behaviors are tied to that identity. The IPTV system needs to identify the user to be able to do this personalization, as well as other personalization, for example of the video content. This is one way the IMS-based IPTV systems can leverage the identity management mechanisms of IMS, as well as the profile management system associated with it. We cover these in other chapters in this book, so we will not go into how they work here – but when creating a personal presentation, you need to read the user's personal profile and apply it to the presentation.

A personal presentation does not only mean personal layout, it also means that the content is filtered according to the user's preferences and interests. That can be as simple as changing the order of columns and rows in a table; or it may mean highlighting different parts of the data (e.g., you may prefer the movies in your VoD archive to be filtered by director instead of lead actor).

Changing the sort order of a table, or presenting data differently, is not a big issue, however, creating systems which change the content according to not what the user has registered as his explicit preference, but as implicit preferences, is more difficult. While there is a default function in IMS-based IPTV to keep track of the users, their preferences and which groups they belong to, there is nothing that enables conclusions based on these groups.

The XHTML profile of the CE4HTML browser is based on the XHTML Strict or Transitional profiles. This means it contains all the required elements in XTML, but also a couple of additional ones, creating a language called CE-HTML, which is what the standard uses. The CSS profile is based on the CSS television profile, with added features from the next generations of the CSS standard, in particular managing image orientation. The bindings between the elements in the document and the scripting elements is done through the Document Object Model (DOM), version 2 from W3C, and the bindings primarily connect the scripts to the AV control point and the notification mechanisms (NotifSocket, which is used locally; and the XMLHTTPRequest mechanism, which is intended to be used over the Internet, when the CEA-2014 server and client are both in the Internet).

We have already discussed the DLNA and its foundation technology UPnP; but the IPTV system will have to go beyond the DLNA (as we will see in later chapters), because the UPnP-based streaming (the only mandatory part of the DLNA audiovisual delivery) is actually quite bad for connections that go outside the home environment. UPnP is not intended for interconnecting hundreds of devices over long distances – from a network topological perspective, that is the number of routers which the signal has to pass through. The RTP, standardized by the IETF and optional in the DLNA, is much better in this regard (as we will see in later chapters).

UPnP is one of the primary bases for CEA-2014, and contains a technique to create remote user interfaces. The idea is that there can be a control point for a user interface. However, that is not quite sufficient, which is why there has to be a mechanism to transport and display XHTML pages. UPnP only provides the basic foundations, as shown by the need for the DLNA.

The DLNA is all about AV transport, although it does not say how the navigation of the AV transport should be done. CEA-2014 is the other side of the same coin; in that standard, the AV profile is optional, and the standard describes a system to create navigation elements in the television set. There is a third part, which is missing from both, and where a body such as the OIF needs to fill the space: where the navigation scripts come from, and how the IPTV services are discovered and served. That is really what this book is about.

CEA-2014 does not leave AV presentation and streams completely by the wayside, however. Even though the AV profile is optional, it is clear that the AV and navigational parts are intended to work together. The AV stream is assumed to be handled using UPnP, and the navigational system contains the control point for it. Figure 3-5 shows how this is achieved.

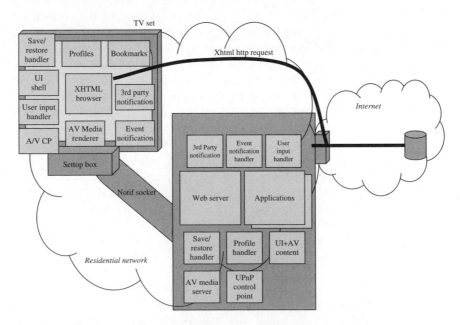

Figure 3-5. Audiovisual presentation with the CEA-2014-A browser.

Since the CEA-2014 system is based on web technologies, it assumes a client and a server. The client is the browser in the consumer electronics device; the server can be an ordinary web server, but it has to have two additional functions: manage notification sockets, and match profile descriptions. Since the server has to serve the devices in the home network, it has to be placed in a location where it can be discovered by the clients. The discovery is done using UPnP, as we have earlier described, but there is a quirk to it: the CEA-2014 standard uses a mechanism to match the UPnP profiles describing the devices. UPnP contains a profile system, and CEA-2014 contains descriptors for additional features, which describe the rendering characteristics of the target device, such as the screen size. When the device is discovered, the server has to match the profile to the rendering characteristics. What that means is not defined, but this could be simply applying transformations to the XHTML document. This could also be done on other characteristics, such as the model of the device, if that was included in the requests from the rendering device (e.g., in the HTTP headers).

Once discovered, the server can provide the client with an XHTML document. The important thing is what happens in these documents. By using features of XHTML that go beyond providing a plain web page with text and graphics, the web page can become an overlay over the video, and the overlay can contain interactivity elements, which can be used to navigate within the video stream, using the remote control. Activating an interaction element (note that saying "clicking on a screen button" is not the same thing) triggers a script which can do something in the client, for example contact the server and make other things happen in the server – or the IPTV system.

The CEA-2014 standard does this using XHTML together with CSS technology, and ECMAscript. XHTML allows you to include elements from other languages inside a document, and that is the intention of how it should be used in the standard. Instead of defining the positioning of text and graphics (which can also be done, of course, but is less interesting for the purposes of this book), the document defines the layout of the screen using CSS (the intent is that the background color in CSS should be "opaque" or, if that is not implemented, "transparent"). Placing a graphic on the screen, and attaching a script to it (by embedding it in the document) can be used to create an interaction button. Tying different scripts to different key signals (which is also included in the standard – since the signals from the different key presses are standardized) will result in a button on the screen that can be activated by the user.

When the XHTML page is received, it will be rendered as any other web page in a browser. But if the client implements the UPnP AV architecture, it will be able to display a video stream. If the CEA-2014 browser is also implemented at the same time, the client will display the browser on top of the video stream; using the CSS transparency property means everything but the areas specifically defined as non-transparent will be invisible. The document creates an object, places it on the screen, and binds a script which enables interaction to it.

There are a couple of anomalies in using this model for interaction. For example, it becomes hard to have multiple buttons on the screen, since the button in focus is the one active, and it is difficult for users to switch between buttons. On the other hand, the linear flow of a television show will probably be better suited to having one (or two) buttons on the screen, which appear and disappear during the use of the system.

One issue is that the CEA-2014 model does not let you synchronize the web pages and the video signals. The synchronization has to be done in some other way, for example using the notification socket mechanism.

Notification sockets standardize something that has been frequently in use ever since the first websites were created. A quirk in the specification of the original HTTP protocol allowed the server and client to have a TCP connection open, and updates to the document in the browser to be sent through it. This works with HTML, where the page rendering starts when the first line is received; but it does not work in an XML document, where the entire document has to be received and parsed before the document can be interpreted. So the underlying assumption of the CEA-2014 standard, which is not spelled out, is that the XHTML document is treated like an HTML document, and that it is rendered sequentially. There is no assumption that there should only be one page per show, but having one page per show and updating it is probably going to be easier than any other way.

XML also enables the use of metadata. This is crucial, as it makes it much easier to categorize and manage content. Metadata can also be automatically generated (even if that sometimes means a tradeoff between speed and quality), and it can be combined with other data very easily, the ease of combination coming from the basic XML technology. It is even easy to create documents which can be used to draw automated conclusions, enabling personalization towards the user to a much higher degree than in the EPG systems that are on the market today.

How does the Multimedia Home Platform Work?

While it is possible to download programs into the set-top box and interact with it, as if it was a PC on the Internet, there is an alternative way of doing this. Just as the traditional model of downloading and displaying web pages has been completely marginalized by Web 2.0, the model of interacting with a program downloaded into a set-top box will be completely marginalized by the ability to interact online with programs that the Internet and IMS enable, for no other reason than that the latencies become more important when you interact with a real-time program. If you just download a game to your computer, you do not really need to interact; and if you play a game on top of the video, the choices you make might change the branching of the video, but they do not really change the program – as in the "interactive DVDs" which have been on the market for a long time.

That said, the ideas behind the Multimedia Home Platform (MHP) are valid, and the interactions which the user can perform are the same as the user interactions, which have been judged most relevant in many user trials. These interactions have also been validated by the voting systems using SMS which programs such as *American Idol*, *Big Brother* and the *Eurovision Song Contest* leverage to enable the user to participate in the direction of the show – and create quite tidy sums of money for the television channels. The problem, as was recently exposed in England, is that the systems lend themselves to cheating when there is a time pressure on applying them. The Open IPTV Forum has also included an application platform in its architecture, which while not following the MHP standard (for good reasons, as will be seen when we discuss the technology), more or less is intended to fulfill the functions, and then some.

However, in the mind of any producer of interactive services, there are warning bells that go off when they are told that the most important applications are government interaction, e-learning (called t-learning by the MHP promoters), and e-health. Never successes commercially, there is no reason whatsoever why those particular services should make anyone buy an MHP-enabled set-top box. As with any service, it has to be the user who wants it for it to happen, not the provider. Shopping via interactive television is somewhat more successful, because users want to buy cubic zirconium bracelets at 2am. The social applications, which have been so successful in the Sky interactive services, are missing from the MHP offering, which just goes to show how hard it is to predict what users will like in advance.

One problem lies in production – the producer must create an open-ended production, where the storyline can change depending on the user's actions. This means more or less creating a game. For less complex interactions, such as the user responding to polls and sending messages, there is a much easier way, and it is part of IMS: presence.

The MHP project classifies the applications it foresees in two broad groups: those which accompany the actual television program, and those which are independent of it. These applications include games and interactivity portals – such as e-learning, e-health and e-government. There have also been attempts of interactive advertising, but MHP has not really taken off, despite its potential.

The interactive applications during a program can be coupled to the user's experience by extending the program experience before or after the show, or by giving users an opportunity to be actively involved through quizzes or voting, and providing additional information, such as statistics. These can also be distributed through unidirectional formats, such as Teletext (familiar to users in Europe, it is a plain text format which is broadcast on the spare lines in the television signal); or there may be applications that are unidirectional from the user to the service, for example voting and responses to advertisements (such as requesting an extended advertisement, or information to be sent to the home).

The final form of interaction is the user interacting with the program itself – bidirectional interactivity – where the user's responses determine the course of the program. This model is more difficult, both for the producer and the user, since it means a much more lean-forward user experience than the lean-back model of traditional television. In many ways, television becomes similar to games, and the games more similar to machinima, stories recorded and generated through games systems.

MHP has its roots in the Digital Video Broadcast (DVB) project, which was created by a number of European companies in 1991; the project was formally launched in 1993, and broadcasts officially began in the late 1990s, with Sweden being the first country in the world to switch off all analog broadcasts in favor of DVB-T, the terrestrial broadcast version, in 2007.

Digital television requires a different set of decoders from analog television, and this was the motivation for the MHP project from the start – since you need a set-top box to decode the signal, why not turn it into a platform for interactivity while you are at it? Especially since there were both deployed standards and proprietary systems, which tried to address the same issues but did not quite fit the bill.

The first version of MHP, DVB-MHP 1.0 became available as a standard from ETSI in July 2000, but there had been demonstrations of the technology the year before. A second version (fixing various bugs) followed in 2001, and this was followed by both a third and fourth version. The major change was that the World Wide Web became unavoidable, and if the MHP working group wanted their standard to be a player in the interactive television world and avoid becoming overtaken by web-based deployments, they had to add an HTML version.

The assumption is still that the set-top box should be the platform for inter-activity, and that it is addressable through a direct or managed connection; the subscription being tied to the connection and managed by a smart card. The same model has been taken up in the digital broadcasts, enabling a subscription model over the broadcast network.

To complicate things a bit further, three profiles have been defined for MHP, plus one option which includes DVB-HTML. The main difference – they can all be combined – is that the most capable profile includes audio and video services, and also download. The second profile enables interactive services, and requires an interaction channel back to the service provider (which the first does not). The third profile simply enables Internet access, requiring an Internet connection.

Applications in MHP are written in Java, running on top of the specialized middleware. It is a Java profile specialized for broadcast applications called DVB-Java, a subset of Personal Java 1.2, and standardized by the Java Community Project. The standard includes several functions that are not present in Personal Java 1.2, which have to do with the interaction, primarily the inter-action with the set-top box and the user interaction. Programs, called xlets, can be started, stopped, paused and resumed just as if they were television programs.

The audio and video functions in the system are based on the Digital Audio Video Council (DAVIC) API for MPEG and switching between the different transport streams, and the Home Audio Video interoperability (HAVi) API for handling video graphics and UI widgets. In addition, it has a number of specialized functions for handling information services, first of all EPG, and also media control such as subtitles, scaling and positioning, and decoder format conversion, which is done using broadcast-specific additions to the MPEG-control Java Media Framework (JMF).

Service selection in MHP is controlled either by using a Java API, or by the DAVIC API. This API enables applications to data (also on different terminals than where the user is currently logged in). To control the file access, there are a number of Java packages which support the regular Java file access (although more constrained, from a television perspective), but also allow for asynchronous file load, caching hints and notifications of object changes.

Java has problems representing graphics attractively, and MHP uses the built-in libraries of Java for graphics; but adds a set of basic widgets for buttons, text entry, icons, dialog boxes and animations. User input is assumed to be done using the remote control. Windows management, however, is done using the HScene package, developed as a replacement for frames by HAVi. It provides overlays and focus, and scopes the availability of applications to input (including creating a kind of user interface based sandbox, which extends the sandbox model of Java). However, Java has a number of problems, and they were not solved by MHP. This comes primarily from the security model, where the MHP model replaces something half-broken with a rigid, unmanageable (but secure) structure.

Channel Switching

A typical behavior while watching television involves the user switching through the tens or hundreds of channels to escape from boring content (often advertising) and getting to something of interest. This behavior is exactly the same in IPTV, but the way it works is completely different from analog television.

In analog television, the television signal is broadcast over radio (or using a coaxial cable). Several channels are broadcast in parallel using separate radio frequencies. When the user changes channel, the tuner in the television set switches from one broadcast frequency to another, making the received signal change and hence the content.

In digital terrestrial broadcast, the channels are also broadcast in slightly different frequency bands, but they are separated using digital encoding. In the same way as in the analog system, the user switches between channels by switching radio frequencies. Since the signal is now digital, there will be a small delay while the signal is decoded at the receiver (there is a delay at the encoder end too, which is why digital television can never be perfectly real time). The user will experience this as a delay between pressing the channel switch button and the channel change actually happening.

In IPTV, there is no need to separate the channels through different frequencies – and no way of doing so either. All content is sent at the same time through the Internet, and separated by virtue of being in data packets with different addresses. All the different receivers in an area receive the same content, which is filtered by their routers, gradually narrowing it down to the individual household, in which the individual receivers filter out everything but the channel they are listening to, capture the packets, and present it to the user.

There are several ways around this. One is to make the bandwidth smaller by more efficient compression. This can be done using MPEG-4 and the variants of that encoding, which helps squeeze more of the redundant bits out of the television signal. A second way is to prioritize the network traffic containing television signals. Another way is to decrease the number of simultaneous channels that the network has to handle, which makes transmission to the edge of the network faster.

The router closest to the user is the least capable in the chain of routers in the network. It is here where the congestion potentially becomes the worst – especially if there are a number of users who want to watch different television programs at the same time. Upstream, where the signals from an entire neighborhood are aggregated, it becomes even more problematic.

The solution to this problem is to send the traffic only once, and create groups that the user can join to receive it. This method is called multicast (when there is a one-to-one relationship between the sender and the receiver, it is called unicast). Multicast only works for signals where the receivers want to get the same program at the same time – for example, watching a live football game – it works perfectly, especially if it is combined with quality of service.

Multicast is different from broadcast – when data is broadcast, it is sent out indiscriminately, that is, not to any particular receiver; when data is multicast, it is sent to a group of receivers. Broadcasting all IPTV channels on the Internet would flood it with unnecessary data (from the point of view of those who are not interested in it, of course). Multicasting, although somewhat more discriminatory, still means that there is no one-to-one relation between the receiver and the transmitter – there is no way for the football fan to pause the game, or for the fly fishing fiend to watch a particular catch again and again (which there is when you use a VoD service, since the connection is set up exclusively between the sender and the receiver, in this case the VoD service and the viewer).

Another major difference between unicast and multicast is channel switching. It is no coincidence that unicast is mostly applied to VoD services, such as viewing movies. When you are receiving a unicast, your commands have to go all the way back to the transmitter to change it. Changing channels has to be done at the head end, since it would be incredibly wasteful to transmit all channels to all users simultaneously, so they could change channels in the same way as in an analog system. It would congest the network so much that nobody could watch anything.

In a multicast network, however, the data stream keeps coming, and you have to join it to get it (actually, it is copied from the interface in the router where it comes into the interface where there is a member of the multicast group who should receive it). Joining a multicast group means simply registering your address with that group, and then whatever is received by the group will also be received by you. This is done using a protocol called the Internet Group Management Protocol (IGMP).

However, if the users had to think about which IGMP address to type every time they wanted to join a channel, there would not be many people switching channels (and possibly not many people watching IPTV either). Therefore the buttons on the remote control, and the step switch which enables channel zapping, must be set to connect to the correct IGMP addresses.

This has to be done as part of the setup of the system, because the channels which should be made available can vary depending on which user is watching (the most drastic reason for this being parental control, where a guardian wants to exclude minors from watching certain content). Which channels you are allowed to watch become part of your personal profile.

If you actively have to log on (in itself a hassle), and then have to select the channels from your profile, this may detract from the viewing of television. Worse, it will alienate the user, since the broadcast experience is that you switch on the television and see the channel you watched when you switched off.

Using the IMS identity model, it is possible to have a "default" identity, which is active when nobody is actively logged in. It is likely that the first time you use the television, you have to tune it, just as you have to do with broadcast digital (and analog) television – in effect, mapping the buttons on the remote control to the channels. In the case of an IMS-based IPTV system, however, the mapping is rather a mapping to a multicast address, using SIP to set up and change the multicast address when required.

Speeding Up Channel Switching

As we mentioned before, there will always be a delay in a digital network. In IPTV, you need to receive a number of packets into the buffer of the decoder before you start decoding; otherwise, you may experience jitter (the network equivalent of an icy road), that is, packets come in bursts, or not at all. Also, if using MPEG-2 and MPEG-4, the decoding cannot start just anywhere in the data stream, it has to begin in an i-frame, which contains the information necessary for the decoding to make sense of the following frames. Typically i-frames are sent with a periodicity of about 0.5 to 5 seconds, depending on the encoding and the type of media (in a seascape panorama, you do not have the same need to change the image as you do in a New York street scene).

There are many tricks that can be played on this, such as speeding up the frequency of i-frames when a channel switch is likely, and the exact method is still being considered by standardization bodies; it is also the subject of battles between proprietary, semi-proprietary and standard solutions, because one of the most frustrating things to a user is if it takes a long time to switch channels. A rule of thumb from human–computer interaction is that it should take less than three seconds to perform any interaction in the network. Trimming each of the delays, it is possible (in labs, at least) to become as fast as a digital broadcast system (which sends its signals over radio, but works in the same way as IPTV when it comes to the decoding of the signal).

If you are using IMS, that will mean setting up a SIP session to handle the multicast stream. This means additional delays if it has to be done every time you change channels. It does not have to be, however. The IMS infrastructure is concerned only when the characteristics of the stream have to be changed, so that a new QoS management has to be set up. As long as the multicast streams do not change, there is no need to set up a new session to handle them. If all other characteristics of the multicast stream (including admission control and charging model) are the same, it is possible to bundle multiple IGMP streams within one single SIP session.

When all users on a network segment have left an IGMP group (i.e. sent an IGMP leave), it would take up too much bandwidth to keep sending the feed (unless they plan to join it again soon). So the IGMP router where the last receiver is located has to turn off the stream from those interfaces where the data is no longer received.

Switching channels in an IPTV system is not very different from switching channels in a regular television set, at least not for the user. Nor is video on demand. As long as disc space is cheaper than network capacity, and as long as the networks cannot guarantee latencies that are below the tolerable levels for user viewing, moving interaction and recording closer to the user will be an interesting option for service providers – since it not only improves on the user experience by speeding up responses, it also offloads the network from handling traffic more than once.

Having a cache in the user's set-top box is one option, and while this can be a boon to response times, it hardly helps the user experience (except, possibly, in the case of advertising). Downloading programs to the set-top box does improve the user experience by letting the user play and interact without having to go back to the broadcaster every time.

IPTV in Japan

As Japan began its reconstruction after World War II and was building up to becoming the second largest economy in the world, there were three keywords which every "salaryman" was working towards: refrigerator, washing machine, television in the 1950s; color television, car, air conditioning in the 1960s. These symbolized the modern world in post-war Japanese society, and television sets were one of the drivers for the growth of the big Japanese electronics firms such as Sony, Panasonic and Toshiba. Despite early experiments, national broadcasts did not start until after World War II, during the American occupation – using the American NTSC standard.

With typical Japanese drive, they were soon in the lead, and the Japanese national broadcaster NHK has demonstrated several world firsts, from high definition broadcast to holographic television. And, having pioneered both the video cassette recorder and the CD, Japanese consumer electronics manufacturers have created game plans for taking the lead in HDTV – from television sets to recording media, in the form of Blu-ray and HD-DVD.

The Japanese have also taken video over the Internet to their hearts, and are second only to the Americans when it comes to accessing YouTube. Home-grown Japanese services have taken over the pole position, though, with an important twist: they can charge for their services. And the content is not home videos, it is professionally produced.

The use of the Japanese services is also much higher than that of Japanese YouTube usage. According to NetRatings, while there were accesses from 11.6 million registered users on YouTube from all over the world in May 2007, the leading Niko Niko Doga service had 1.3 million accesses. They typically spend 2 hours 10 minutes on the site, more than double that of YouTube's 1 hour 2 minutes. Competitor video-sharing sites Yahoo Doga (run by SoftBank) had accesses from 5.7 million registered users in May 2007, and GyaO had 4.9 million. Users spent 44 minutes on GyaO and 27 minutes on Yahoo.

The reason for the popularity of Niko Niko Doga is its interactivity. Users can comment on the videos as they are playing, and share their comments with other users. This is not true for the mobile version, however, and that service has been growing by the same leaps and bounds as the fixed-line service. From 3 million registered users in September 2007, the fixed-Internet service has grown to 4 million in November, a trend that continues to grow.

However, when registration and usage are free, the number of registered users is less important than how you can make money from them. And here, Niko Niko Doga has an important twist. Run by mobile games and contents company Dwango, it has 144,000 members who pay a premium of 525 yen per month for quality. Usage is more than double that of YouTube.

The 4 million registered users are not lazy, either. With 3.2 million comments per day, and 59.49 million page views, the 1.43 million unique users watch 15.67 million videos per day. The target is even more ambitious: 9 million registered users, and 0.5 million premium members. International expansion is also on the cards for Niko Niko Doga – it has just started in Taiwan, so far with a modest 20,000 members.

The most popular content is Korean soap operas, a genre the Japanese have a love affair with since housewife heartthrob Bae Yong Joon created a stir with *Winter Sonata*, a tear-jerking love story about a couple in trouble and their undying love. So popular are Korean soap operas in Japan that there even are charter tours to the shooting locations of the most famous ones.

What has the potential to change the IPTV world completely is the Japanese leveraging their strong position in consumer electronics. Still the world leader in the development of television sets and video recorders, the Japanese manufacturers have been looking hard for ways to keep ahead of their low-cost Chinese and Korean competitors – and they may have found it in a unique VoD service, named acTVila (pronounced actovila).

Japan has changed the world before, by listening to what consumers want. According to a recent study by Ericsson's Consumerlab, the message is loud and clear: give us televisions in the living room – where we also have the network connections. Already, the flat panel television is the currently unconnected device users most want connected to the Internet.

Long a driver in the DLNA, which aims to make it possible for home devices to connect and share data easily and seamlessly, Sony is also one of the main drivers in the acTVila consortium, together with marquee names such as Sharp, Panasonic/Matsushita, Toshiba and Hitachi.

Available in a free and premium version, the service is based on a simplified browser installed in the television set. When connected to the Internet (in addition to – or instead of – the regular antenna), the television automatically looks up the acTVila portal, which is also available over the Internet. Not all television models do this yet, but when they do, the user can choose from a large number of free videos, and become a subscriber, with access to the premium content.

One reason for doing this now is that the Japanese market for televisions is being renewed – the Japanese analog television network will be shut down in 2011. As in the rest of the world, this is a strong driver for IPTV services, and almost all television sets will be delivered with an Internet connection in 2008.

As of this writing (November 2007) there were 100,000 users – two months after the video service started (the basic version started already in February), but the goal is much more ambitious: 70 million acTVila-capable television sets in 2011, with 10 % of all users being paying subscribers with access to the premium content.

Today, the video content includes Japanese television series, and anime classics such as *Gundam* or the infinitely popular *Sazae-san*. There are also international standbys such as CSI and recent movies. In the portal part of the service, there are television listings, news, weather, games, and other content which bears a striking resemblance to the early i-mode portal. In addition, there is shopping – so far only to view the goods, the ordering is done with the mobile phone or the PC. Also, navigation services, a hit in the busy Japanese cities (without street names), are available for the acTVila users.

The Internet connection in the Japanese television sets has other uses. Sony has created a system to show "widgets" – exactly the same as the popular Web 2.0 technology used by Google and Facebook – in two models of their television sets. Connecting to a server on the Internet, they give users access to information services such as Yahoo, weather forecasts and e-mail.

Video on demand from television sets and expanding use of services for user provided content means lots of traffic in the network, even though IPTV, in the sense of traditional television distributed over the Internet, has yet to take off in Japan, despite a penetration of fiber to the home that is second only to Korea. The Japanese broadband connections are the cheapest in the world, and leading operator NTT targets 20 million Fiber To The Home (FTTH) by March 2011. The Japanese government has launched an ambitious plan to turn all the 25 million broadband users in Japan into fiber-optic users by 2010. This puts entirely new requirements on the network.

With broadband becoming a reality in Japanese homes, users start expecting it to work as well as the telephony network – especially when the two largest suppliers are NTT and KDDI (the second biggest fixed-line operator). For customers, voice over IP (VoIP) has long been a reality, and the quality of service for OAB-J numbers (the regular number series) is dictated by Japanese law.

However, this has raised a number of problems for NTT, which they decided to address in their NGN trials by solving all the QoS problems for NGN in one stroke – and created an open network to boot. Realizing that fiber is only the first step, NTT announced in 2004 that it would build a next generation network, with IP everywhere. Field trials started in December 2006.

Based on the ITU-T standards being worked out in TISPAN, the NTT NGN features high security and reliability. By virtue of being standardized, it is also intended to be open, and interfaces have been released to other actors. Interoperability is being shown in the NTT showrooms in Osaka and Tokyo (fully booked several months in advance), followed by large-scale trials with customers.

The most important feature of the NGN being created in Japan is the quality of service. Using IMS to control the traffic, NTT assumes that four types of communication will dominate in the FTTH networks of the future:

- Bidirectional communication, such as VoIP.

- Unicast communication, such as VoD.

- Multicast communication, such as IPTV with HDTV resolution.

- PPPoE connections, for ISP service setup.

In the showrooms, users can enjoy HDTV video conferencing and videotelephony, as well as demonstrations of telepathology and e-healthcare, showing the stress that the Japanese government places on technology to solve the escalating health problems of its rapidly ageing population (10 % of the Japanese population being over 75 in 2008).

For video on demand with high quality (movies instead of YouTube clips), quality of service management makes it possible to download videos at the same time as using the Internet, without problems such as pixelated artifacts showing up in the transmission of the stream. The technically most complicated area is the encoding and decoding of the video signal. Using the telecom grade H.264/AVC codec, it is possible to shrink HDTV to less than 10 Mbps. This makes sure the network is not overloaded, since even in a fiber optic network, the capacity is not infinite.

In an NGN network, users can get all the advantages of IMS-based IPTV services, such as those developed by Ericsson and currently under standardization in the Open IPTV Forum. These include viewing of missed broadcasts, trick play (back, forward, etc.), and interactive services.

In Japan, the Ministry of Public Management, Home Affairs, Posts, and Telecommunications has used the NTT NGN to validate retransmission of programs broadcast over the digital network using IP multicast. The intent is to cover those areas of the country where the digital broadcast network may not be available from the start, or conditions are unfavorable for other reasons. This has required tweaking the coding of the television signal (from MPEG-2 to H.264), as well as solving the signaling of the service information – both of which NTT has done as part of the support work for the ministry. It is not likely, however, that traditional television will be re-broadcast over the NGN from the start of commercial service.

The infrastructure of the NGN conforms to the international standardization, using IPv4 as well as IPv6, IMS in the core network for control of services, and a number of different suppliers of equipment (including television manufacturers) connecting their products into the network. NTT has also realized that the equipment required will be of a totally different magnitude than today's equipment. Terabit routers will be the norm, and NTT is working on studies to create a "carrier-grade" server platform with Japanese manufacturers. Based on the Advanced Telecommunications Computing Architecture (ACTA), it is intended to run Carrier Grade Linux, an open-source operating system which Ericsson has participated in developing. Ericsson is also active in the ACTA standardization.

IPTV in the Mobile

In Japan, mobile television is literally everywhere. Most new mobile phones have television receivers (popular models include Sharp's Aquos, and SonyEricsson's Bravia models – spinning on the popular television brands). Television receivers are even appearing in other devices – electronic dictionaries being one example.

Mobile television in Japan, however, is still the same as the terrestrial television – using the same technologies. Analog television broadcasts will be switched off in 2011, and in preparation for this, digital broadcast has already begun. As part of this offering, the broadcasters send out their programs on the regular network – but also in a lower resolution format in one segment of the digital multiplex, hence the name "one-seg".

It is these low-resolution signals, originally intended for use in EPGs and similar, that the one-seg receivers use. On the small screen of the mobile handset, the picture becomes crystal clear.

As yet, broadcasters are the main beneficiaries of this trend – together with end-users, who get a new source of entertainment during their famously endless commutes. Operators are so far reduced to being the ones who sell the handsets with the new capabilities – which can have its benefits, the Aquos mobile with television being an early hit for Softbank after their takeover of Vodafone Japan two years ago.

Regulatory changes may yet make different services possible, and the operators are preparing for services. But paradoxically, mobile television is likely to be nothing more than a niche application for operators in Japan – even though it is ubiquitous.

Chapter 4: Designing Interactive IPTV Applications

That users can have a direct influence over a show is a difficult concept for filmmakers. They are used to telling a story without discussion (if there is any, it will take place afterwards). The story does not change because of comments from the audience. Where a traditional television program uses the language of film (loosely based on that of theater), the interactive television program has to base itself on different formats – in particular, standup comedy – which allow for interaction with the audience. When users can interact with the television program, this is when services become interactive. But it is possible to take this approach much further.

The bigger picture: how to use interactivity

One-on-one interactivity over television broadcasts would seem a waste of bandwidth, but the audience can be bigger than one when the interactor is someone who has performed something exceptional within a group – for example, won a prize in a competition, or the winner in a race. However, having someone call a television show and showing that person on the screen does not change the way television is created, instead it replaces the need to send a television crew to the scene with a single reporter equipped with a mobile phone.

Why IPTV? Interactivity, Technologies and Services Johan Hjelm
© 2008 Johan Hjelm

> **(Continued)**
>
> *For a producer, and for the creative team devising new television formats, it is much more complicated to understand how to use the "red button". Up until now, the only opportunity that more than one person has had to interact directly with the television show is through voting, which usually takes place using premium telephone calls or SMS. While voting can change, or even determine, the outcome of a program, there are other ways in which users can interact with television shows.*
>
> *Another type of interaction – which is also one on one – is mashups. Here, the television show is combined with something else, for example a map or subtitles. The other media stream comes from a different source than the video signal (which may be received over the air), and the combination creates a completely new program. Mashups are not the primary focus of this book, but the same technologies employed in creating interactive television can be used for video mashups.*
>
> *Interaction requires three things: a way to capture the user's interactions and send them to the broadcaster; a way for the broadcaster to change the program; and a way for the result to be visible on the television screen. In this chapter, we will look at the last aspect: how to make interaction visible. In the rest of the book, we will look at the other two aspects.*

In this book, we will not make a deep analysis of the cinematographic implications of user interaction, that is left for others to do. We will simply assume that the technologies described in this book are used, and show how they can be combined to create the user experience. The actual program design is not something that will be discussed at any length either. However, just like the standup comedian does not improvise his routine entirely, but composes it from blocks of ready-made jokes, which are combined together to create the show, the script for the interactive show can be created in the same way, and the different possible blocks selected based on the participation.

There are already some shows that work this way, although they tend to be filmed beforehand, with little real interaction from the remote audience (limited, e.g., to the ability to call someone to get an answer). These are usually quiz shows, and this was the first interactive format to be pioneered on television (although it has its roots in radio). *Who Wants to be a Millionaire?* is probably one of the easiest examples.

In creating an interactive show, there are a number of options for gathering and handling information. The main purpose, however, is to make the audience in the studio, the remote audience, the host of the show and the participants interact and create a different experience based on their participation. We will not discuss the actual program production, since that does not really change compared to other live production formats (in particular, talk shows); the host will still receive his instructions from the teleprompter and the audience will still receive the cues about when to laugh and when to vote. The important thing is the scripting, and the role in the production team most affected, which is probably the producer. This is only one example, there are other possible models of this type of production, which can be used to create an interactive show.

Dynamic Creation of Interactive Television

Another important difference in designing an interactive show is that the creation of the show is dynamic, rather than static in the traditional way (apart from the actual program production), and that there is a framework for how the show is produced, rather than a traditional manuscript. The framework consists of the technologies discussed in this book, and their applications; the dynamic part is the composition of the script (and the advertising, although not part of the program). Not only are shows distinguished by their story and the studio design, but also the design of the interaction and the graphic profile of the textual presentation has to be an integrated part of the production. It is also possible to take the interaction from mobile telephones into account when using the IP Multimedia Subsystem (IMS); this is something we will discuss a little bit later. The important thing is that the design of the application is not only a script and a visual, it is a program design as well, and the three parts have to interact when the production is created. As with all multimedia projects, the more all the participants from the different parts are involved from the start, the better it will work out.

From a software designer's perspective, this means working together with a number of other people to create the program. Modules can be reused between different programs, and, as a designer, the more productions you participate in, the more you can reuse the knowledge. Ideally, there should be tools for designing the software and reusable program libraries, which could be combined using a graphical tool, but there are no such tools or libraries available at the moment. Therefore, we will discuss the software design in terms of the interface and protocols. An interactive TV application is different from traditional TV programs, since the flow consists of several smaller flows linked through events, as shown in Figure 4-1.

In a recorded interactive show, each event is a self-contained item, the transition points between each item enables the user to make the dramatic choices and determine the sequence. This means that the user does not have that much freedom to create the story, since the plot is given by the segments the user is only able to interact by connecting the segments. It is different in real-time shows, such as a quiz show, where the idea of letting the questions be affected by the replies of the users is not new. However, it is not yet possible to interact more often than every few minutes, and include the users' demographics, so that the show can be changed depending on who is watching it.

It is not only the programs that can change when the user interacts with them, advertising can also be adapted, both to the user's interaction and to the

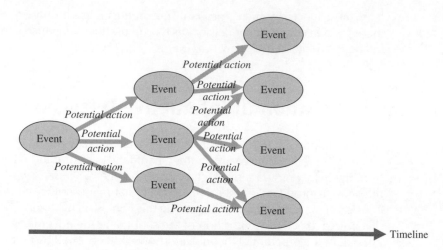

Figure 4-1. Building an interactive story script.

demographics of the viewers – and the latter can be done automatically. There are a number of different ways of designing the matching in the background, and an important aspect here is that the design is not dependent on real-time interaction (a couple of seconds latency is probably sufficient for the processing). However, a real-time interaction (where there is no latency at all) puts much higher requirements on the system, and makes it harder to build.

There is another constraint that can be designed into the program, especially when using IMS. In an IMS system, you can get feedback from the network on how much bandwidth is available, and which QoS parameters are presently set. This can be used (by application servers, which can get this information through the ISC interface) to change the system parameters. In particular, the number of multicast groups can be changed dynamically, especially if different multicast groups are used to handle different advertising demographics; as the bandwidth available decreases, different multicast groups can be merged dynamically. This could potentially also be done with programs, for example, when different multicast groups are used to display different aspects of a program (such as different camera angles in a football game). If the traffic situation worsens, the number of camera angles can be cut; this can be automated, but the automation has to be under control by the producer, so that the editorial integrity of the program is not compromised. This means that service providers have to cooperate in a different way from what they are used to with the broadcaster, especially when they want to provide interactive programs. How this is done depends on their other relations – if the service provider and the broadcaster are the same, this is easy; if the broadcaster and the service provider are different companies, the relation has to be negotiated, and this is likely to be different every time two companies interact.

In a system that does not use IMS, this is much harder to do, even if it is technically possible. The interaction depends on how and what can be retrieved from the routers of the service provider, and this will vary depending on which system is used. If the parameters can be supplied using a standardized format, such as Diffserv, it is possible to extract and use them, but it requires a great deal of work and additional logic (which comes free with IMS). If multicast is not used to manage the different aspects of the broadcast, there is of course nothing that can be done in this regard; if each user gets a personally tailored unicast, this means that the service provider can do nothing but manage the encoding to perform optimizations. While this may be fine if you assume the service provider and the broadcaster are always different and never cooperate, a situation where they do cooperate will help create a more interesting user experience.

Interactive user experience

"Interactivity" is a term misused in the industry, occasionally even meaning that the user can change the channel, as if that was a means of interacting with the show itself. In this book, "interactivity" means that when the user performs an action, something happens to the program being displayed, and that something is transmitted back to the user.

Interactivity is often taken to mean an individual user changing an individual show, this would be very close to a computer game (not that this is necessary a bad thing – computer games have a lot to teach the content industry); but the cinematographic storytelling in a game and a movie are completely different. In a game, there are a number of branching events, which can lead the story in different paths. In a well-written script, the same modules can be reused, and the branching trees merge together into an ending, which may be varied depending on the effect the interactions have had on the characters. It is easier to apply this model to interactive television than to computer games.

There is also a different way of interacting: users acting together to change the show. They may act in concert, or they may act as individuals and the responses collated based on their interactions. When users vote, they act as individuals, but their votes are tallied and affect the result – in this case, the show. The individual and the collated models of interactivity are so far the only ones that have been explored in the industry, for lack of an effective means of users coordinating their actions.

Integrating Interaction in the Script

The model we will use in this example will collate the responses of individual users, and use their responses and the backgrounds of the interaction to change the script of the show, based on a pool of questions and actions of the host that will make the program different – although it will still be a quiz show, with basically the same outcome: a winner will be appointed, and go on to the next level of the tournament; contestants will win prices; and the audience will have been entertained. And not to forget that advertisers will have received a much more attentive and appropriate audience than what they would otherwise get.

The script for this show goes something like this: the show opens with the host and a panel of participants. The interactive component is not active when the show starts. The host poses the first question, which is selected based on the previous programs (or an assumption of the demographics of the show, if this is the first program). The question is a multiple-choice, where the audience has the opportunity to respond using the colored buttons on their remote controls. The panel tries to guess the answers as well. When the question is posed, there is a break (either for advertising or editorial), and then the second question is posed – before the answer to the first question is presented. During the break and the posing of the second question, the responses from the audience and the panel are collected and collated (possibly with different weights, depending on the show premise); the answer to the first question is then presented, and then the third question presented, possibly followed by a break. This is repeated until the show ends.

There is a twist to this program, however. Instead of having a catalog of correct answers, which the user has to provide to progress, this quiz show is based on opinions. The responses of the users will be aggregated to form an average, and the respondent who comes closest to the average opinion will be the winner (it can be made even more complex, for example, by dividing the user opinion according to demographics, but since this is only a model description, there is no need to go into this here).

The show thus becomes interactive in the sense that the responses of the users – and their demographics, as well as the opinions – are used to steer the questions posed and also what the responses will be, since there is no "right" answer. To present one example, the question could be "Which is more delicious: vanilla ice cream, blueberry ice cream, strawberry ice cream, or chocolate ice cream?". Alternative responses can be created based on the audience demographics – for example, a younger audience can be assumed to be more aware of the flavor of cookie dough ice cream than an older generation. The trick to this show is for a script writer to come up with questions where the opinion of the audience is more correct than a response which can be looked up in a dictionary.

The basis for this is what we will discuss later: the profiles and the metadata. Metadata characterizes the content in different ways; profiles characterize users. If users are taken as individuals, the comparison of the user profile with the profile of the content expressed in the metadata can be used to make a selection of which content should be displayed, such as when a user gets personalized advertisements when using video on demand. The metadata has plenty of other uses, however. In the example above, it will be necessary to categorize the questions using metadata, but since the scope is given, the metadata can be designed to fit the show presentation (questions should be characterized in terms of the demographic and topicality, for example if they fit the season; it is not necessary to categorize them in terms of the origin of the ingredients).

At the end of the show (or for each question), one winner is appointed from the audience and from the panel. If there is a consistent winner from the audience, that person can be contacted and participate in a one-to-one competition with the panel, either in this or the next show.

Using Profiles to Adapt the Show

User feedback is always valuable, especially if it can be put in relation to user attention. When users are interacting with a service they are obviously paying attention, especially if they are paying for the privilege. Users who are actually doing something online are hence much more valuable than a generic demographic profile of the probable viewers of a program – because you know they are there. Even if the privacy concerns make it impossible to give out users' personal details unless they have explicitly approved, the aggregation of user profiles can create a common profile of the group of users watching the program, which can be leveraged in determining which advertisements are more appropriate for these users. It can also be used to create an automated auction for those advertising opportunities where users are interesting to a larger group of advertisers.

It is probably necessary to put in a word of warning for producers of such shows. Users are not stupid: if they find themselves at the receiving end of a commercial presentation, they will switch channels. So making the questions and advertising interdependent are not a good idea, however it may sound like a commercially attractive proposition. It will make the show less interesting, and scare users away. If the show and the advertising are independent, but targeted at the same demographic, then the audience attention is likely to be held and the response to the content better.

One effect of the widespread use of personal video recorders is that users are more likely to skip the commercials, and video on the Internet has enhanced this effect, as well as speeding up the viewers' consumption – they tend to be less interested in the responses of the show host, and more in those of the show guests; and they want a quick summary, not a lengthy lecture. A great deal could be written about the cinematography of YouTube, but this is not the place for it.

When users respond to a question online, they typically expect either to be the winner or disappear in the crowd of losers. Winners can be appointed through different algorithms, usually "first past the post", that is the first one who gives a correct answer is the winner. This is typically not revealed until after an interval of some time, which may be used to display a commercial – but the reason is not to display the commercial, it is to enable more users to respond. If users found that they had already lost when they responded, they would soon get tired of responding. On the other hand, when users are voting, there has to be a certain time to allow them to react (and when based on SMS, for very popular programs, time for the servers of the operators to collate and distribute the answers).

In this case, however, it is not possible to crown one user the winner before others, since the result depends on the responses from the audience. So there has to be a period during which the audience responses are collected, and the average computed. During the same time, in another process, the participating audience demographics can be collected from their profiles, and used to create a common profile, which can be used to select questions and advertising, the latter for example using an automated auction system.

When users give responses, they can be collected into a pool and used to generate a common profile. This profile can then be used by completely different systems with completely different algorithms for completely different purposes, especially if the answers provide several options. For example, if users have four different options, the person who comes closest to the average response can be appointed the winner, especially if there is no "right" answer (which, however, is the premise of all current quiz shows).

There are a number of requirements that such a show poses on the system. First, the users have to be able to give answers, and they have to be identified (since one user is to be appointed the winner). Second, all users have to have a profile containing demographic information; that information has to be collected when they send a response, and the profiles aggregated. Third, there has to be a way of characterizing questions (and advertisements) in such a way that they can be compared against the aggregated user profile, and the aggregate used to select an appropriate question (and advertising). The questions have to be possible to characterize using metadata.

There are several ways in which the aggregation can be optimized. For instance, a fixed vocabulary of key words can be implemented in SQL, which speeds up processing. That would mean a traditional a traditional database system (ingesting a number of XML documents, and using the ingested documents to compare entries in a database). Programming databases and describing the different optimizations that have to be done to the system (e.g., normalization of data tables) is not something which we need to go into in this book, since there are many who do a better job of it.

The software design of the system is only one part of the design. There is a second part which is equally important, if not more so, and that is the user interface design. The combination of the two is what makes the show work as an interactive show.

The user interface to the interaction has traditionally been a Java program, which is downloaded into the set-top box and exposed as a widget on the screen, which the user clicks on to interact. There are problems with this. Apart from the problems of latency in downloading Java applications, there is an issue with the overly complex security model of the existing standards, as well as the fact that it is rather hard to program the graphics of an application in Java (it is much better for back-end application, despite its stated front-end goals). Here, JavaScript widgets, such as are becoming widespread on the web, are a much better solution from a programming perspective, being more lightweight and containing the appropriate programming primitives for on-screen interactions. The back-end connectivity can be used to send SOAP messages (XML documents in HTTP), but there is nothing which says that JavaScript cannot connect to a SIP stack equally well as the HTTP stack required to send SOAP messages. This is discussed in Chapter 8.

To display the widgets, there has to be support for JavaScript in the device used as the end-user display. Normally for IPTV, this would be a television set, and the JavaScript support can be built into the television set or the set-top box, if one is used – it is likely that there will be television sets with JavaScript support in the future, as these are already available in Japan (as discussed in Chapter 3). The placement of the interaction elements was also discussed in Chapter 3, and while the standards are not quite set, the use of CSS is a simple way to put the interaction widget in the right place on the screen.

CSS is independent of the absolute coordinates on the screen, instead it makes the system place the elements on the screen in relation to each other. In a coordinate system, the relations between objects are described in relation to the coordinate system; in a system like CSS, the object placement is described by the relation between objects, windows for the object placement also being objects in their own right. This makes it possible to design interaction objects, and place them on the screen, with the same flexibility as on the web. However, there is a constraint in the handling of interaction objects (such as widgets and menus) in an IPTV system: they should be placed so that they do not interfere with the program itself. This means there cannot be a static placement, the objects have to be moved according to the presentation (and the producer has to make sure that there is a reasonably constant spot in which interaction objects can appear, otherwise users will not be able to identify them when they are supposed to interact with them).

Design of Interaction Objects

How to design interaction objects, and how to place them on the screen, is something that is traditionally the domain of the user interface designer. Human–computer interface (HCI) design is a science in its own right, but it often has problems with regard to aesthetics, especially graphics design. The important thing in our TV program example from a user interface design perspective is to entice users to participate, and make the participants interested in continuing to participate. The design of the user interface to the interactive TV application has to interact with the script of the TV show, so that the widgets shown in the application give the same message as the show host, and entice the viewers to interact in a way which interests them.

Such interaction, however, is outside the traditional domain of HCI design. Although completely within the domain of user interface design, it is more like creating a script for a sequence of actions than creating a traditional user interface. The user interface designer has to learn to work with the script writers and the software designers to ensure the user interface of the interaction makes the user interact appropriately with the system to present the expected results.

As luck would have it, there are guidelines that can be leveraged for designing appropriate interaction systems. Two of them come from the two television companies with the longest experience in interactive television: the BBC and Sky Interactive. A third set of guidelines is published by the American CableLabs, which works at designing interactive applications (among other things) for the digital cable systems in the US, which have a backchannel in the same way as a pure Internet connection (some cable networks, indeed, use the Internet protocols for their networks). The BBC, especially, being publicly funded and also interested in driving experimentation, has created a set of guidelines that can be used as a basis for the creation of the user interface of any interactive service.

There are a limited set of things which can be done to enable users to interact with a television program, since the interaction has to be integrated into the program without disturbing the viewing. It becomes even more complicated if the interaction is also to be available to people who are not using the built-in backchannel of IPTV to interact (e.g., users who send SMSs from their mobile phones to interact). Here, the program has to be designed so that the program which is sent over the terrestrial broadcast channel (where interaction is out of band, i.e. done through a different medium, for example the mobile phone) has to create the same user experience as the program which is delivered over IPTV.

Unless, of course, the Internet program is sold (or presented) as a premium service. Then it makes complete sense to have more possibilities available through the IPTV service than the traditional broadcast services. It also makes sense to divide the audience into two groups: those who participate in the premium service, and the rest. In the example program, the two groups could compete against each other, or the non-premium users could guess what the premium users would answer (something which might motivate them to become premium users, which is probably desirable from a service provider's perspective).

This creates a number of requirements on the user interface to the service, which are not covered by the BBC or Sky Interactive guidelines. The BBC has a mandate to make all its services available on an equal basis to all users in the UK, which means it has neither the motivation nor the ability to present a differentiated service. BSkyB could do it, but does not have the tight relationship with service providers which might be necessary to create this (since it wants to be independent from the different service providers). In the US, the different cable-TV providers could do it, or telecom companies such as AT&T, but this might require them to provide their own programming. Some of them do, but it would require a different organization from what they have now.

Human–computer interaction science is much more than a set of rules, however, there are a number of rules which can be applied to the creation of IPTV applications, derived from the way a television set produces the image and the way people perceive color and shapes.

How to Handle Colors

The first guideline when it comes to color is not to forget that there are about 15 % who cannot see particular colors at all. There are two types of color blindness. One type means you can see some colors, but others appear as gray shades. The other type means you cannot see any colors, only shades of gray. For this reason, unless the group of users is guaranteed to have perfect color perception, the design should be done using shades and contrasts as well as color. There are, however, people who have problems seeing (and hearing) even normal television programs, and "design for all" or "appropriate design" is a major subdiscipline of HCI. That is not something we will go deeper into here, but designing so not only people with perfect vision and hearing can enjoy an IPTV show can easily be done using the guidelines which bodies such as the W3C and others have produced (although they discuss the television as a means of consuming web content, they do not consider interactive IPTV applications).

There are a number of other errors which web designers make surprisingly often, such as putting red text on blue background (the most difficult combination to perceive); and not using combinations with the highest contrast (black on yellow). In general, television screens are much more limited than computer screens in terms of color and have a much higher gamma value, which will result in a much higher contrast and saturation display. The colors need to be toned down and desaturated when taken to the television screen, if they are to look good. Many of these problems disappear with HDTV, but since it is likely that many broadcasts will be done in standard definition and with standard color scales designers have to be careful about color.

This holds especially if there is any likelihood that users will view the display on traditional cathode-ray monitors, since the movement of the scan line across the screen can cause moiré patterns, bleeding of strong colors on neutral backgrounds (it will look like the color is flowing), and flicker of pixels which fall on the scan line. These problems largely go away when plasma and LCD displays are used – but if there is any likelihood that there will be users who have only cathode-ray displays, this has to be taken into account in the design.

Changing the color from one strong color to another is also something to be avoided. The contrast can cause overlapping strips to appear, or create a double-exposure effect (such as when looking at a red shape, and looking away to see a green one). With text, there can be glowing edges around the brightest objects. In general, the hottest reds and oranges, and the purest whites and blacks are likely to cause problems.

There are a few issues with screen size, which have to be taken into account as well. The traditional television screen has a 4:3 relation between width and height; but HDTV and most new television screens have a relation of 16:9. This means they are wider than the traditional television set, so content in standard definition has to be shown with black stripes by the sides. There is also a risk of distortion of the image when displaying it in a display proportion for which it is not intended.

Text is also likely to be a problem, as text does not display well in standard definition. On a cathode-ray television screen, letters do not display well (due to their sharp edges); in general, displaying lots of text on a television screen is not a good idea, as users are neither used to or motivated to read text on the television screen.

Generic Interaction Models

There are three generic models of interactive television applications – see Figure 4-2 (as outlined by CableLabs):

- prompt, load and run;
- load, prompt, and run; and
- load and run.

These three models make an important assumption: that the application has to be loaded onto the execution device (e.g., a set-top box or a television set). The first two models assume connectivity. The last model, however, assumes that most of the interaction happens in the local device, and while this may be relevant for certain types of applications, popular interactive television applications do not require loading – a JavaScript widget is loaded with the presentation page (whether this is overlaid over the video or included in the video stream), so there is no need to load. Indeed, loading and waiting are two major turnoffs for users, as plenty of studies have shown – we are used to television being instant, so waiting for things to start is something associated with a computer, and computers have different assumed characteristics than a television set (although technically, nowadays they are pretty much the same).

These three models work in completely different ways from a technical perspective, and the prompting of the user is also different. The first model loads only enough data to present the trigger (and the associated program), and loads the data only when the trigger is selected (or continues to load data in the background until the trigger is selected). In principle, this is how a browser-based overlay works. It brings up the trigger quickly, and ties it to

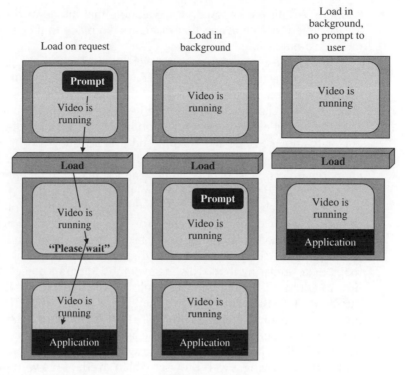

Figure 4-2. Models for interactive television applications.

the interaction; if the latency in the network is high (which is an assumption in the OCAP system), or if the roundtrip delay is big, server-based execution becomes a problem, since it will lead to a perception of latency for the user. When this model is used, it must be used with a protocol which allows for quality of service, especially if there are multiple users using the service.

The second model loads the application, and then presents the trigger. The loading can be done in advance. This model, however, has a problem if the assumption is that the data is loaded at the same time as the program. If it does, there will be a delay from the user switching over to the program until the data is loaded (typically, a set-top box is not able to hold more than one program at a time – at least, not the present models). This model has a huge problem when the user switches between programs, since – unless the data channel is wide, or the downloaded program very small – the system will never catch up; the interaction elements cannot be loaded until a few seconds (in worst case, minutes) after the user has selected the show – and then, he may have got tired and gone away.

A third model is to load the application in the background, and the interaction takes place only in the background without user involvement (or explicit user involvement). This model can be used when, for example, there is data that has to be provisioned; or where there is a user selection before something happens. In this book, however, we assume that the networks and systems are designed in such a way that this can take place in the server, which is more cost-efficient.

This said, the applications are not web pages. The remote control is not a mouse. If interaction is coded to the color buttons, it does not matter where you aim – the blue button means the same wherever it is in the screen. There are just a few choices possible (more if you use the phone and ask people to call different numbers depending on how they want to vote). There is no cursor to move around the screen, and a good thing too. If there were a cursor, the user would have to think about where to place it. Sudoku and chess regardless, most people do not enjoy thinking to be entertained. This is a constraint on the choices, and something the designer has to live with.

Both CableLabs, BBC and Sky Interactive focus on the placement and design of the interaction elements. The placement of the interaction triggers are often done using Photoshop, taking the images from the video and putting the interaction elements into them, tying them to links using image maps or a similar method. This means embedding an object, which has a link, to another action in the video itself. The different platforms that exist for interactive television solve this in completely different ways. The ones which are most successful in terms of usage are those that have a tighter connection with the video stream – although these are the ones hardest to program. The standardization, however, is driving towards interaction which is easier to program. Just having a web browser which overlays the web page over the video is not sufficient – there has to be a better connection between the video and the interaction. This does not have to be technical, though, as witnessed by the strong interest of the users to interact by telephone, when prompted. The lesson is that there has to be a tight integration between the video and the interaction – both technically and editorially.

Triggers, bugs, menus and bridge graphics

In practice, there are just a few interaction elements that can be easily applied to create a service: triggers, bugs, menus, and bridge graphics.

Triggers, or prompts, are the elements which result in a user interaction. Triggers can be widgets, they should not be present all the time (like an icon on the screen), but appear when they are required. They should also not appear in front of an important part of the video. While it is possible to do this automatically, this is the domain of the producer, and the designer has to be flexible enough to enable his design to fit the program – it should not be placed on top of the speaker's face; and it should also not slide around the screen to avoid being placed in the "wrong" position. The BBC requires the trigger graphics to always appear in the upper right corner of the screen; the producer and director can then design the video so that this corner is not important in the action (e.g., things should appear to the left instead of from the right).

A bug looks like a trigger, but is not clickable. It is a small icon or logotype appearing in the corner of a program (not to be mistaken for the station identification logotype) that shows who is sponsoring it, or promotes a coming program *in a graphically pleasing way, or something similar. The only limitation is what is legally allowed, and how much screen real estate there is to place it.*

Bridge graphics can be designed in the same way as bugs, but they are often designed to relate to the next event, since they tell the user that something is about to happen, and that he should wait for it. This is necessary if you have a program which has to be loaded (and very irritating). Even on computers, where people are used to waiting for the system to perform its functions, hourglass or wristwatch icons are an irritation. Bridge graphics should not be designed in the same way as such computer icons; although they could conceivably borrow the visual language from computers. Bridge graphics and bugs can work together and replace each other, and be used as a means outside the program to move the user along to the next sequence, for example telling them to wait during a commercial break before it starts.

The most important part of the interaction graphics, however, are the triggers – and what follows after the trigger. When a trigger is activated, two things can happen: it can lead to an action (e.g., selecting a different camera angle, or different point of view, or longer commercial); or it can bring up a menu. Menus are a mainstay of human–computer interaction, with a history going back to the beginning of graphical presentation of computing (if not earlier). They present the available choices, and enable the user to make these choices.

There are a number of rules of thumb that apply to menus – which have been well substantiated in research and practice. The first is to avoid them. Paradoxical as that may sound, a menu presents the user with multiple choices, and this is directly contrary to the way television programs (and films) are designed. They interrupt the flow, not in the least the flow of thinking, and this means the user experience is lessened (since the entertainment is disrupted). There may be ways of using menus to enhance the user experience – but that would mean designing something very similar to a game, and that is not what interactive television is about, since it means the user leaves the watching (lean-back) mode, and goes into an interactive (lean-forward) mode. The trick in designing any interactive application is to enable the laidback viewing and enhance it – and menus have the opposite effect.

So menus should be used carefully, only when presenting the user with a set of choices that do not interrupt the current viewing (e.g., before a news program starts, selecting summaries of the Olympics, or only highlights from certain types of matches, and so on). This holds regardless of the design of the menus – whether they are only graphics, or graphics and text, or only text. If there is a set of icons which can be used, that is preferable to using text in menus, for a simple reason: the user is not sitting close to the screen.

Designing Menus and Text

When presenting the user with an item on the screen, the user must be able to read it from two to three meters away (the typical viewing distance). Designers of computer graphics must be careful – the computer offers much higher resolution than the typical standard definition television screen, and the user's eyes are closer to the screen, so windows can be smaller. The only reason home videos work so well on YouTube is that the windows are so small.

To start with, cathode ray televisions create problems here, too. A shape with long, straight, vertical edges may blur if combined with other shapes and colors (especially those which cause "bleeding" in the visual perception). Letters tend to have such shapes, and they can blur into each other, especially if the font is too small. Light on dark is better than dark text on a light background on a television screen.

Designing television menus is not dissimilar from designing PowerPoint slides, because the font should not be too small (24 points is probably a minimum), and there should be little enough text that the user can read it in one sweep of the eyes over the screen. Ninety words per screen is a maximum according to the BBC, but the design must also consider that these words are not passive information – the user is expected to make a choice, and has to think about them. Chopping up the text into "chunks", which can be read easily, helps the understanding. The brain works by breaking up the information in our environment, and de-selecting the pieces which should not be worked on further; so helping the user to understand which parts of the menu are significant (e.g., by only including the meaningful parts) will help understanding and hence help the user to perform the correct action with a minimum of thinking.

In other words, the wording is as important as the graphic design. The more the user has to think, the more work it is to make a choice, and the less relaxation and fun is the television viewing; which is exactly counter to the way that interactive television should be designed. Navigation has to support the "lean-back" experience of the television viewer, which is very different from the "lean-forward" experience of the PC user.

The latter is harder than it sounds, because contrary to what happens during most PowerPoint presentations, the user is distracted by what is going on in the background – unless the video stops until the user makes a choice, which causes other technical problems as we discussed in earlier chapters. Indeed, the successful Sky Interactive format uses a markup language which forces the design into "cards", similar to PowerPoint slides. That technology is severely dated, however, and can comfortably be replaced by the technologies now being standardized by the consumer electronics industry.

The design around the framework, how the service designer has set up the user interface, becomes very important since it becomes a problem if the menu has to include actions towards the system. If there always has to be a "back" choice, instead of a "return" button on the remote control, there will be an additional line in the menu that the user has to consider – consciously or unconsciously – and this will slow down the user's understanding, and hence the ability to interact, and hence the way the application can be designed, because the interactive television application has to be designed on a timeline. If you have to include actions toward the system, having two choices within three seconds will feel difficult for the user, since the thinking time may exceed this. If there is no interaction within eight seconds, the user will switch (that is according to the BBC; other research says three seconds – so the shorter, the better).

Selecting typefaces can also make the reading easier – or more difficult. There are typefaces which work well on television (typically sans-serifs), and they tend to be those which do not work well in print. Then, the broadcaster may have special requirements on the design – they may have graphical profiles that the designer has to follow, which normally includes the typeface; these typefaces may be specially designed. This depends on the broadcaster, and has to be investigated as part of the design process.

A lesson from this is that the graphics – including the text – has to be designed so that it can be removed and changed. In practice, this means exposing it in a separate plane, which probably means an overlaid browser page. There are two reasons to do this. The first is internationalization: if the program is to be presented in a different language (e.g., a television show from Quebec being translated into Spanish), the menu choices must follow this. Using graphic icons helps here, of course, since that means no translation of the text, but then, the icons – and the bugs – have to be reusable, and the bugs have to be designed so that they are not dependent on the current time. If a program is re-broadcast, it makes no sense to tell the users that they cannot call a phone number, or that they cannot select a choice in a menu or click on an icon. It is easier to remove it.

That means the menus – and bugs and icons – should be generated depending on the broadcast occasion and location, and also on the service provider's preferences. Unfortunately, manufacturers of consumer electronics equipments and service providers have not yet been able to sit down and agree on a minimum set of actions to be built into the remote control. The Consumer Electronics Association, which has standardized the browser for consumer electronics equipment, has taken a first step, but the service provider angle has not been considered.

One reason for this is that there is a fight behind the curtains of the interactive television industry: Who should own the user experience? Having a consistent user experience is as important, maybe more important, than having a user interface that is well designed. People can forgive the occasional blooper if they feel they know what they are doing, and to do that, buttons and menu selections should have the same meaning wherever they are. Is this because you are using the BBC, or Sony? Or even Telewest (a British cable television provider)? Who should be in control? That is not yet settled, and there does not seem to be any academic research that could help. So far, however, the broadcaster has the upper hand: without them, there would not be any interactive services, so they get to decide.

One such consistency issue is the remote control. Today most remote controls come with color buttons; the only one that has a consistent meaning is the "red" button, which activates the service (and then only in some countries). Being consistent, first within the program and then within the service, will decrease the efforts, and hence the enjoyment, of the user. The same goes for the user of numerical keys. If the user does not remember what to do, it would be useful to have help text; but only if that paused the viewing. The timeline cannot continue while the user is trying to figure out what to do, then the interaction opportunity will be lost, so help text, while it sounds like a good idea, is actually bad. If you need to press "help" in an interactive television application, the design is wrong.

HCI rules for interface design

In the HCI community, there are several well-researched rules for what an interface should do. It should, for example, support the user with knowledge about where they are and what has happened; ideally also what will happen when they press a button (something that can be accomplished by having different types of buttons for different types of interactions, e.g., one type for voting, one type for donating, one type for buying, or whatever interactions there may be). The viewer, after all, only gives the television show three seconds to tell him what will happen. Consistency and predictability are de rigeur, in other words. So the user interface has to be intuitive, understandable without thinking, and not involve too much reading. That means it must relate to the cultural mental models and metaphors of the audience: a British interactive program will be very different from one used in Japan. At the same time, users do not like to be given too few choices, or being forced into a structure of thought; if you give users a navigational structure, they have to be free to navigate within the constraints of that structure. This means the designer has to understand the inherent semantics of the design, and that is much harder than it sounds, especially since many designers do not share the users' perception of what the navigational metaphors might mean. Freedom to navigate also means freedom not to navigate; to leave, or not participate. Having that choice is important; if the television starts behaving like a computer, and hangs because the viewer did not press the right button at the right time, television viewing will start declining rapidly.

The linear flow means there is a limited time to interact (as we noted in Chapter 1, the best trigger is the presenter asking users to interact). That, and the audience having less technical know-how than the users of Internet, means the punishment for breakdowns is much higher than in an Internet application, where the user can pause and come back again if there is an error message. In interactive television, it either has to work or you have lost the user. If you are charging the user for interaction, this may mean a significant loss of income. So everything has to work. That means testing is more important than it is in Internet applications. Luckily, there is a lot that interactive television designers can learn about testing from the Internet.

Testing Interactive Applications

Web designers have a tradition of "shoot and go" design, where a service is created and released to the users as fast as possible, without any testing to see if things actually work. In the world of telecoms, however, the tradition is completely the reverse, testing is done to verify every crossed t and dotted i. Often, the testing takes more time than the development, and bug fixing is a major effort. But then, the service is guaranteed to run with less than 1 % probability of failure.

Interactive television has no method of testing, since the services are created within a tested framework (if the MPEG encoder and decoder works, and the transport works, the video will work). There is no need to test an individual video. In the same way, testing individual designs for interactive programs will not work either, since the effort of testing will be disproportional to the development effort (which, in turn, has to be proportional to the effort of traditional production, if the program is to work as an economic proposition).

As long as the framework is not established, it will not be possible to test it. Assuming, however, that there is a framework and that it has been tested (which is probably a condition for its use), there will still have to be a test for the individual program, so that it is usable. This will require human inspection. Even though there are automated methods, which can be used to identify objects placed on top of each other, there is no method to determine whether the placement may be appropriate (if an object is to be framed to enable interaction by clicking on it, for example).

A designer can look through the program and see if there are any examples of an object being in the wrong place. But before a program is aired, the only traditional way to get feedback on the storyline and the storytelling is to show it to a focus group or a closed audience. This is routine in the movie industry, where a film can be completely changed depending on feedback from the audience. It is also routine for most television productions. However, the feedback required to change an interactive production is not the same as that required to change a television series. Also the collection of the interactive feedback can be made much cheaper than the collection of feedback concerning a storyline.

In the software industry, in particular the Web 2.0 part of it, there is a method of user-testing applications, which is "quick and dirty", and this is the name usually applied to this type of testing. The advantage is that the testing can be done before the application is finished, which makes it easy to change parts of the design that do not work (e.g., the design of the interaction widgets and their placement on the screen). It is based on the imagination of the users, and it has to be adapted to television slightly, but such testing makes it possible to see very quickly if something is wrong with an application, and how it should be corrected.

When developing an interactive application, especially mobile applications but also web pages that allow for more interaction than just following links, there are two parts to the testing process. The first is a quality assurance part, simply making sure that all the links and other interaction elements work, and give the expected result. This is surprisingly often ignored, with resulting user dissatisfaction. It may be acceptable when designing web pages, where the expectations are low to start with, but it is not acceptable in interactive television, where the users may even have paid extra to become part of a premium service – and the service provider could then become liable if it did not deliver a working service.

Quick and Dirty User Testing

Therefore, quality assurance is the first part of the testing of applications. The second part is testing the user interaction. These two parts can be done independently, and be combined when the application is installed in the interactive television system (another reason to make the design as independent as possible from the actions that it calls). Just as many early websites were tripped up by the PC pixels being square and the Mac pixels being rectangular, many interactive television applications, which are not tested in the target environment, are likely to be tripped up by the television pixel being slightly broader than it is high. This kind of amateurish mistake may work when designing websites, which can be corrected between times when they are viewed, but it can be a problem in IPTV applications, which essentially only exist during the short time the program is aired – unless the applications are used in a VoD application.

Following the "quick and dirty" methodology of mobile and website testing, the service is constructed as a set of paper cards. Each card represents a screen – something easier to do in interactive television services than on the web, since the linear flow of television means that the interactions are constrained within the current scene. Moving through the cards leads the user to walk through the application as if they were watching it on the screen. Of course, the experience of watching a video is very different from watching a sequence of still pictures, but as long as the interactions are tied to scene changes, screen changes will give the same effect in terms of user feedback.

Creating the cards is easy; it actually helps if they are not too professional, since the simpler the design the easier it is for the user to provide feedback. Paste the interaction elements on a photo representing the scene, and put it in front of the user. Record everything the user does as a consequence of the interaction. If they are doubtful, and especially if they make a mistake, the design has to be fixed. The important thing about this method is that it does not give you an answer as to whether the design is right; but it does tell you when it is wrong. Trying the proposed design with five users is sufficient to isolate problems that are not the result of the reaction of an individual user.

As mentioned above, the problem with this method is that it does not automatically tell you what is wrong; you have to change the design, and investigate again, until the user does not make any more mistakes. In practice, this is where the designer can have a lot of help by human–computer interaction design, since even though it is not a method to correct mistakes, it can be used as a knowledge base to inform you what types of errors will lead to mistakes, and how they should be corrected.

A special issue is that the broadcaster may want to approve all the graphics in the program. The approval process can work in different ways, but the consequence is still that until the design is approved, it may have to be continually redesigned. This is why it is so important to follow the style guides by the broadcasters (if any exist), and why it is important for anyone who wants to design for more than one broadcast channel to make the widget design interchangeable. The placement and the result of the interaction is the same, but the look and feel of the widget is different.

In practice, the design is also one of the major differentiators between programs, and part of what is licensed with the program concept. For an interactive program creator, the tradeoff between following the interactive style guide of the broadcaster and creating an own image is difficult. There is no way to provide generic guidance; each producer has to make his own choices.

Making Mashups in IMS-Controlled Interactive IPTV

Mashups, in Web 2.0 terms, are applications that merge different data sources into a visualization, creating a new application. The first mashup was a map of the San Francisco Bay Area with apartments for lease overlaid on it – there are many mashups that perform similar services. Add a timeline, and a mashup would make a perfect interactive television program. As a matter of fact, it would become a traditional visualization of two datasets over time, which is something much more familiar than the static instantaneous view of the mashup (think "apartments available for lease in the Bay Area every year since 1849"). In traditional television terms, these mashups are always displayed within the scope of a program (e.g., Al Gore showing an animation of how the sea levels will rise over time); the novel aspect that mashups can add to interactive television is an animation that can change depending on user input. How much it can change, and what the changes may mean, are the interesting parts.

There are two models for making mashups: horizontal and vertical (see Figure 4-3). The Web 2.0 mashups are horizontal, the mashups which interactive IPTV enable are vertical. There are three differences: display, delivery and integration point. To confuse things, these models can be combined (mashed up, if you will) in various ways. And then, there is the major difference: timing – a function of the delivery and the display.

To start with the integration point. This is where the different data sources are merged to form the new application. In Web 2.0 mashups, the assumption is that this takes place in the client, because there are two or more data streams that are supposed to come together, and that come from different services (which, incidentally, can be exposed in the same client). For example,

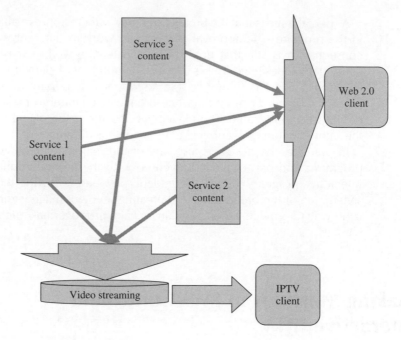

Figure 4-3. Mashup models.

the apartment listings and the Google maps were both intended to be viewed in a browser; with a little programming, one could be superimposed on the other.

If that is to work, there has to be a programming environment in the client that can expose one dataset on top of the other, and add interaction with visualization. The client must also have the capability to receive or retrieve the two datasets (one of them can be video).

There are several problems with this from a television perspective, as well as from an interactive television perspective. For example, it assumes that there is a fairly capable programming environment. This is available in certain types of set-top boxes, but not in most. It also assumes that there is a means to receive or retrieve the data; this is possible in an IPTV environment (but not in others), since there is a connection to a managed overlay on top of the Internet, and it can be used to connect a browser to many different sources, as well as the television signal, at the same time. But the more data, the more memory and processor capability required, apart from the correct programming environment.

In the vertical model, integration happens in the server. This is a more traditional deployment of computer systems – taking two databases and merging them together is nothing new, and nothing that any programmer with some experience will need to think twice about. The only problem is if the data is not available in formats that can be easily translated into each other. Comma-separated data, which can be imported into a database without any work, is best; XML is also easy. Screen-scraping, that is, deconstructing an HTML file and retrieving the actual data from it, is worst. Not only is the consistency and data quality hard to maintain, it also introduces a delay, since there is more processing required to get the data out.

There is an advantage to the vertical model as well: the application server in the IMS can access a number of data sources which relate to the service usage through the ISC interface, for example the user profile and presence information. From a software design perspective, it may be more feasible to have a separate server that handles the creation of the merged datasets and the animation; this can then be sent to a separate streaming server for delivery to the user. The exact design will depend on the system under construction, but the principle is different from the client-based mashup: the data is mixed before it goes to the client. Only the interaction elements are added.

That has impacts on delivery. If a mashup is to be meaningful, it has to happen within the same timeframe as other interactive television applications – which means three seconds. This is not only the time for the retrieval of the data from the different sources, the calculation of the overlay and creation of the display, but also for the command to go from the user to the server, and the visualization to go from the server to the user. Retrieval can be speeded up through caching or pre-fetch (if the user retrieves apartment data from Palo Alto, he can move either to Menlo Park, Santa Clara, or San José next – so the data from those cities can also be fetched). The data can be stored in a cache so the next user can get it (or the same user can get it the next time) without fetching it from the origin server again.

The actual mashing up of the data can be done using a database system, and this is typically faster than using the browser in a set-top box. The database system is specialized for this type of application, and it can process multiple threads and multiple users at the same time. The set-top box has little memory, a small processor, and a limited environment; except for the specialized video processing circuitry, the cheapest possible components are used to hold down the price. The same is true when the client part is integrated into a television set.

Another bottleneck that can be removed through having the retrieval in the server is the network. A server can have massive bandwidth and perhaps even direct connections to the origin servers, minimizing the latency and getting the data faster. Clients are, by necessity, a few more hops away, which means a few milliseconds delay regardless of how well designed the network is. They not only have lower bandwidth, they also may share the connection with other users, which means lower capacity and potentially higher latency. Even if it were theoretically possible to apply quality of service to the mashup communication (which is difficult, since it typically is retrieved over HTTP, and you would have to look inside the data packets to find out that they were important – which means delay), it would cost resources and make the link to the service more expensive for the end-user.

The third issue is the display. This breaks down into two different aspects: how the display is done technically; and how much interaction is required.

How do you interact with a mashup using a remote control? A mashup assumes the user can give input about the data streams which have been mashed up. To interact with the mashup application, the user has only the color and number buttons. Interaction must be limited to those – which means the designer has to think quite a bit about how the application should be designed. What does it mean to press the blue button in a mashup – and how do you visualize it in a way that does not require any thought from the user? This is really difficult, and may require verbal presentation to make sense of the interaction, which means the mashup has to be designed as a combination of video or audio segments, interactive elements, data retrieval, and visualization. And if it is to be overlaid on a television program, it has to be designed to follow a timeline.

The timeline is what causes the most problems for the browser-based mashup model in interactive television. There is no inherent timeline in most Web 2.0 mashups – they exist only as snapshots of the dataset at an instant in time, that instant mostly assumed to be now. If there is no interaction with the mashup, it is very similar conceptually to the still images that exist in television, such as weather maps; if a short times series is added, the effect is the same as the visualizations that already exist (e.g., weather patterns, changes in sunrise and sunset times during spring and autumn). If this fits into a different timeline, such as a news presentation, it means the program has to be halted while the user interacts with the mashup. If the mashup is meaningful, it would mean changing the story, branching out into a different storyline, something which is not technically hard to do (just load another video in the background), but hard from a cinematographic point of view (what about continuity?).

We have already discussed timing delays in delivery; combined with the difficulties of interaction and the importance of the timeline in television, the mashups that are suitable for interactive television are those which can present a snapshot (and a snapshot which the user can come back to, creating familiarity by re-use). Having a number or small mashups that appear as part of the lineup of a television program are likely to make users more interested in coming back, especially if they can be combined with the editorial content in a way that enhances it. If the user's profile can be leveraged as well (easy to do technically, but hard when it comes to real-time viewing), the result will be something which the user can come back to mentally and interact with (just as morning shows repeat the weather map a number of times).

There is no limit to the different types of mashups that are possible, this only depends on the data available. Stock portfolios, currencies, electricity prices, other types of data that have a real-time pricing can be used, especially if they can be combined with other types of data (e.g., in countries heavily dependent on hydroelectric or wind power, overlaying the wind strength or level of water in dams with electricity price and weather trends). The limit is the data sources available, and the imagination of the designer.

When it comes to data sources, there is a caveat that has to be added. No data is free to use without permission. Sometimes, this permission is implicit, published along with the data (e.g., when data is published under a Creative Commons license, or when out-of-copyright data is published). Databases are not protected by copyright in the US, but in Europe they have a special type of protection. And, regardless of the legal protection, getting data is the result of an agreement. Individual users may claim fair use, but as soon as they publish the results (and especially if they intend to make money from it), the rules enabling a private individual to make and share copies with other individuals do not apply. Anyone publishing a mashup has to make sure the databases are used with the proper agreements. This is the same as discussed in Chapter 12, since although the ownership of databases typically is different from the ownership of video, the same rules apply.

User-Provided Content

It used to be very complicated to create content for television, but three things have happened through the constant improvements of the Internet, computers and video technologies: it has become easier to record content, produce content, and distribute content.

There are two types of content a user can create for interactive television: video and interaction. Interaction elements can be overlaid over any video (using some technologies); in some examples, such as "crows" and subtitles (a voice track with comments, or a text with translations), these are interdependent, but not interlinked, with the video. Interaction that is interlinked with the video is, for example, triggered by the announcer. That interaction is typically in a controlled format, since it can only lead to a few well-controlled results.

Users can provide any type of content, but here, too, there are two interdependency dimensions: if the user-provided content is independent, or if it is dependent of a television program. When uploading a video to YouTube of her baby dancing, the user is providing a piece of independent content. It does not have anything to do with any television show. When, however, the user uploads a video of her baby in a *Star Trek* costume singing the anthem of the United Federation of Planets, there is a strong dependency on a television show. A continuation of this is when the users write new (imagined) episodes of their favorite televisions show (fan-produced content), something that lately has also taken the shape of videos produced in the image of television shows.

There is a dilemma for the copyright owner here. On the one hand, if people produce content which builds on the show, they loved it and are committed to it. On the other hand, that means they are infringing on the copyright, and this may damage the story, merchandising, and other aspects relating to making money from the show. Apart from obvious examples such as pornography, shutting down a user-produced show, created by people who have committed to it to the extent that they are identifying with it, will create enormous amounts of badwill, of which there have been several famous examples in recent years. Better to create a model where the process is under control, such as letting the fans submit scripts, which are read by the scriptwriters for the show, and vetting the fan-produced content to put it on an "official" site.

Technically, however, creating a derivative show (one that builds on the same concepts) and distributing it is not very different from creating and producing the show itself. The actual filming and post-production is of course differently managed, but once the file is encoded, the distribution system does not care whether it is produced by two guys in a garage, or Cecil B. de Mille. The technology is the same. Only the system to manage the submission of the content is different. Systems for PC-based viewing such as YouTube and its competitors are technically nothing but huge video on demand servers, but from a social perspective they are something completely different.

User-provided content can be a derivative video, that is, one that builds on an original concept. This is what is problematic from a copyright point of view. If the user-provided content is not video, but creates something different which is attached to the show, then this is a different matter. If the only thing that is distributed is the subtitle (or crow), and not the show itself, there is nothing the copyright owner can do or say about it. A good translation is an independent work. So is an interaction. If a user is clever enough to create an interaction sequence on top of the program that enhances it, then he is free to distribute it to anyone, as long as they get the actual program through legal means.

Technically, however, this is a bit more complex. The first question is what the interaction means; the second is where the mixing takes place, and how it is done. Finally, there is the matter of how the application data and the video are distributed to the viewer.

First, what the interaction means. If the interaction is created by the broadcaster, voting, for example, means making one of the contestants win. But if the show you are watching is a replay of the original *Andy Pandy Show*, what is possible? The show cannot be changed, since it was transmitted 50 years ago. What would voting mean? What other types of interactions could there be?

Shows as popular as *Star Trek* have their own base of fanatical fans who are willing to shell out several hundred dollars for a specially designed box with the first edition DVDs. These fans are willing to communicate with each other, especially when watching the same episode of the show. An IMS-based IPTV system makes it possible to build applications which let you see who in your (extended) list of friends are watching what, and get an alert when they start watching a particular show or episode. Such implicit interaction does not have to be run by the television station (or the VoD provider); the fan club itself could run it.

Another example of interaction based on old programs using the technologies described in this book is creating statistics. In some sports, such as baseball, there is a large number of different statistics that can be collected about each player and each match. If someone is fanatical enough to want to find out all that can conceivably be known about a certain star player, there may be particular statistics that he will want to see when the star appears on the screen.

Technically, to show statistics on the screen at the same time as a certain star appears requires three things: that the statistics are available for display; that they can be mixed in with or be overlaid on the video signal; and that the star can be identified so that the statistics can be displayed appropriately.

The first constraint is the mixing with, or overlay on top of, the video signal. If the receiver has a web browser that is used for the interaction, that is, displays the triggers as Javascript icons on top of the video, then the mixing takes place in the terminal. Interaction and video stream in principle do not meet until they are combined in the display. Interactions are decoupled from the video. That this will happen is the assumption in this book, since the consumer industry is firmly behind the CEA-2014-A standard.

There are two alternatives. The first is that the mixing takes place in the display, but is not done using a browser. The triggers are mixed into the video stream in the display device (set-top box or television set). The execution environment in the display device, for example an MHP or OCAP Java environment, executes the programs which result in the content. There are two problems with this approach. The first is how the programs get to the display. If the content provider is not the service provider (which is the assumption in the case of user-provided content), then the content must get to the display some other way than along with the media stream. This means retrieving it from a different source, and that in turn implies a client capable of this – an HTTP client, probably, if not a web browser. Theoretically, there could be a registry for user-provided content run by the service provider, which users could register to and get the updated content through a SIP Notify as a result of a SIP Subscribe, or as a SIP Message. That would require using this for the notification mechanism in the IPTV system, but if the content is larger than a few lines, this is not a good approach – SIP Subscribe and Notify, and especially SIP Message, are not intended for file transfer. The SIP-based FLUTE protocol could potentially be used, but there are still two issues which have to be addressed: the discovery of the service, and the registration of the user to receive information.

If the service provider runs a registry, the discovery becomes easy: anyone who has produced content can register, and everyone else can get it. This is no different from other similar systems. However, if there is no registry, the discovery has to be done out of band from the IPTV system – that is to say, there has to be a web page (and a browser to read it) with links on which the user can click, to either download the service (i.e. the statistics) right away, or get a notification when it is updated. This assumes there is a web browser in the device.

The second alternative is that the mixing of the content takes place at the service provider. This is possible if the user has a way to register with the service provider that he wants the content, and a way of discovering it. The easiest way is if the service provider runs a registry, and the users add the services to their profile; this can be done automatically if they have a way of registering. An enterprising service provider could, for example, do this as a revenue-sharing service, where the users pay a small fee for the use of other user's content, and a proportion of this is shared with the user who created the content. In IMS-based IPTV systems this is relatively easy to set up (it requires setting rules in the rating engine, and registering the charges and payments in the user's account – something done automatically anyway). This type of approach has been very successful in mobile systems. The mobile network operator 3 in Italy, for example, ran a service (which design-wise is very similar to YouTube) that let users take short films with their mobile phones, and pay a few cents for viewing it; when someone viewed your film, you got paid a percentage. The people who produced hits, with millions of views, ended up rather wealthy; the other users had fun, and the service provider got paid, so everybody was a winner. The same approach can easily be applied to a system to manage user-provided content in an IMS-based IPTV system.

In the second case, the content is mixed in with the video stream and delivered as part of the video stream to the viewer. This means no interaction with the content decoupled from the video stream is possible, but it does solve a rather tricky problem: synchronization with the video stream, which can be done in the server. To figure out when to expose information, the meta-information that the system generates (e.g., MPEG-7) can be used to identify when a certain player comes on, and an application in the server look up the user's profile and puts in the appropriate statistics.

The synchronization problem applies to all services that work with content from a service provider without actually being part of it: subtitles, crows, statistics and anything else that the user may provide that needs to be synchronized with the media itself.

If the synchronization with the service is not done by mixing in the content in the video stream (essentially in the application server), this becomes much more complicated. There is, especially if it is receiving an MPEG-2 stream, no way for the client to identify what happens in the video. The only synchronization that is automatically available is when the video started and ended. When a client wants to provide a service on top of a media stream, the media stream has to be the particular version for which the service was developed. Otherwise, the synchronization may or may not work, because there is a risk that it has been changed.

While it is possible to add content to a video in this way, it means that the content has to have its own synchronization information, and it has to start when the streaming ends, and run uninterrupted until it ends – assuming that the content has not been modified, for example by inserting advertisements. And if the content version is not precisely the one for which the user- provided content was created, the synchronization will not work either. The service provider cannot provide any synchronization information – how could there be any, if the service provider is not aware that there is an additional service and the video itself was produced 40 years ago or more?

Chapter 5: Monetizing IPTV: Advertising and Interaction

Without advertising, would television still exist? This may be close to the truth in many countries, where commercials finance the television stations. In other countries, there is public service television, which makes its living from taxes, license fees or from viewer contributions, however, public service stations are now beginning to feel the pressure of lower taxes, and sometimes include advertising.

Advertising tries to make use of the viewer's attention, and today, television advertising is seen as a major irritant among users – precisely because it interrupts their attention. On the other hand, viewers do appreciate certain types of advertising.

The bigger picture: managing advertising in IPTV

Advertising is, simply put, a way to capitalize on the interest created by the editorial team to send a different message to the user. There are borderline cases, such as infomercials, where the content is paid for by someone, but the "normal" form of television advertising is a way to send a message to many users at the same time.

> *(Continued)*
>
> *Television used to be ideal for this, because broadcasts can be immediate with the same message reaching the entire group of receivers. However, as television has fragmented into multiple channels, so has advertising started to suffer. Advertisers have increasingly started to experiment with more interactive formats, such as advertising on the web; but the very nature of advertising is to interrupt and direct interest somewhere else for a short while. The personal nature of the user experience on the web makes it less suitable for advertising that is directed at changing the user's perception of a brand (as opposed to informing them about an opportunity, which the small-print advertisements do – and Google is proof of success that this is a viable way of using the web).*
>
> *Video advertising is effective for certain types of messages. But it is notoriously hard to measure, and while it may be possible to measure the effect of sales the day after an advertisement was shown, the broadcaster has no clue as to whether it was in a particular program that the advertisement was effective, or even whether it was in his programs at all. This is where measurement agencies come in, trying to gauge the interest of users by sampling the potential audiences.*
>
> *In IPTV, there is an opportunity to directly measure what the users are watching, and match it to demographics. It is not particularly hard to anonymize the results, so even though the service provider may know exactly what each user was watching, they do not need to tell the advertiser – only that "males in the age group 30–50 with an income over $500,000 per household" watched this show. This type of measurement can help advertisers place advertising in a much more accurate way – it can also help them place advertisements in real time. Shopping channels, where the user interacts with the channels by buying things, change the programs to reflect the sales. Extending this to placement of advertisements is not particularly difficult.*

There are many ways of classifying advertising, but for the purpose of this book, it makes sense to divide it in two types: that which expects an immediate reaction; and that which does not.

An immediate reaction here does not have to mean a direct user response; after all, advertising has a purpose and that is to sell products (or services). No sales, no income, no company. So, rather than having the user press a button and give feedback, most advertisers would rather the user hauled out his credit card and bought something. Not all products are bought from impulse, of course – advertisements may be used to build up to a sale.

The first major category expects a reaction; the second is building up to sales. In the supermarket, it is far more important for the advertiser that you remember their brand when you are about to make a purchase. Advertising is usually seen as one homogeneous mass (probably because it is usually handled by one homogeneous industry – the advertising industry), but there are vast differences in selling shampoo versus selling cars. For one, shampoo is an example of one type of product which most people buy every now and again without prompting, simply because they use it every day and it runs out. When the user goes to the store, they will buy a bottle of shampoo, and if you have advertised successfully, it will be your brand. However, it takes a lot for users to go to the store and buy an extra bottle of shampoo, if they have a full one at home. So for the shampoo company, it is much more important that the user remembers its name when making a purchase.

The second type of advertising is intended to make people buy something, and the best way to do that is to engage them. The thing you want them to buy may be anything from a car to a cereal bar, but the distinction is that it is not something that they would regularly buy. Most people do not buy cars (or washing machines, or vacation packages) particularly often, so making them first identify with the purchase, and second making them go ahead with the purchase, is really the purpose of this kind of advertising.

By the way, since it is handled by different specialized companies, advertising is often seen in isolation from other types of marketing material. Direct mailing is very different from television advertising, but when people have become hooked through television advertising, and registered to receive information, they will be receptive to direct mail. The coordination of this is what marketing departments are for, even though this is often forgotten in the industry dominated by the specialist companies fighting to show off their creativity.

What happens when advertising becomes interactive is that it starts to interface with the direct mail and other sales material. This changes the requirements on the advertising itself. It also makes advertising more similar to direct mail. Hopefuls within the industry will tell you that a consequence is that the $200 billion spent on direct mail every year can now be added to the 450 billion spent on advertising. Of course, this is a pipe-dream; but the reality is that if advertising can deliver, there will be more money available. It may come from higher prices paid by the consumers, since they value the product more; it may come from lower costs in other parts of the marketing chain (direct mail is expensive, even though it is efficient as a marketing medium). So any figures must be taken with a big pinch of salt; the fact is that there are still only experiments in the industry, nobody can tell you what the successes will be like. Of course, that means the fun of trying out what works is still available, and that makes the industry all the more interesting.

That said, what are the new things offered by IPTV when it comes to advertising? As you may have understood by now, it is three things (the same things as the industry generally offers):

- interaction

- personalization

- reporting

In this book, we will go through how these work technically, but for starters, we will look at what they mean in terms of constraints on the advertising which can be made available and used in this medium – and what requirements the different types of advertising place on their use.

There is a famous quote that "50 % of all advertising is wasted, but I do not know which half". Advertisers are happy with a response rate of around

1 to 2 %, but they have huge problems in measuring it – how many people actually watched their advertising, and who responded? IPTV can solve those problems.

IPTV enables automated reporting, since the users are having an invisible dialogue with the service at all time. They subscribe to a session, which means anything they do as part of that session can be collected and reported. How this can be done we will look at in later chapters, but this works less well in some solutions than others (by now, you should have figured out that this book is rooting for IMS as the solution). The important thing, however, is what type of statistics you can gather about the viewer response (and how you can secure it so that their privacy is safeguarded). Part of this feedback comes from the ability to personalize the viewer response (actually, it uses the same mechanisms). Personalization is automatic in IPTV, since it comes as part of the package with the subscription management.

The user interaction is the icing on the cake from the advertiser's perspective. Interacting is only required when you want to engage the user, and that is only necessary when the user wishes to participate in a dialogue with the advertiser (or other users, which probably is more likely). The television set is superb as a mechanism to provide viewing, but it is lousy for interaction – so here, again, IMS does something that is not possible with other techniques: it lets you redirect your session to a more appropriate device. And it can save money for the advertiser, since you do not have to have a huge bank of telephone operators to perform the interaction; if you can direct the user to a chat session, the chat can be performed by a program, for example a "chatbot" (an intelligent agent configured to respond to chat messages), and the cost is much less than for a human. If the dialogue is limited enough, users will be satisfied. So it really matters how you set expectations.

Advertising the brand

If you already have a user, independently of whether that user is using a car, a cereal brand, or a certain kind of shampoo, they have made a choice. They chose that brand over other brands. Today, the market is more crowded than ever, and if the user always goes to the store and chooses the products based on price, there is no way to make them a recurring user – other than minimizing the price, which is not a good way to run a business, since it leads to pressure on the margins and in the end, a business very vulnerable to loss. The best business is where you can get users to pay a premium for your product, come back willingly for more, and stay satisfied.

Most products do not do that in themselves. Few products are so good, given the existing choices, that people want to pay a premium. There are some exceptions, but mostly a product is the result of its brand. And the brand is much more than the product: it has properties that have nothing to do with the actual thing. For example, if a brand of sesame oil is produced by a craft process, it will be good, but maybe not so good that it means buyers will pay 30 % over other brands. But package it in handcrafted containers, and make sure the brand has a 150-year history of satisfied customers, and those customers will come back for more,

> *paying a premium of more than 30 % (this is a real example from Tokyo, but they do not advertise on television).*
>
> *Most brand advertising is long-term oriented. Most advertising that requires an action is short-term. Of course, when you request something you want it to happen in the short term; when establishing a brand, the activity can be long term (even if most companies do not think in 150-year terms). What has happened with the Internet is that the distinction between the brand building and interaction has become blurred. To maintain a dialogue with users is now part of building the brand.*

If you have a satisfied user community who are coming back, why advertise at all? For two reasons: attracting new customers, and adding values to the brand. If the customer community is attractive, other customers want to join it, and if the brand has high values, they will want to continue using it, and will be more satisfied for doing so.

This is the theory. What about the practice?

In the world today, there are several thousand projects that involve interactive advertising. Most of them are experiments, and not repeated. Most interactive television projects happen in countries where the penetration is low, and the usage is a novelty. However, there is one place where interactivity is commonplace, at least to a limited extent, and it makes sense to look at how advertising has evolved there. That is the UK.

In the UK, when television went digital, the cable companies and satellite providers decided to put in a backchannel. This is a telephone line in most cases, but it is sufficient for what it needs to do: trigger a change. Remote controls come with a specific button, the "red" button (from the color) that enables users to request a change – which is preprogrammed by the broadcaster, or in this case the advertiser. The term has already entered British advertising vocabulary.

There are a few ways in which advertisers are using this button. The first is telescoping, which does not work in linear programs (such as ordinary television shows), but work perfectly when the user has requested the video. When pressing the button, the user gets a longer version of the advertisement, which gives more information. Of course, the producer of the advertisement is sorely tempted to make the advertisement a cliffhanger to make the user go to the interactive ad advertisement, and it also makes clever filmmaking possible since the viewer can get what is essentially a parallel program, which can be anchored in the program they came from, while promoting the message. Telescoping works since the advertisement can squeeze the demanded video forward in time while it is playing.

The second way in which the "red button" is used is to go to a specialized website (or user area). In the BSkyB system, the interactivity is based on WAP, a technology originally developed for mobile phones, which make it possible to provide pages to the user which are more graphics-intensive and has small scripts included. Making websites that work for television is quite feasible, but it requires a browser in the television, which most do not have yet (except in Japan). It puts quite different demands on the design of the website compared with traditional website design, though (which is discussed in Chapter 11).

The third way the "red button" is used is to order things. In the first wave of interactive television, at the end of the last century, the idea was to enable people to buy things using the interaction button. Disregarding the fact that users find it highly intrusive that you can buy a sweater like Jennifer Anniston's while watching an episode of *Friends*, selling things using the television has taken off in the shopping channels, but not in other programming. When watching a shopping channel, the user is mentally prepared to buy (and will gladly do so using the "red button"), but when watching a soap opera, they are not interested in buying, they are interested in being entertained.

But you can order things without buying them, and here is where the red button becomes useful, in particular as a feedback tool. If you can target advertisements reasonably well (and you can with IPTV, although not on an individual basis), the percentage of users who order additional material is a good measure of how well the commercial did in leading them on to the purchase. Taking the user to a website is cost efficient, but a package in the post is more efficient, since it also serves as a reminder of the interaction outside the time spent watching television. Certain types of products, especially those which are capital intensive, benefit more from this than others; but when introducing a new product, it can be a great way to hand out samples only to those who are interested in them.

Whether buying Jennifer Anniston's sweater, or getting a free sample of a new shampoo, viewers need to fill in their address and personal details; and this can be done automatically using IMS (as well, it should be acknowledged, through a number of other techniques – but not as easily).

Advertising used to be unidirectional. Users used to be expected to watch passively. But increasingly they are expecting to be engaged in a dialogue, even though they may not have any intention to buy, or be recurring customers. The good news for marketers is that each time you engage the viewers in a dialogue, you are getting an opportunity to let them identify with your brand, and eventually make a purchase.

Interaction with the television set often leads to a communication in a different medium, mostly the mobile phone. Combinations of different media become especially appropriate if they can be leveraged to enter the user's personal sphere. The mobile phone is more personal than the television set by definition – the mobile phone is in your pocket, the television set in a semi-public space in most homes. But the message must invite a dialogue. It is not sufficient to send a message like "thank you for using the red button to order the interactive package. You will receive it in the mail shortly. If you have any questions, please do not hesitate to contact us." Instead, make the user respond. The message has gone to his personal device because he requested it. Invite him to a prize draw if he answers three questions about the product (also an excellent way of figuring out how much people actually remember about the product). And redirect the session to his PC (or trigger an invitation the next time he logs in).

The problem for advertisers (actually for anyone with a corporate message) is that you cannot just present the message and expect the viewers to receive it. You have to listen, and this is where the convergence happens, since the dialogue has to be carried on in several media at the same time. That is only half the story. You can shape the dialogue, by focusing the possible responses. This is hard to do without becoming boring, and if the user has a feeling that you are not listening, he will go somewhere else. But if given the right questions, a "bot" can give the right answers.

Voting for advertisements

Previously, we talked about the opportunities for interaction in IPTV, especially the opportunity to vote. Voting is the most prominent opportunity, but there is one thing at least that broadcasters can learn from the voting that already exists. It is a significant source of income for broadcasters (even more so to the service bureaus, less so to the telecom companies), thanks to most voting being premium SMSs. "Premium" means the price for sending them is 3 to 10 times the price of ordinary SMSs. And the broadcaster only receives a small percentage of this (how much varies, but on the order of 10 %).

This works for premium SMSs, so why not make an effort to do the same for any SMSs associated with the IPTV service? This will require some smart negotiation and setup, for example setting appropriate triggers in the charging system and getting the messages to the right numbers, and so on. The negotiation comes from getting paid by the telecoms operator (who basically should be very interested in increasing the revenues from non-voice services). It may also be interesting to use the SMS to redirect users to a premium WAP site, that is a mobile website which the user pays more to access (in the same way as they pay more to use premium SMS).

This can be done today, but it requires a lot of work. If the IMS provider has both a mobile and fixed arm (or if they have agreements with them), the same profile can be reused, and the user's communications directed to the appropriate device (the richest communication, the most personal, the one that pays most – whichever is most appropriate).

Another readily available information source – at least in some countries, and some television channels – is teletext pages. In Europe, text pages, containing program information and news, are broadcast along with the television signal . In Japan, portal pages are broadcast along with the program. In the existing interactive television systems, these pages have to be accessed through the menu system, which means it is virtually impossible for the user to get the information at the same time as the advertisement is shown. However, in interactive television, you can put a small interaction icon in the screen, and it can take the viewer to a text page, which can contain more information (if this is supported by the broadcaster).

Probably the best use of the interactivity button is for the user to get coupons. These can be downloaded to the website, but this is clumsy; it is much more effective if the user can press a button and get a coupon on their mobile, which they can redeem by showing it. This is fairly simple to set up, provided you know the mobile phone number of the user; the coupon can be given a unique number, which can be connected back to the user profile, for example (if this is possible given privacy concerns). But it may not be that important to tie the coupon to individual users, because while coupons in many countries are used as a method to attract customers by giving them a discount, the best use of coupons is to get feedback on the effectiveness of an advertisement. After all, the important thing is to sell, not to give the user a discount. One unique feature with coupons in IPTV is that the coupons can be tailored to geographic areas and time slots, which means that for example a hamburger chain can trial a number of different designs at the same time, and by measuring the number of coupons redeemed can determine how effective the different designs are. This of course means connecting the IPTV system to both the POS system of the store, and the messaging system of the operator; but this is a rather small investment to be able to micro-track the effect of advertising.

If you are looking at television to build brand, sponsorship is an option, but it has to be carefully thought through. Sponsorship associates the brand with a service, and that can be a very positive experience; it can also be a problem, as Deutsche Telekom became aware when the cycling team they sponsored were convicted of doping. Sponsoring existing programs is expensive, since advertising salespeople know how to charge; sponsoring an emerging media property can be cheap, since it is essentially an experiment. The association has to be thoroughly examined; is it more appropriate for an insurance company to be associated with a baseball statistics service or a fantasy football page? Or a page with healthy recipes? How does it relate to the brand, and the message of the brand? And what does the sponsor get for his money – does it include an icon in the television guide? All this has to be considered.

In this perspective, IPTV cannot be seen in isolation. Television is already a part of the marketer's toolbox; now it is a part of it technically as well. Advertising, and other communication, has to be designed to work together. The chatbot has to have the appropriate vocabulary based on the advertisements; the advertisements have to have the right information on how to get the user further; the user has to be able to receive and give appropriate feedback. And even if a chatbot is maintaining the dialogue, how are the responses used? All this has to be thought through, but this is a technical book so it would be out of our scope to go deeper into it here.

An IPTV Toolbox for Advertisers

The IPTV system gives designers a toolbox to work with. How they get the appropriate result is up to the designers. In addition to the possibilities created by IPTV itself, there are (as we have discussed) a number of other services which can be used to improve the effect of advertising, both short term and long term. Table 5-1 lists some of the technologies we have discussed in

Tool	Use	Audience	Connection to IPTV	Connection to IMS
Telescoping advertisements	Enabling users to get more information by extending the commercial	Prospective purchasers	Use with VoD	Session, profile
Ordering	Enabling users to get items or information by pressing interaction button	Prospective purchasers	Use with advertisements	Session, interaction message, profile
Chat	Engaging users in conversation to give them more information, or solicit it from them	Existing customers/ prospective purchasers	Use for user communication with existing customers	Supported as communications service
Presence	Provide information about the users' activities, to enable adaptation now or in the future	Existing customers/ prospective purchasers	Use to track users, statistics	Integral part of IMS
Use of profile information to select appropriate advertisements	Provide appropriate information	Existing customers/ prospective purchasers	Use demographic information of users	Integral part of IMS

Table 5-1. The interactive IPTV toolbox of the advertiser.

Tool	Use	Audience	Connection to IPTV	Connection to IMS
SMS to mobile phone	Provide confirmations, entice dialogue	Existing customers/prospective purchasers	Trigger interaction with advertiser	Not part of IMS
Premium WAP site	Provide additional information	Existing customers	Provide additional information	Charging system part of IMS
Redirection of session to website	Provide additional information	Existing customers/prospective purchasers	Provide additional information	Requires session management from IMS
Redirection of session to chatbot	Entice dialogue	Prospective purchasers	Initiate interaction	Can use IMS messaging
Interactivity icon redirecting to text page	Provide additional information	Prospective purchasers	Initiate interaction	Can use IMS session redirection
Redeemable coupon	Entice purchase	Prospective purchasers	Communicate with existing users	Can use IMS charging system
Sponsored additional information	Build relationship	Existing customers/prospective purchasers	Communicate with existing/future users	Can use redirection to website, etc.
Icon in EPG	Build relationship	Prospective purchasers	Initiate interaction	Can use IMS session redirection (e.g. to trigger a video)

Table 5-1. (Continued).

this book, which can be used to strengthen the message of the advertiser – plus some additional technologies, which you will have to read about in another book, but which will help build the total user experience around the advertisement.

The IPTV Advertising Design Project

While IPTV is still not in place with all its possibilities according to this book, and while the market is very small, there have been a number of projects from which you can learn – and draw conclusions about. In many ways, the early Internet advertising projects can also give some hints about what will happen. One of the first fundamental conflicts is between the creative and engineering people.

For an engineer, or even more a programmer, creative people are a pain in the behind. They do not know what they want, they have imprecise ideas about it, they care more about how things look than how they work. For a producer or a director, the engineer or even worse, the programmer, wants impossibly precise instructions – and they are not willing (or able) to change them as you change the program to reflect a different slant. Creative people probably think engineers are from Jupiter; engineers that the creative people are from Saturn. To the project manager, when the project has progressed for a while, both of them could be from Uranus.

As usual, the issue is setting the expectations right, and creating the kind of documentation that is useful to everybody, while making sure the project progresses in a good way. There are some tips, which can help avoid a split in the project, and bring the two together.

The first, and this may be difficult if you work in an advertising agency, is to focus on measurable goals, which are rooted in sales of the product. Advertising is an investment, and it has to be tied back to sales, otherwise the customer will not return – especially if there is a recession, and marketing money is scarce. But if you can prove, and prove easily, that the advertisement campaign led to sales, then you have it made. So look at the toolbox, and think about what you can use. Get the engineers cracking about developing solutions.

Then, talk to the creative people. Find out who the target group is, and how they should be approached. Are there constraints which have to be met, such as are users likely to use IPTV only at certain hours. What are their ideas for addressing the target group?

Before the engineers and creative people have gone too far in their own directions, bring them together and have a brainstorming session. This has to be done while both groups are thinking about what to do, but before they have formulated an opinion about what should be done, so that they will be defending a position. This is nothing unusual in terms of project management, and it is very much a matter of feeling out the right timing. But don't wait too long.

When you have the brainstorming session, bring a customer representative; and make sure you have a bulletproof project specification with clear goals and scope. If you do not, there will be guesswork, and impossible to set expectations on all hands. Then, brainstorm campaign ideas.

After the brainstorming (and the party afterwards, which is really important for social cohesion), make sure the engineers and creative people formulate measurable, manageable goals for themselves. If they do not have goals which are clear, and fit into the overall goals of the project, there will be a lot of to-ing and fro-ing about when the work is done and what it was actually supposed to be. Make sure they understand that the goals should be maintained, but that they can change the system in the course of the project, specifically at the milestones (which should not be too far apart for this to work).

Create project groups which include both engineers and creative people, and start them working on the ideas from the brainstorming. If you bolt the interactivity or measurement on top of an existing advertisement, it will feel pasted on at best (at worst, it may ruin the advertisement). If you try to build an advertisement around interaction buttons it will feel clunky. Neither is good; there has to be a dialogue.

Make sure there is a point where the project has to display the result, preferably with the steering group, in a relatively short time – before anything else the storyboard and draft must be finished. It is much harder to change something when you have started working on the real code or script; but if all you have a storyboard and a UML diagram, you can move things around without hurting. Create a couple of versions, if you can make project members do it. Make them as different as possible. You will need them later.

When the steering group (or the project members) has approved the project, give them some time to start working, but before they are completely done, make sure there is another review. Changes to a finished project are difficult; changes when it is 80 % done is not as difficult, and it is possible to understand what the result will be.

Then, when it is 95 % done, put it in front of a focus group and see how they react. Focus groups can be involved very early (there are techniques to make sure users are involved in design, but since an advertisement is about telling the story and asking the user to react to it, it is not likely that you will find those methods as useful as if you were developing software). Make sure the project members understand that they have to adapt to the focus group feedback – and repeat it a couple of times, with iterations through different versions.

If the focus group feedback is positive, finish the project as soon as possible. If the focus group is negative, you have just found out that the client does not know his customers (or at least not the focus groups), and you have to go back to the drawing board; since there is now a crisis, use that to create a sense of crisis, look at how you can dismantle what you have got, and how you can rebuild it in a way that meets the approval of the focus group, so you can get the customer to approve.

Splicing Advertising into the Media – Or Putting it in the IPTV Set?

If the broadcaster does not have control of the stream, then there is no way to verify how a distributor (such as a cable television company) inserts advertising, and how much the broadcaster should be paid. That model is actually prevalent in video on demand, where the distributor has bought the rights to the program, and is able to sell advertising into it. Since the distributor also has the user profiles, it can charge more by personalizing the advertising. On the other hand, a sales organization is needed for advertising; and if one does not already exist, it is costly to build, and it may make more sense to ally with a broadcaster who already has a sales division. So there is good reason for the distributor not to alienate the broadcaster.

There is an alternative to the broadcaster maintaining control over the stream, of course; and that is to create a trusted reporting mechanism. This can be done with IMS, as we touched upon in Chapter 4. But trust is not only a matter of mechanics, and it will not happen unless accepted by everyone in the value chain. Importantly, however, you can create a reporting mechanism using IMS, without changing the distribution model.

Today, there is no automatic reporting mechanism. Advertisers have to trust third-party agencies (which in themselves make a big deal of being trusted), who collect data from a random sample of the appropriate demographic, and come up with an average for how many users in a given segment watched a certain show (or at least had their television sets on – there have been some trials with cameras in the measurement equipment, showing that people do the weirdest things in front of their television sets, and sometimes things which certainly do not focus attention on the television). Getting real-time feedback, however, is important for advertisers, since this is the best way of being able to modify the advertising as the effects start to show (unless you are building brand, in which case it is probably better to repeat your old messages as often as possible).

The splice commands in the in and out points are not normally carried through to the end-user. If they were, the advertisement insertion could be done there, from a cache in the user's device (set-top box, home computer, video recorder, or similar). It could also (potentially) be used to carry interaction events, but this is not done today, either.

Another thing which bears mention is that in the market which is most advanced in the distribution of advertising on television – the US – there is an ongoing "arms race" between the end-user and the broadcasters, fuelled by devices such as the TiVo (which are trying to switch sides, to help the advertisers, however). When switching to advertising, producers typically shut off the program stream for a few milliseconds (or even tenths of a second) before switching on the advertising. This "fade to black" can be discovered by the video recorder, and it can use it to jump forward to where the program continues – essentially eliminating the advertising. There is not much that the broadcaster can do about this, except avoid fading to black. Some companies have tried to force IPTV standards to prohibit the ability to cut out advertising, which is just an encouragement to hackers and will not help – since once hackers start mucking around with the data stream, what is there to stop them from doing things which are even less desirable, from a broadcaster's point of view?

Inserting Advertising

Advertisements are a way for the broadcaster to leverage the attention of the user by selling it to someone else – in this case, the advertiser. Advertising interrupts television programs to insert their message. There are different kinds of commercial messages, of course, like infomercials and other paid-for programming; but these make for a different kind of user experience, where the user voluntarily looks up the content and treats it like a program. Users care a lot less than journalists about objectivity, and they have no problem distinguishing commercial messages from those that have a different sender (or, sometimes, the other way around).

So if the premise of advertising is that it should interrupt the program, how do you do it? And how do you do it in a way so that it interrupts the program at an appropriate place? Also, in some countries, you have to make sure there is not too much advertising (e.g. more than 30 seconds of advertising per 4 minutes of program); and you have to manage the advertising anyway, so the user does not get bored and leave. The consequence is that the broadcaster has to manage both the stream and the advertising. For broadcasters, there is another big advantage to this: they maintain the control of the stream.

Today, there is a given stream that you have to insert advertising into it – and you have to deliver a complete stream from the headend to the end-user. There are many ways in which this process could be optimized, but the way in which it works today is that the broadcaster gets a program from the producer (or aggregator), inserts advertising into it, encodes the stream, and sends it out. This is not likely to change just because you introduce a nice new format for users to interact with television – such is the conservatism of the industry – and the enormous momentum required to change all the receivers out there.

When the broadcaster receives the advertising stream, the advertising locations can either be pre-inserted, or the broadcaster can insert it. The mechanics of doing so have been defined by the Society of Telecommunications Engineers in the US, and documented in a report known as SCTE 35 (which is also an ANSI standard). The mechanism defined, simply put, is based on the mechanism used in analog systems where a tone, which a human cannot hear, is inserted into the video stream to trigger the insertion of the advertisement.

In the old days, this meant switching from one tape to another (and running it at the appropriate time). This, in turn, is based on mechanical cutting and pasting of tapes (originally, of films): you have the program, you cut it off at the cliffhanger (where the hero is hanging off a cliff and you are sure the audience wants to stay), and then you splice in the advertisement (or advertisements); and then go back to the program (to find the hero had a helicopter standing by all the time).

The trick here, when television goes digital, is to automate the cue tones. Even better if you can make the automation a function of the cue. This is where the standardization goes much further than defining the tone itself, and how it should be inserted. The solution is to enable the databases of programs and the database of advertisements to interwork, so that the advertisements are automatically picked up and inserted in the right place. This is where the profiles we talk about in Chapter 6 are so important. The important thing about the mechanisms, however, is that they have to work regardless of whether the broadcast is done from a real-time event, or whether it is a VoD unicast.

The standard defines "splice points" in the MPEG-2 transport stream. The splice points make it possible to switch the MPEG elementary streams (ES) from one source to another. Today, splicing is normally done in a specialized device (essentially a switch), which merges the streams together and forwards them to the encoding device. There are two types of splice points in the standard: out points and in points. "In points" allow the system to switch out of the current bit stream and into another; "out points" are points where the bit stream can be exited. In practice, this means that an advertisement is triggered by an in point in the program bit stream, and the advertisement has to end with an out point. Both out points and in points are imaginary from the presentation point of view. An out point and in point can be co-located, and this must be so in the case of exiting the program stream to introduce an advertisement.

It used to be that splice points were represented by tones, but it is no longer correct to talk about cue tones, since the signal is digital and the audio and video portions may actually lag each other in transmission – and that would mean advertisements would be spliced in the wrong place, if the only clue was an audio tone. Instead, there is a way to insert an identity in the individual MPEG elementary stream, which defines what the splice point is about – and also serves as inventory control.

SCTE 35 provides three important functions through the same mechanism: making clear the identity of the individual objects to be combined in a data stream; when there is a timeslot available to combine them; providing a reporting function to the advertiser and the charging system of the broadcaster. The advertisements are inserted where there are signals presenting slots for inserting advertising, as shown in Figure 5-1.

This is done by creating a time-based Unique Program Identifier (16 bits long), which can be used by the advertising insertion system in the commands that are defined in the standard, to define the slot where the insertion can occur. The identifier can also be used to convey information about what the program is, so the metadata about it can be looked up, and then the system automatically can select which advertisement is appropriate to insert in that slot.

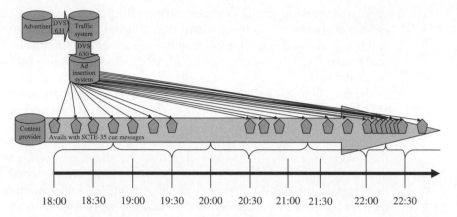

Figure 5-1. Inserting advertising in a media stream.

Since the assumption, traditionally, is that there is a dedicated switch which switches the streams, the system takes all the streams which are to be composed into the MPEG-2 distribution stream, and combines them. This is done based on a table representing the different elementary streams which are combined into the program. This table provides information about what is happening in the program (based on the table of the elementary streams and their descriptors), which is fed to the splicer. This table lists splice events – in and out points. The table may be sent in advance, and it is possible to cancel events in the table, which is good if you have sent a splice event in advance and something happens in the program that changes the scheduling (e.g., overtime in a football game).

When a splice event occurs it can be picked up by the system preparing the distribution stream, and advertisements can be inserted according to the criteria set by the broadcaster and the advertiser. This is based on metadata, which we talk about in Chapter 7. The process has to happen fast – too fast for the user to notice. This makes it difficult to personalize advertising in real-time streamed events (as yet), and easy to do in video on demand. This is where industry leaders such as Tandberg Television has focused, creating a system which reads metadata, user profiles and other information, and uses this to splice in personalized advertisements in video on demand – the advantage being that the user has to log in to the VoD system (and usually does not mind doing so). This makes it possible to get the user profile, select the appropriate advertisements, and compose the individualized stream within a time that is too short for the user to feel that he is waiting for something. The mechanics are the same from a broadcaster's point of view.

The splice commands in the in and out points are not normally carried through to the end-user. If they were, the advertisement insertion could be done there, from a cache in the user's device (set-top box, home computer, video recorder, or similar). It could also (potentially) be used to carry interaction events, but this is not done today, either.

Chapter 6: P2P, TV on the Web, VoD and (n)PVR

Video is more than television, as countless video stores around the world witness. And as countless Friday nights around the world also bear witness, renting videos is a way of enhancing the home life for millions of people around the world.

Video on demand came long before broadcasts over the Internet, of course, and it was the trigger behind the first boom in networked television, as long ago as the 1980s. Technically speaking, downloading video is much easier than receiving a real-time stream. So easy, in fact, that millions of people are doing it at home, and millions of others are making content available both through peer-to-peer file-sharing networks, upload sites such as YouTube, and through other sites. Upload and download are two sides of the same coin and technically very similar, but there are two major differences: the user interface, and the network.

The bigger picture: video on demand

Before the VCR, unless you were rich enough to own a video disc player, there was only one way to watch television: when the program was broadcast. With the

---

(Continued)

VCR, users gained a way of deciding when to watch the shows they recorded – and to buy and rent shows and movies they could watch later. Passing through a number of technical iterations, the business model of the video store has been trying to move online for some time; since the last television boom in the 1990s, in fact.

Providing video on demand is not particularly difficult from a technical perspective. Create a big server, hook up a number of network connections to it, find a way to make people pay, and run with it. There are two hurdles, however. The service provider has to pay for the content; and the server has to be really, really big if the model is to be successful.

Instead of having one really big server, many small servers can be used instead, that is, hundreds or thousands of servers. Connecting them directly to each other, while maintaining control over the directory, is how peer-to-peer file sharing works. This has been around since the 1980s, but now broadcasters are getting interested, especially those publicly funded, since it is possible for users to view programs again, without the broadcaster having to provide the resources. The issue is how to maintain control over the distribution – and there is no way of getting paid. For that, the service provider still needs a central server, or a trusted agent in the peer of the user.

Peer-to-peer file sharing became infamous during the 1990s, because it allowed users to upload shows and programs they had not purchased the rights for, and make these shows available to others. While this is a copyright issue, the service providers to some extent are liable if they have control over the directory. This forced the original Napster to shut down; but it is also the foundation for YouTube, where the service provider has relinquished the control to the users.

Getting paid for providing videos on demand still requires more than a big server. Above all, it requires that you have a way to track the usage. This is the subject of this chapter.

Video on demand, the next big thing according to Bill Gates of Microsoft, has been around since the 1990s. It is basically very simple: the user requests library information, gets a catalog page, makes a selection, and receives the video. This is the basics of all VoD services, but there are a number of twists. Users providing content for download, video libraries of films going back to the 1960s, and users recording videos which are stored online feel extremely different from a user perspective; from a technical perspective, they are not. A video library which can receive uploads, and deliver downloads, is not even state of the art in today's data centers, it is just a server with very big discs and large network connections.

From a network perspective, there is a major difference between VoD and other types of IPTV: there is a one-to-one relation between the receiver and the sender. The user has ordered content that is delivered to him, and him alone at that specific time. And there is an acceptable delay in the delivery, which can be used to create a personalized version of the content, applying the user's profile to the selection of advertisements and other aspects of the content.

When distributing a video from an event in real time, it is not sustainable to have a one-to-one relation between the server and each user; not only would this require a very large server with a tremendous number of network interfaces, it would also create a very large load on the network. When dimensioning networks, the telecoms provider traditionally calculates the probability that users will require the use of the network, and dimension the shared resources in the network accordingly. If there is a 100 % probability that everyone would watch, the network would be so loaded that it could not be used. For real-time events, the IPTV provider instead uses a different technology called multicast (where the stream to be presented is distributed to addresses which the user can connect to in the same way as water flows through the mains, initially at high pressure in a broad pipe, then in smaller and smaller pipes as they branch out and finally reach the home of the end-user). Of course, in the case of the Internet, the water shares the same pipe as the gas and electricity, and there is also a back channel which has to be given space (and it is not sewerage). We will talk more about how that works in Chapter 10.

There is a version of video on demand, called near video on demand (NVoD), where the broadcast of a movie starts at certain time intervals (e.g., once every 30 minutes). This is much more efficient from a network point of view, but much less satisfying from a user point of view. If the user wanted to watch the movie at a certain time, he would have gone to a cinema. The alternative is that it is downloaded in the background to the user's personal video recorder, and this requires that the client can handle background downloads – and is always switched on, something for which there is no guarantee.

A special variation of video on demand is creating a way for the user to record content, which is under the control of the system. This can be done in a personal video recorder, which is part of the set-top box, or it can be done in a server in the network (which may mean setting markers in the stored data where the user has requested recording, instead of creating a separate store of data for the user). Getting content from a network Personal Video Recorder (nPVR) is very similar to getting it from a video library, with the exception that the user is the only one who has access to the content – it is not a shared video library. The RTSP protocol, which is most suited to manage remote video viewing, can also be used to record content – the "networked video" device can be used both for viewing and recording content.

One advantage of the networked PVR is that it is always on. In the home, users can switch off the PVR, which would disable the download; so would switching off or removing the router in the home. It also means there is a constant traffic load on the access network to the user's home, which may be a problem if the user wants to send and receive other traffic when not watching videos (since part of the network is always occupied by the download – this is familiar to anyone doing even less than medium-sized peer-to-peer (P2P) file sharing). This is less of an issue if the home network access is fiber-based; but it will constrain the home network if the user is only using an ADSL connection.

Figure 6-1 shows the basics of a video on demand system. This works in the same way regardless of whether the content comes from a service provider or from other users; and regardless of how many users there are. There are a number of constraints which come from the different components. These constraints become more important as the number of users increases.

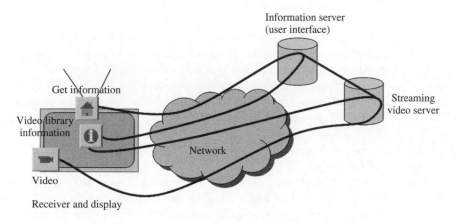

Figure 6-1. Video on demand.

The first constraint lies in the user interface, representing the commands and information available to the user. The signaling can be of two different kinds: controlling the video that is currently being played (e.g., fast forward, back, stop and play); and selecting a video to be played. This is fairly straightforward, and the most common way is to represent the commands as buttons, either on the screen (where you may have to move a pointer to them), or as buttons on the remote control, in which case there has to be a protocol between the renderer (the television set) and the remote control. The protocol receives the signal from the remote control, and translates this into a command in the appropriate protocol which makes the video go fast forward when the fast-forward key is pressed, and so on.

The information available to the user is of two kinds: the video information itself; and the information about the video. We will look a bit more on the video information itself in Chapter 10, but from a user interface perspective the important thing in VoD is the information about the video, and this is also the part that is easiest to change and manipulate. It is also the area where the IPTV service providers can differentiate themselves. The *Spiderman* movies are the same whether you buy them from provider A or B, but in one case the guide pages may contain information about the actors and the director, and in the other case there may be cool teaser pictures from the movies, depending on how the service provider is trying to sell the content. This is also where the opportunities for personalization are greatest, as we will see later.

How the information about the available content is presented is very important for the user's selection of the content. There are a number of ways to present this information (based on metadata, which we will look more into in Chapter 7); regardless of the presentation format, the basic technology for the presentation is the same: an XML page is taken into a presentation tool and displayed to the user with clickable links that trigger signaling.

The clickable links can be several different types of URIs, that is, trigger several different types of protocols used for the "signaling" to the content server. In HTTP, probably the most familiar case, clicking on an HTTP URI will trigger an HTTP GET directed to the resource, which is referenced in the URI (referenced here means the URI contains its address). There can be several other types of URIs, such as RTSP URIs, SIP URIs, and proprietary protocols whose addresses are URIs. The URI is a generic mechanism used to address resources.

There is no standard that says "build a VoD system this way", although the combination of HTTP for getting the page, RTSP for signaling, and RTP for the content download come close. There are a number of alternative ways on the market, most notably using HTTP for signaling; and in this book, we will also see that there is a need to provide an additional encapsulation of both the signaling and stream, which creates a need for an additional protocol. You could do the same thing using HTTP, but this creates more overhead, does not enable control over quality of service, has problems with real-time signaling in constrained environments, has security problems, and does not enable the reuse of a generic user profile for the subscription (and the payment of the services). Each protocol should be used to do what it is good at, and while this means more work for the implementer of the client, it means the system as a whole works better. For many developers, the only concern is with the own component; for the service provider the concern is with the system providing a good user experience. This means a more complex environment, but one which works better in combination.

The architecture of the system is the subject of later chapters. Let us now look at the user experience. The user interface to the directory of the VoD information is an XML page, retrieved through HTTP. This is probably the most agreed generic design requirement in the entire IPTV industry, but is not quite as clear-cut as seasoned web developers think.

For starters, the format of the information page – and hence its design – is not necessarily a traditional web page. It can be a set of objects presented in the viewing window based on JavaScript, and present aspects of the information based for example on what other users are watching. This gives designers much more freedom than the traditional design of a list of content, ordered in different ways and presented in different fonts; but there is no reason to let the storage of the content in a database force the presentation of it. Relational databases may be tables, but if the VoD system is intended to help the user find interesting content, and consume more of it, then there are better ways of presenting the information about the movies that the user can download. That may mean going beyond the browser, but it certainly means letting the VoD server re-organize the list of content according to the user's interests, and that means knowing more about the user than the fact that he has an identity and has logged in.

The format for VoD content guides is not standardized, although there are some guidelines in the TV-Anytime guidelines (and metadata for films is standardized). However, there is a problem that applies to both the EPG for VoD, and the VoD itself.

In most VoD systems, the service provider tries to get a direct connection between the video source and the client that plays the content. If the user is expected to pay for the content, there has to be both a security management and trust relationship to enable the payment (and the trust in the payment); if, however, the video that is being played is only a video from the Internet, there may not be a requirement to get paid – it may be advertising which the user downloads, and in this case, the requirements will be somewhat different.

Getting Paid for VoD: Advertising

In the case of advertising, when there is a legally binding offer in the video, the content provider wants to make sure that the content is current (so that the user does not go to the store with a coupon for the end-of-the century sale of 2000); if advertisements are embedded in old programs, they either have to be replaced, or a disclaimer has to be included. As with all disclaimers, it is very hard to make this clear that it is not legally binding. If some of the advertisements are valid, and some are not, then the invalid ones have to be marked with a disclaimer. This means a lot of effort (since the service provider not only has to have functions to manage the video, but also to track the advertisements); it probably makes more sense to use an automated advertisement placement system, which removes the advertisements and replaces them with current ones, which can also be adapted to the user.

As we discuss in Chapter 5, there are two types of advertising that are likely to occur in IPTV systems. The ones that are most likely to occur in a content listing for a VoD service are those which try to convince the user to watch some other type of content than – since most users are not looking for a particular movie when watching content, they are looking for a relaxing experience, and maybe a type of movie. Here, there is even more potential for applying a recommender engine to select those advertisements (and couple them to the type of video that the user is most interested in), but this has so far not been applied by any of the many companies providing VoD services, mostly because they lack the means to collect the user's profile and preferences information.

Another reason for tracking each advertisement is that advertisers want to be able to count each individual impression of content, and that is hard to do when the content is downloaded once and then viewed independently of the connection to the content provider. According to the guidelines concerning broadband advertising on the Internet from the Internet Advertising Board (IAB), for example, content should always be counted (as an advertisement impression, i.e. as something which the user has viewed) only when it is viewed by the user, not when it is downloaded to the user's computer (which is easy).

There are three places where content can be manipulated: in the client, in the server, and in an intermediary proxy. In theory, the manipulation can be done in either the client or the server; in practice, it will be done in a combination of the different nodes in the chain of delivery, with different parts of the change of the content done in different places. For example, while the server may deliver an XML document, the proxy may change it to XHTML, which can be presented in a browser; and the client applies the formatting instructions, including instructions to remove certain parts (e.g., if the user has no interest in seeing the column of director information, that can be removed in the client).

How the content listing is created and how it is presented are not really related, although they can interact in various ways. In particular, this becomes important when the content is personalized.

A web page can be created as a template, and populated with elements from a database. This is often done on web pages today, but it is not quite as usual to provide personal information. However, when presenting television listings, this can both make the user experience more interesting and easier to monetize (since you can tell who has watched which advertisements, they will be more valuable to the advertiser and can be priced higher).

When presented as a web page, regardless of how it is composed, the content listing will have to follow the same rules and conventions as web pages follow on the Internet, otherwise users will not know how to interact and work with it. When presented in other ways (as independent widgets) the user interface is not constrained by the rules of the content as a page. Instead, there is a large amount of research and experience from the discipline of human–computing interaction (HCI) that can be applied, sometimes in different ways from what a designer would do –the important thing to remember is not that it looks good, but that the user experience is good.

When content is presented as a web page, there are conventions for how to insert advertisements (personal or generic). Searching for content is also easy to do in a generic way. When presented as a widget, however, content listings work in a different way, and how to insert advertisements when the content is presented in a widget, or how to do searches through a widget instead of a list is not understood, nor are there conventions for it. But it opens up possibilities to search for what looks like something else, rather than something that has a similar name; and it opens up possibilities to insert advertisements as a background, rather than as a banner in a page. This can be much more efficient from the presentation perspective, as well as from the perspective of the advertiser.

If the content library of the service provider is reasonably large, it is too big to present in one page, regardless of whether that page is a web page or a specialized display widget. Somehow, the content to be presented has to be selected. Listing it alphabetically, or by date, is a very simple way, and almost completely useless if it means the user has to browse through several screens to find the content he wants – there is a well-known risk that the user gets tired and does something else, which means loss of income for the service provider. A good rule of thumb is that the user should be able to find anything he is looking for in "three clicks and three seconds". It is not probable that the service provider will be able to earn more from advertisements on the four pages the user watches before getting tired than from the video he could have sold to the user if he had followed the three clicks rule, especially since this both builds the user profile and can be further monetized by inserting advertisements around it, which can be personalized and video advertisements, which generally cost more and hence sell for more than banner ads.

To do this, content has to be preselected, and the content, which the user is most likely to buy (or rather, the content which the user is most likely to buy which the service provider is most interested to sell), has to be listed first. Preselecting content is where the user profile becomes useful, especially when combined with the profiles of similar users, in order to compute the most likely content the user will want to view next.

The technology to do this is called recommender engines, and they are a well-known area of content management, with considerable amounts of research and development around them. They work especially well with XML. There are various ways of implementing recommender engines, and the algorithms for what will provide the best recommendations are not clear – the (online) video store Netflix did run a competition to find a recommender engine, which would provide recommendations that resulted in higher sales and better recommendations, by matching the users' profiles and known preferences against existing databases: the competition ended without any first prize being given out (there were several second prizes).

Building a recommender engine is relatively easy (at least a simple one), the complex and difficult part is collecting the user preferences and user information. A recommender engine has to have good data to work with, which can be compared with the user preferences to become meaningful (see Chapter 7).

Creating a list based on the user's historical behavior is rather easy; creating one based on current behavior is harder. Here is where IMS becomes useful, since it makes it possible to connect information about what a user is currently viewing with information about other users' preferences, and so provide a user with content that is more relevant in the context of his friends. To do this without IMS, you have to have both a session, which contains the user information as well as the program information; and a method to communicate the profile information and match it with the program information. You end up with a system that is very similar to IMS, but more complex because it is not optimized to do this.

Getting Paid for VoD: Charging for the Service

When the user clicks on the link in the content list, however that has been created, the video is downloaded. but, before the user is allowed to become a viewer, there is something crucial that has to take place: the user has to pay for the right to view the content.

VoD systems are rarely free. Users pay a fee per video to watch. While this model is borrowed from the successful video rental business, the lower cost of the online distribution is not quite enough to cover the expenses. Although the model has the same costs and incomes for each rental, it ignores the crucial fact that in the video rental business, the late fees used to be a bigger money earner than the rentals. Since the physical infrastructure is gone, the revenue from actual rentals is not quite enough, therefore the volume needs to be higher. The industry is not quite there yet.

Having disabled the primary source of revenue by changing the distribution mechanism to a method where it is impossible to be late returning a video (and even impossible to return it at all), the industry has managed to short-change itself. The non-physical distribution of IPTV should mean less actual costs, and less delays in terms of the viewer getting the requested medium (no discs involved, no trip to the video store or cinema – where the margins for popcorn are more than double those of the tickets). On the other hand, the margins are already razor thin in video stores, and will not get thicker in video rental online. The provider of the VoD content library needs to monetize the attention of the user while he searches the content listings, and as luck would have it, there is a ready-made group of advertisers who are looking for that user at that very moment: the entertainment companies.

While users have got used to regarding content from VoD sites as necessary to pay for, they do not see content from the web, or even worse from P2P file-sharing sites in the same way. Here, the distribution model has become associated with the content being free (which may be true in the sense that the user is not paying for it). However, if the content rights owner were to decide, it certainly would not be free. To get some money back, the service providers resort to selling advertisements. This may well be an extremely profitable proposition, if you have a community which is driven by its own users and where there are rights clearances between the content provider and the service provider to ensure that the content is paid for (in the case of user-provided content, it may be sufficient to have an agreement to provide the content on the site without compensating the user). This is supposedly how Google intends to make money from YouTube.

However, if a service provider wants to charge its viewers for watching the contents of the movie library, there has to be a mechanism to charge for the use of the content, and there are several preconditions for its use.

To start with, if you want to charge someone, you need an unambiguous way of identifying them. If bills were distributed at random, the companies sending them out would not last long. There are various ways of doing this: in

IMS, it is done by the client having a cryptographically secured identifier (usually embedded in the USIM, the User Secure Identity Module; this is typically a SIM card, although it can also be a software token, provided to the user when they originally subscribed to the service). The same module can be used to identify the subscriber both when it comes to the connection and when it comes to the use of the service – and also for additional services that use IMS, for example voting or presence.

Alternatives to IMS

The alternative to using IMS is to embed the cryptographic token in the client. This probably requires creating a special browser, which has to be tamper-proof, since otherwise there is a large risk that users will steal service by copying the keys in the secure module. This is a major problem in satellite television today, where the decoding is done based on keys, which are embedded in a smart card; the card is both the service identifier and the encryption key, but it can be copied and the copy can be used in other places. This is, perhaps perversely, not illegal in some places.

However, the mobile industry has long had the same problem, and has a long history of continuous improvement in this regard. Cloned SIM cards were a problem in Europe for a long time, when cards were copied and sold to eastern Europe (just after the fall of the Soviet Union); but this was possible to counteract, and the IMS user identity module is much more secure and – if embedded in the SIM card – cannot easily be copied.

There are several fees, and several service providers, involved in the charging for the IPTV service. To start with, you have to have an Internet connection with sufficient bandwidth. How that is set up and charged for is out of the scope of this book, but the same mechanisms as with IMS can be used to charge for connection and content – and to identify the user. If IMS is not used, other mechanisms have to be created; charging is not too difficult, since it only involves verifying that the subscriber requesting the content is the one receiving it, and that the subscriber receives all of it. The same mechanisms can be used to charge for real-time consumption, but when watching a television channel it is only a matter of a subscription (paying for the usage) rather than paying for the consumption of the actual content item (which would be pay-per-view, and where IMS can also be used).

Since there are often several service providers involved in providing the IPTV service, the service provider fees are added together to create the final fee to the user. How this is charged depends on who is fronting the user. If the telecom provider, who provides the connection to the network, also provides the IPTV service, the telecom provider can either have a licensing agreement with the content provider (this can be a bulk agreement); or it can be an agreement to pay for actual consumption; it may be combined with hosting services (the service providing hosting the content service for the content provider; or the content provider hosting it for the service provider), or it may not. There are simply too many variations to say exactly what it may look like. It is probably unusual that the IPTV service provider is the one who sends out the bills, but this could potentially be done. It is all a matter of two things: registering the user's requests and fulfilling them; and rating the services provided under the agreement with the user.

Most people associate charging and billing, but they are not that tightly connected. A service provider could potentially provide information about the usage (and that it has been verified) to several IPTV providers; this would be done as charging data records (CDRs), which can be adapted the requirements of each service provider. When receiving it, the service provider system has to apply any business rules which they may have (e.g., "half price before 6pm") and create the bill. In this sense, charging becomes yet another service, which can be provided to many service providers.

There are a few preconditions for charging the user, which have to be taken into account:

- unambiguous identification;
- subscriber agreement with service rating agreed;
- mechanism to request service;
- charging system mechanisms:
  - rating system (providing information about how much each service costs)
  - fulfillment verification mechanism (making sure the user has received the service).

Any charging system conforms to the same generic model (defined by the OMA). In that model, there are four different parties:

- the customer (who gets the service);
- the Merchant (who provides the service);
- the Issuer (who provides the customer with the means to pay for the service);
- and the Acquirer (who provides the merchant with the means to receive the funds which are payment for the service).

How the charging is done, and how the settlement is done, may vary between different businesses. In principle, the Acquirer and Issuer are banks, which interact with the Merchant and the Customer; the Customer triggers the service, and the Merchant delivers it – and the rating information. Rating information is crucial, since it is here that the cost of the service is defined. IMS describes a charging architecture for setting up an IPTV service – the IMS charging architecture. The idea is that the charging system captures the events, and sends them to the billing system where the rating information is applied to the charging event. This is done using the Diameter protocol, which is a protocol defined by the IETF specifically for managing this type of service.

IMS provides an offline Charging Trigger Function (CTF), which is located in a network element such as the CSCF, and which interacts with the Charging Data Function (CDF) over Diameter. The CTF provides a way of collecting accounting metrics at trigger points (which are specified by the charging model selected). It then forwards the data to the CDF. Within an IMS session, there are charging sessions, which start when they receive a starting trigger, have a few intermediate triggers and then an end trigger, which are tied together to generate a chargeable event. The CDF uses the data to construct Charging Data Records (CDRs), that have a predefined format and content. A separate function, the Charging Gateway Function (CGF) acts as a gateway between the CDF and the billing system, correlating different CDRs from different CDFs, managing persistent storage, error management, etc. Rating and account balance management are grouped together in a different function, the Charging Control Function (CCF).

In IMS, the basic principle is that a service does not start until the charging is authorized. This means checking that the purchased subscription is valid, or that funds are available in the user's account (e.g., for pay per view). In a traditional billing system (including the traditional billing systems of cable-TV providers), the charging is centralized around the invoicing (the billing). This means that the billing system gets and updates the account data, collects and aggregates it, and prints the bill. It also contains the business logic for pricing (applying rating and discounts), and grouping of charging data. If the operator is to be able to terminate the service, the account management becomes important, as does the rating. A traditional rating engine for the telephony world is a standalone application, calculating the information that goes into the CDRs. The actual price is not calculated until after the event, which may mean that the user is charged more (or less) than expected.

In a real-time charging system the rating is integrated with the session management. It is a prerequisite for the use of the service, and it is applied as the user consumes the service. There has to be bidirectional communication between the charging system and the CSCF (or other network element). In an IMS charging system, the data and the logic is instead collected into the user's account, where it is managed. Since the information is not done in a separate (often batch-based) system, but close to the session control, the user can get immediate feedback on whether the account status allows for the selected action (e.g., if there is enough money in the account to pay for the pay-per-view the user has just ordered). The credit authorization is done directly, and this has direct effects on the session management: if the user is not authorized, there is no session. Users can also receive alerts and warnings during the session (e.g., a message that their account balance is low). If the user runs out of money, the session can be automatically terminated.

The account management for the charging system in the IMS standard also allows for multi-user accounts. In the IMS-based IPTV system, where the default identity is the family identity and there can be several public user identities tied to this, multi-user accounts are possible, and they can be charged in real time, but there can also be dedicated accounts specifically for promotions and other bonuses. Since rating means that the price can deduct from an account equally well as add to an account, it becomes relatively easy to reward the user for certain behaviors if they are given triggering points – for example, paying the user for participating in a survey.

When it comes to the fulfillment, there is less work done – because this is service-specific, and because it is harder to determine when the user actually received the service, or not. For IPTV, there is a different problem: even if the data stream was delivered, was the data quality sufficient? And is delivery only a matter of network quality?

If the user receives an IPTV data stream without errors, the service provider has a good claim to having delivered the service. The quality of the programs is, alas, not part of the measurements. But it is a very unusual network that can deliver a service completely without errors – although there are ways to manage errors (such as TCP for connection-oriented services; and forward error correction (FEC) in UDP-based RTP). Errors create glitches in the viewing – artifacts on the screen in terms of pixelations or frozen images, and disturbances in the sound.

An IPTV system can measure two things: that the data stream has been correctly delivered; and that the MPEG grammar was correct. If there are errors in the MPEG file, or if there were errors in the data stream, there will probably be issues with the reception. This can be logged –unfortunately, the logging is up to the individual IPTV service provider to require, and the systems manufacturer to provide. And then, there is the issue of what is sufficient quality – this is entirely subjective. Users also associate the content and the quality, and the sound quality with the image quality, making the measurement equation very complex. How to measure the user's quality of experience – without being intrusive – could probably be the subject of a different book in itself.

User-Provided Content

The technique for downloading content is completely independent of the nature of the content. Downloading a three-minute trailer from a movie company's website is not, in principle, different from downloading a home-made video created by a couple of girls in a garage. The constraints on the system are the same, and when the video becomes a number one hit on YouTube, the mechanics of providing it are no different from providing the trailer for *Spiderman IV*. And, in principle, providing a library of home-made rock videos is no different from providing a library of silent movies.

Users can provide content in two ways: it can be recorded from a transmission or another recording (such as a DVD); or it can be recorded by the user (with a video camera, or the user's own computer). Whichever method is used, the result is a video, which is stored in a file. This file can be shared with other users in various ways. One way is to upload it to a central site; the other way is to upload a description of it, and make it shareable. The technologies are slightly different, but there are additional constraints that apply regardless of the means of production and the distribution technology. The first constraint is technical, the second is legal.

The Network and User-Provided Content

The first constraint has to do with the network that connects the user to the video-sharing mechanism. This is less of a constraint when using YouTube, because that is more than a directory of available files, it also hosts the content itself.

When a home network is connected to the Internet, the connection can be made through several types of "access networks". These range from dial-up connections and wireless connectivity to optical fiber (where the connection can theoretically be up to terabits per second, although only two people have that kind of home network connection). Some types of networks have a built-in constraint: they are asymmetric.

In a home network, where the different devices are connected using Ethernet, there is no difference in the speed by which content can be uploaded and downloaded – Ethernet has the same speed in both directions. This is also true for some home access networks, such as optical fiber, but it is not true for the most widespread connectivity form today: Asymmetric Digital Subscriber Line (ADSL).

ADSL is designed to run over telephone lines, and while the magic is in the receiver and transmitter, the capacity, which can be shared over the connection, is not infinite (and much lower than what is possible with an optical fiber). Since ADSL was designed for users to download content, rather than provide content, the system has a much lower capacity in the uplink (where the user provides content to others) than in the downlink (where the user gets content from others). Hence, YouTube is a better technical solution than peer-to-peer technology, at least over ADSL.

Peer-to-Peer Versus Central Server

Both television on the web and peer-to-peer are user-requested media: the program is sent until it is requested. This is different from broadcast and multicast, where the program is sent regardless of whether it is requested or not (there is a hybrid, where the program is sent in a loop which the user hooks into – so-called "near video on demand"). Since the program is not sent until it is requested, it is rather similar to video on demand, however, the technologies used are different, even if the result they give is basically the same, that is, they both send content in a stream over the network. In the case of the peer-to-peer file-sharing systems, there are problems with the limited bandwidth of the network uplink.

Streaming and downloading

The original model of delivering data over the Internet was to download a file. This means starting the download, getting the information about the file, receiving all the content, and then stopping. The File Transfer Protocol (FTP) was designed to do this, and helpfully – for its time – has a download channel and an upload channel for status messages. While useful in managing bad network conditions, the number of checks required, as well as the constant traffic, makes it a bad idea when you just want to view something.

HTTP took one step forward. It gets a file and then throws it away. The original is kept at the origin server, and a copy displayed at the client. In HTTP, the file is downloaded without the checks required by FTP. Since the file is stored, you can view it again.

But if the file you want to see is very large, why not start the viewing when only a part of it is downloaded? And why not throw away the parts that you have seen, and then continue watching? That is how streaming works. Streaming data means getting a stream, not discrete chunks of it. The media is sent in a continuous flow from the origin server to the receiver. It can be compressed (MPEG-4 is particularly good at this), and it can be either an existing file which is streamed, or a file which is captured as an event takes place – live streaming.

When all receivers have a one-to-many connection with the sender, the stream is broadcast. There is no selection, everyone gets the same thing. When only one user gets the content from the server (the situation in video on demand, normally), the stream is unicast. When the content is sent to a limited group, it is called multicast, and this is how live streaming normally works on the Internet. A streaming protocol, such as RTP, is connected to a set of users, who are grouped (e.g., by connecting their addresses together) using Group Management Protocol (GMP), and then they will all receive the same data stream.

In HTTP, streaming can be realized simply by not closing the connection between client and server and continuing the download, which is not how the protocol was intended to be used. The reason this may not be ideal is that it is very sensitive to network load – and it also imposes a high overhead in the network. In other words, it may be detrimental to your neighbors.

This is easy to prove. To figure out the required storage size from the streaming bandwidth and length of the media, for a single user and file, multiply the storage size (in megabytes), which is equivalent to the length (in seconds) with the bit rate (in kbit/s) and divide by 8388.608 (since 1 megabyte $= 8 \times 1{,}048{,}576$ bits $= 8388.608$ kilobits). One hour of video received at 300 kbit/s (this is a typical broadband video) will then give $(3600\,\text{s} \times 300\,\text{kbit/s}) / (8 \times 1024)$, which gives around 130 MB of storage required. If the file is stored on a server for on-demand streaming and this stream is viewed by 1000 people at the same time using a unicast protocol, you would need:

$$300\,\mathrm{kbit/s} \times 1000 = 300,000\,\mathrm{kbit/s} = 300\,\mathrm{Mbit/s} \text{ of bandwidth}$$

Of course, using a multicast protocol the server sends out only a single stream that is common to all users. Hence, such a stream would only use 300 kbit/s of serving bandwidth. Far better for all your neighbors.

Another issue is intermediaries. In HTTP, the protocol is designed so that you can have a chain of servers, with the request automatically going to the first server in the chain, rather than the server where the file was originally stored. The design is intended to hide these intermediaries, so that the user believes the request always goes to the first server – but the traffic does not have to go over the network every time. When HTTP was designed, disc space was cheaper than network capacity, and even if that is still true, it is on the way of turning, so that the intermediary storage of data is not as attractive. Still, it makes sense from a latency point of view.

The intermediary server can capture the response (and the request), and send it again when the request is made the next time (in which case it acts as a cache); if the server redirects the request (or the response), it acts as a proxy. A cache usually does not change any of the message (headers – which contain the information about the request or response itself; or body – which contains the actual contents, in the form of HTML, or a file). A proxy, on the other hand, interacts with the message, changing the message headers. The headers that are changed are at a minimum those concerned with the timing of the message (which are updated to take the actual transit time into account), but the proxy will interact much more with the headers in the request and response. This is not noticeable in HTTP, but in SIP it has been taken to a completely different level, and this is one of the cornerstones of IMS.

HTTP assumes that there is a client and a server (and a number of proxies acting on behalf of the client and the server). A server is a piece of software in this regard, not a hardware box (although there are of course box providers who build computers optimized for running server software, and call them servers). When the web was designed, almost 15 years ago now, there were still terminal-based systems around, and the PC had not really taken off (the original web server and browser were written not on a PC, but on a Unix-based Next computer). Computers on people's desks were not very capable; and servers resided in air-conditioned rooms with raised floors, although they were no longer tended by men in white coats or programmed with punch cards.

It makes a lot of sense to divide the functionality between the part presenting the information to the user and the part interacting with databases; this requires very different computational operations and different optimizations of the system performing the action (to interact with a database, you need a lot of shared memory and fast discs; to present content, you need a good graphics engine and lots of memory dedicated to this process).

Computers have come a long way in 15 years, and now a desktop PC is more capable than the Next computer where the web was created. Provided with fast network interfaces, a PC has a fast enough disc and enough memory to be able to distribute files to other computers, if given appropriate software. In a peer-to-peer-based file-sharing system, the users connect to the next available computer, which has the appropriate resources, and gets the files it wants to use from there. Conversely, it makes its own resources available, sharing what is available in its own system.

Streaming, especially of a live event, assumes that there is a canonical source: the content comes from one single server, and is received at many different clients. To buffer the content at an intermediary can help reduce the bandwidth significantly, especially if the content is to be viewed later. This is the premise for peer-to-peer systems, when used for streaming data.

P2P in the European Broadcasting Union and EU

When talking about "peer-to-peer file sharing" the "file sharing" part often gets dropped, and "P2P" becomes synonymous with file-sharing systems. The theory behind the interconnection of large numbers of computers has been investigated very thoroughly, especially since there are two major variations. In the home network, using UPnP, the computers are all peers. SIP is also based on peer technology, and does not make assumptions about computers being servers or clients (which makes it somewhat hard to build appropriate proxies). Peer-to-peer technology means only one thing: all computers are equal, and they transmit to other equals. The problem is that while computers may be equals from a theoretical point of view, they have different bandwidth and storage capacity. Especially when receiving large files, it would be nice if they could help each other out in order to receive content.

When a user installs a client in a peer-to-peer file-sharing system, making their hard discs available for storage, the distribution of the files that can be downloaded would be random if there was no method of control. In addition, if there is an identity associated with the client (as is the case in the Skype system, for example), there may be a need for centralized authentication and authorization to ensure that the users are who they claim they are, and have a right to do what they are doing (which often is associated with payment). Peer-to-peer systems, for that reason, in some topologies are coordinated by "supernodes", which maintain directory systems and direct requesting nodes to those which have the files they want, and which are near them in the network. While peer-to-peer systems are often touted as a way around bottlenecks in networks, the supernodes risk becoming bottlenecks in themselves.

This does of course have plenty of legal implications, which is one reason the peer-to-peer technologies have got a bad rap. However, it is not an inherent feature of either file sharing or peer-to-peer technology, it is just a feature of the way the software is designed. Some content providers actually use peer-to-peer technologies to provide content which may not have a sufficient audience to merit being put on their servers. This requires the content provider to have control over the allocated area of the user's hard disc, which is unlikely to happen when the user is using a PC, but more likely to happen when providing set-top boxes to the users; the set-top box is supposedly never switched off and so always available; and the agreement with the user can make sure the allocation of the disc space is allowed.

The first variation is that the directory of the content is shared between the peers in the same way as the files (this can also be done in various ways; the most advanced systems divide the file into small pieces and distribute these among the computers in the network, with redundancies to safeguard around computers withdrawing). When the directory of the content is distributed over several computers, the access is not guaranteed and it becomes very hard to keep the directory current if computers come and go in the network, which is likely to happen in a system based on voluntary effort. The alternative is to have the content centralized, but verify it as the computers come and go. The centralization means that the content directory can always point to a place where the content should be available. When this type of file sharing is used, this may mean sharing things the user did not put on the hard disc himself, since the resources in the network of peers – which you become part of when you download the peer-to-peer file-sharing client – are allocated by a directory server, making use of the resources the user has provided without his explicit consent.

Instead of having one humongous hard disc, and hundreds if not thousands of connections to it, peer-to-peer technologies spread out the storage and access to the stored resources among the participants in the network. This is exactly the opposite of how YouTube works, since that system assumes that there is one central server from which all video clips are served. This also allows for control over the content and the presentation, which is precisely what the proponents of P2P technologies want to avoid.

The method of presenting content in P2P systems is rarely more sophisticated than the file listing in the operating system. To make the user see the content as part of their own computer helps them see downloads as a simple option for getting content free. One special case of this is when you access the stored content from a cache – which may be in your neighbor's PVR.

BBC Internet media player

The BBC, Japanese Softbank, and Chinese television have all developed P2P file-sharing systems to ease congestion in their networks, leveraging viewers' resources – and in the case of the BBC, controlling the content distribution as well. Their intent is not to make people distribute content illegally, rather, it is to make sure that as many viewers as possible can get their content with a minimum of infrastructure investment. What they do is to overlay a multicast infrastructure on top of the existing network, instead of implementing it in the address management system – which you have to do with IGMP.

The BBC media player is probably the most high-profile of the different players around. It controls the distribution of BBC programs (which are supposed to be available for free to people in the UK only). And it saves the cost of creating huge banks of servers – otherwise, the BBC might become a competitor to Google when it comes to infrastructure. The BBC integrated Media Player (iMP) uses P2P technology to reduce costs. In this case, the P2P technology is used only for downloading media files. It was developed by a company called Kontiki, who claim the iMP increases quality and at the same time reduces the distribution cost by approximately 80 %. However, there are serious issues with the digital rights management (DRM). P2P players – as with Kontiki – can coexist with DRM, but also offer extra security for the content. Other examples of broadcasters using P2P downloading are Time Warner, BSkyB, OMN and NBC. In all those cases – as with file sharing without the permission of the content owner – the users provide a piece of the infrastructure, since their memory and network capacity is used for the re-broadcast of content.

Peer-to-peer broadcasting of live events is harder than VoD. If there is latency of the delivery of the stream to the viewers, there will be delays, which mean viewers will not see the program at the same time. Or worse, the delays may result in lower quality – or resending of content, which means further delays in the viewing. The broadcaster has to trust the network. However, if the media streams can be distributed over a large number of computers in parallel, the streams can be combined to provide a live stream at all the receivers. This works, and the Eurovision Song Contest, the Chinese New Year celebration broadcasts, and the Word eSports Games 2006 Master cup were all transmitted over peer-to-peer systems in parallel with the traditional broadcasts.

Softbank of Japan, which started out as an investment company, but soon expanded into becoming the leading Internet connectivity provider and has since branched out to mobile telephony (becoming the fastest growing operator in Japan), and owner of the Fukuoka Softbanks Hawks baseball team, among other things, is not a company to shrink from challenges. Its offering of a peer-to-peer-based system for television viewing is nothing but audacious. Notoriously hard to manage from a traffic perspective, yet immensely popular among Internet users for sharing of multimedia files (copyrighted or otherwise), peer-to-peer technology has a bad reputation as a malicious technology. However, when used in a managed way, it can be leveraged to provide services with very little investment for the operator.

That is the intention behind Softbank's BBbroadcast, a system which uses a peer-to-peer overlay on top of the IP network to enable PC users to view and redistribute content in near real time. Using its Softbank Hawks baseball franchise as the test vehicle, baseball games (against Nippon Ham Fighters and Lotte Mariners) have been sent to more than 120,000 simultaneous users.

The number may seem low, but it is exactly the target for BBbroadcast. Softbank intends it to be a service providing "long tail" content, making things available that are not broadcast on national television (such as Softbank

Hawk games, since the team is based in the southern island of Kyushu). "Regular" IPTV services would be provided from a server (such as the games of Tokyo-based Yomiuri Giants, if they were available).

The BBbroadcast client does not replace IPTV, but enhances it. Incorporating DRM, it does not create a problem with illegal file sharing. The overlay peer-to-peer system overcomes some of the issues of the uncontrolled nature of peer-to-peer. Still, it is an unproven technology, not yet established as a popular feature among Japanese users.

Like its counterparts in other parts of the world, it uses a mesh technology to establish the peer-to-peer network. Traditional P2P systems act like a tree, where one user sends and receives only to one or more other users. By splitting the streams and adding links across the tree, the transmissions become much more secure against disruptions (such as one user switching off his PC in the middle of a download). Instead of a tree, you get a grid as shown in Figure 6-2.

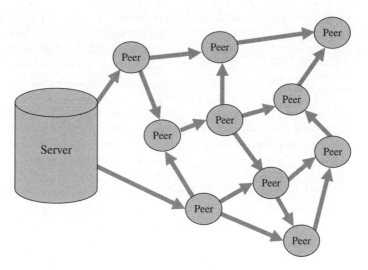

Figure 6-2. Grid structure and P2P broadcast.

The European Broadcasting Union (EBU), which manages the Eurovision Song Contest among other things, is taking P2P live broadcast one step further in an EU-funded research project, P2P Next, which intends to work as a test bed for a number of television stations that want to distribute their content to users outside their traditional coverage areas. The project, which has just started, has a fairly conventional setup: the user goes to a portal, signs up, installs the software, and can watch the program. Since the project started in 2008, it remains to be seen how popular it becomes. Much depends on the broadcasters – if the programs are only the traditional fare, they limit their audience to expatriates and students of their language. Including interactivity would be a nice twist – but the industry is not there yet.

Chapter 7: Digital Rights Management and Next-Generation IPTV

In IPTV, there is no "thing" to show you have the right to consume content. There is no physical token, such as a book or a CD. Nevertheless, content is licensed to use on certain conditions, and as a user, you have to follow the contract you have signed. Think of it as a lease, instead of buying. But add interactivity, and the equation is complicated. Instead of protecting the rights to the program, the service provider will be making money from the users interacting, and so be more interested in getting their programs to as many users as possible – the famous Gillette model of "giving away the razors to sell the blades". This turns the way the television industry has traditionally thought about content management on its head.

The bigger picture: handling rights digitally

From time to time, YouTube and other video sharing sites have to remove videos from their directories. These are videos where the rights are owned by someone else than the uploader, and where the upload was not according to contract.

(Continued)

It may seem strange that there can be a contract for television, especially that received over terrestrial broadcast. After all, it is sent out to anyone interested in watching. So how can there be any rights attached to it?

"Copyright" is a legal term, and there are certain rights that a creator of the work (which may be a television show, or this book) automatically gets when the work is created. Those rights can be traded (as the author of this book has sold the right for publication to Wiley, the publisher, for example).

A better way to think about copyright is that you receive a lease. You, the user, pay a sum of money (in the case of terrestrial broadcast, you either pay a license fee or you pay with your time watching advertising). But there are conditions attached to the lease you have received for the content, which you have to conform to if you want to see the content. For instance, to read this book, you had to buy it.

How this works, and what the broadcasters can do to track and enforce it, is the subject of this chapter.

Generally speaking, content rights is the hardest part of the IPTV industry – because it is not a matter of technology, it is a matter of education. Both users and producers need to understand what rights they have, what rights they can claim and sell, and how to safeguard those rights. When you buy a DVD in the video store, you have bought a copy of a film, and at the same time a physical token for that film. With the Internet, it is now becoming possible to distribute multiple copies of a film – without making physical copies.

When you have a physical copy, you can sell that to somebody else, or give it away. However, if you rented it to someone else, and set up a business doing so, you would probably have a visit from the film company sooner or later. If you read the fine print on the DVD case, you will find that you are not allowed to rent the film that you bought. You cannot do whatever you like with the video. Video stores actually rent the films from the film companies themselves (and are smart enough to pay less than an average Joe buying the video, so do not feel sorry for them).

Information is different from other types of goods (and most services) since it can be shared by many people at the same time, and copied without degrading once it has been digitized. However, the old rules still apply, copyright, in all countries, is automatically created "as a result of a creative expression". There is no need to register, no need to put in a copyright mark (although this helps if there are disputes, as we will discuss later). Since there is no registration for copyright (which there used to be in the US, but no longer) other types of rights management, which can be registered, are also used, such as design patents (which only apply for 20 years, but do not override the copyright).

Two rights in copyright

There are two types of rights associated with the copyright: the economic (or exploitation) and the moral. All countries in the world, bar the occasional rouge state such as North Korea, are party to the international conventions that make the rules, and contain means for enforcing them. National laws have been adapted to this, for example, extending the copyright for written works to 70 years after the death of the author, when the work then falls into the public domain (which is why Disney is so nervous about the copyright for Mickey Mouse expiring – in about 30 years, since Walt Disney died in 1966); and making the World Intellectual Property Organization (WIPO) the final call for resolving disputes.

Some countries have laws that include the moral and economic aspects of copyright, but those rights exist regardless of how the law is written. The easiest way to think about them is that the economic right is the right to sell the information, and the moral right is to have the form it has been given preserved – if someone publishes a story you cannot change it without asking, regardless if it has been published on the Internet. The sticky point is if you start changing the story – at what point does it become a new story? This is even more an issue in video and other graphic forms – can you take Donald Duck and make him a hero of an adventure story?

The courts regularly judge cases of copyright infringement – a photograph used to create a graphic print, a story plagiarized from the Internet. There is, however, no percentage or rule to automatically apply when it comes to copyright, that is, to say what is infringement on the original. The judge and jury (or equivalent in the countries with different systems) have to determine what is infringement, and what is not. In practice, this is not too difficult – it is as US Judge Potter Stewart said about pornography, "I cannot describe it, but I know it when I see it".

If we take away the pieces of plastic, and the video exists on the Internet, the same rules still apply. There is no special legislation for the Internet, and there is no anarchy there either (as was claimed in the beginning of the century). The same rules apply as when you buy something in a store, but slightly differently.

There is a set of basic rules that apply to the management of copyright. The first is that copyright expires 70 years after the death of the author. This not only applies to books and other written works, it also applies to any work where there is an "author", in the sense of someone who makes an original contribution. A large part of the preparation of any video production is to sign up the copyright holders – because not only is there an author's right, there are also rights of executors. When a singer sings a song written by someone else, the recording of that performance also becomes a "work", for which the singer has copyright. When the copyright of a work expires, the work passes into the public domain.

Creating a professional production means signing up all the copyright holders – with due compensation. The production company holds all the copyrights for the production, and if infringed, can request a court to enforce them. This is the basis for the entire television and video industry. Singers and composers usually sign over their rights to a copyright management organization (there is usually one per country), which is tasked with collecting from the radio stations when they play a song, for example.

Exceptions to Copyright

All this may seem complicated, and there are two exceptions: fair use for individuals, and copyright that has been assigned to the public domain. There are a number of things you are always allowed to do with a video or other digital work, which fall under the rule of "fair use" – which, it should be noted, applies to individuals, not to companies. The rules do vary a bit between countries, but you are always allowed to cite, parody, and make backup copies for home use. Often educational use is also exempt. In most European countries, what is fair use is outlined in the law; but in the US and many other "common law" countries, it is a matter for the courts. File sharing over peer-to-peer networks is something which is covered by the rule only if you share with one or two friends; if you share your files with thousands of unknown people, it is a clear copyright violation in most countries. This is of course a gray zone – are you allowed to send a copy to a friend? And is the friend allowed to send it to other friends? And so on. The answer is generally yes. You are not allowed to take a copy and put it on your website, or put it in the company newsletter. There has to be a personal relationship with those who receive the content. If the receivers are people whom you do not have a personal relationship with, the sharing falls outside the bounds of what is allowed. Of course, there is a gray zone here too, but few people have more than 15 close friends anyway. More than that, and it starts to be suspicious. And as soon as you make money from selling someone else's copyrighted work, you are stepping over the line.

Derivative works

You are also not allowed to change the work. Here is another gray zone, because you are allowed to change it if you create a new "work". So, for example, if you work very hard in Photoshop and replace Darth Vader's face in Star Wars with your own, you have not created a new work, and George Lucas would probably feel you had damaged his creation. However, if you combine Star Wars *and* Star Trek, *and paste yourself in as Darth Vader, your boss as Captain Kirk, and invent a new story – then you may have created something new, which at least could be considered a parody. Depends on how good you are.*

These rules apply to any video, any book, and music as well. How much you are allowed to "steal" is a hard question (when Madonna sampled Abba, she asked – and paid). If in doubt, ask the rights owner. Even though the idea is that the rights

> *owners should have put in considerable effort to create something new, and this is why they have a say in how the work is used, the rights holders today are more often corporations. That does not mean it is right to take the video, or book, or whatever and do with it as you like – the corporations have bought the rights from the creator of the film, or book, or whatever; and they have a right to say what to do with it. That is the meaning of the international conventions that regulate how you can use video and audio, and any other work that falls under copyright. It can seem stiff and clumsy, but as with any other law, if you do not like it, talk to your parliamentary representative. You cannot change it yourself.*

This is true even though (according to some industry figures) 75 % of all content is now created by amateurs. Note that this does not say anything about how frequently this content is accessed; there is a well-known distribution of content which says that 20 % of the sites get 80 % of the hits (it may be even more extreme; this is a mathematical relationship called a power law or Zipf distribution). This is true for the individual videos in sites like YouTube, too. But since so much content is created by amateurs, how the copyright for the content can be managed becomes an issue. There is no difference between the photos of your children dancing and Harry Potter from a copyright perspective.

So, when you get a copy of a video, it is more like a lease than a purchase. When you lease a car, you are not allowed to do many things you would with your own car (e.g., fill the trunk with manure regularly). The reasons are different, but the effects are the same: you can use it, but there are some things you are not allowed to do. It does not matter that it cost 99 cents on iTunes.

When you start making money from a production, however, the rules change. An individual, acting as a private person, can share videos in a limited way; but once the work is commercialized and someone starts making money from it, the honey has to be shared out among the copyright holders according to the agreements struck when the production was made.

There is an exception to the rule: signing away copyright to the public domain. Seventy years after the death of the author, his work becomes free for anyone to do anything with – without restrictions and constraints. This is true for any work. There are some additional ways this may happen, for example, when the work is done by a publicly funded agency in the US.

When a work is assigned to the public domain, no rights to the work are reserved. Of course, if you create a new work from previous works (e.g., doing a montage, collage, or sampling), this becomes a new work, and is protected in the same way as any other created work. An author can still put his work in the public domain while he is alive, and even add some conditions to it. All an author has to do is make a legally binding statement that the work created is in the public domain, and then he has in effect created a gift to the general public of the work.

Attaching Strings to Copyright Gifts: Creative Commons

A gift can come with strings attached, and not just those tying the parcel. When making a gift of a work to the public, it is possible to give away some rights, but not others, and put conditions on how the work can be used. This is because, when making the gift, the author retains both the moral and economic rights to the work, and the gift is a donation of parts of those rights. There is no constraint on how much of the part can be given away, and how. If an author wants to give away only the right to freely use the word at the top left corner of every page in a book he has written, that is possible – but not very practical. So a group of "cyberlaw" experts (cyberlaw and intellectual property experts James Boyle, Michael Carroll, and Lawrence Lessig, MIT computer science professor Hal Abelson, lawyer-turned-documentary filmmaker-turned-cyberlaw expert Eric Saltzman, and public domain web publisher Eric Eldred) founded an organization to create a way of regulating the gifts of rights to the public, called Creative Commons, in 2001.

Today, this is an American "charitable corporation", in practice a non-profit project to ensure that there is a set of legally coherent terms applicable worldwide, which can be easily communicated about how an author is able to give away the whole or pieces of a work.

Creative Commons has created a set of licenses for giving away content, and these licenses are non-revocable – once you have made the gift, you cannot change the terms. It is possible to stop distributing the work under the Creative Commons terms (in effect, saying "this is not a gift any more"), however, if someone obtained it under the Creative Commons terms, the gift has been given and cannot be withdrawn.

Just like any other form of copyright, the rights occur the moment the work is created; and the author does not have to register anything, just state that the gift is given and it is done. There is no central registry (although in the US, there is a copyright office that can accept registrations of copyright, which could be useful in cases where there are doubts about the dating of a work). There is no formal requirement to sign any documents, and nobody keeps track of the licenses except the rights owners. To complicate that part, it is possible to create non-exclusive licenses, so the work can be given away to some people or parties, but not to others. The easiest way to understand this is via an example: a work can be given away freely to anyone who wants to use it, for example, for educational or other non-profit purposes, but if the work is to be used to make money, a different license will apply.

When a work is licensed under the Creative Commons terms, it becomes available under the terms of the license to anyone who comes in contact with it. Theoretically, that means anyone who encounters a piece of the work, whether through downloading it, getting a paper copy of it, getting it from a friend, or whatever means, gets a license. In practice – just as in regular copyright management – it is not practical to manage copyright at that level.

Since the license applies to anyone who comes into contact with the work, Creative Commons has created a "deed", a document which can be affixed to the actual content and presents the terms of the license. This is available in three different forms for the different kinds of license that Creative Commons provides:

- a human-readable form (the "commons deed");

- a legal code ("lawyer-readable"); and

- a machine-readable code, encoded as metadata in RDF which can be encapsulated within the document itself.

The Resource Description Framework (RDF) is a way to create statements that a computer can interpret. In this case, the statement is about which rights apply to the document. The legal code is the actual license; the others are (formally speaking) references to it.

There are four key elements to the Creative Commons license (shown with the following icons in the license terms, to be easily identifiable):

 Attribution: others can copy, distribute, display, and perform the work, but they have to give credit to the author (and owner of the copyright, if they are different) when doing so, in the way defined in the license terms.

 NonCommercial: that the work is not going to be used to make money (a work under Creative Commons can be licensed for noncommercial use by one group, and still be licensed to others for commercial use – the license is non-exclusive). So if a work is marked with this symbol in the Creative Commons license, and you have found a great way of making money from it, you have to approach the copyright owner to get a separate commercial license.

 NoDerivatives: it is forbidden, under the terms of the license, to make works that take the current work and change it (e.g., by sampling a song by a choir for a gangsta rap record; or painting moustaches on a portrait and call it "bearded lady", putting it in your online gallery). It is free to distribute copies, as long as there is no change of the work.

 ShareAlike: any derivative works created must be licensed on the same terms as the original work. So if the original license was NonCommerical, all works, which derive from it, must be licensed under the same terms (which means that you will never be able to sell posters of the bearded lady, unless you buy a commercial license to the original from the photographer – but you can give it away on the Internet).

Note that these conditions go together in different ways. There are a number of standard conditions that Creative Commons has defined, but there are multiple other conditions possible, as well as impossible (it is not possible to combine the ShareAlike license term with NoDerivatives, for example). The licenses that Creative Commons has created are:

Attribution means others are free to use the work (even commercially) as long as the author is credited.

Attribution ShareAlike means others can change the work to create new work (even commercially) as long as they give the author credit, and share the new creations under the same terms as they received in the original work.

Attribution NoDerivatives means the work is licensed for redistribution, both commercially and noncommercially, as long as it is not changed, and as long as the credit for the work is given to the original author.

Attribution NonCommercial means the work can be used freely by others, including changing it and using it in various ways, as long as they give a credit to the original author and they do not make money (note that this does not mean just selling the derived works, it can equally well mean performing it in public and asking a fee for watching the performance).

Attribution NonCommercial ShareAlike means the work can be used, changed, and freely shared as long as the original author is credited, and as long as any works that are derived from it follow the same terms – i.e. are noncommercial and attribute the author (both the original, and the author of the derived work).

Attribution NonCommercial NoDerivatives means that the only thing others are allowed to do is to download the work and share it with others, as long as they give credit (i.e. link back) to the original author, but they cannot make money from doing so.

The author selects the appropriate license, uses the engine on the Creative Commons website to generate the appropriate metadata and other code, and puts it in the work. It is perfectly possible to choose two different licenses, which gives the public a choice between which of the licenses applies. This is tricky because it is about legal matters, and the author has to be careful to select the right sequence and order of the licenses, which is why the Creative Commons licenses are so helpful. It is also helpful to understand what can be done with the work in order to create a new one. As usual with lawyers, there is a disclaimer, so it is by no means certain that any and all possibilities are included in the license terms. Note also that it is possible to license only part of the work, for example a web page but not a book, or an audio recording but not the book that is being read from. This makes it tricky to select the appropriate license terms, but it also means that anyone who uses such as work has to be careful to make sure they are covered by the terms of the license.

When you give someone a license to something – especially if that is done without ever meeting the other party – there is always a risk of misuse (i.e. they do things which you did not allow according to the license). The problem is that there is no way for the rights owner to control what happens once the work starts getting widely distributed. There are two ways to handle this: try to control the distribution; and trust the licensees.

The second way is how the Creative Commons license works. By putting the mark of Creative Commons on the work, and then trusting that it will not be misused, the author (or other rights owner – you could, potentially, buy a work and then put it under a Creative Commons license) is giving up the rights to enforcement as well as the rights to derivative works, which the license provides for. Since Creative Commons was explicitly created as a license especially for the digital age, that makes perfect sense.

If the author, on the other hand, discovers that a licensee has misused the document, he can sue them in court; or request that the name of the author be removed from the work (since he may no longer want to be associated with it). The license terminates automatically once someone breaks it, and from a legal perspective that will hold in court; if the author has licensed the work solely on the terms that it cannot be made to make money, and someone sells it, the license expired from the first minute they put it on the website with a price tag on it. Note, only for that user, not for everyone else. Basically, the author puts his work on the Internet, and his faith in his fellow users, by saying "I trust you to follow the license; take this work".

Legal Constraints on User-Provided Content

The legal constraint is concerned with the rights of the content creator. This is something we will come back to in depth in later chapters, but it is sufficient to say here that just because you have possession of a piece of content, you are not allowed to do anything you like with it. Users can provide content in two ways: create the video themselves; or record content created by someone else – "ripped" from a DVD or captured from television (increasingly, television shows provide content on the web as well). However, very few users are capable of providing professionally produced content – although this has changed with the emergence of video cameras that are able to capture content professionally at consumer prices, as well as editing programs, which make the management of the captured video easier.

There are a large number of sites showing user-provided content. As hosts for postings of others, the providers are exempt from responsibility under the laws of many countries; however, there is a fine line between the responsibility of the provider and the freedom of speech, which running a video bulletin board enables. When in doubt about the law, the court has to decide. Of course, it is useless to sue someone who cannot pay (unless you want to create fear among the population, as the Recording Industry Association of America (RIAA) has done, when suing single mothers without substantive means of support).

YouTube, the prime example of television on the web, is really just a very big video on demand system for short video clips, based on the same technology used to serve web pages. The reason it has become so popular is the same as Napster in its day became popular: users published content for which they did not have rights (but thought they did), for others to use. Google, quite a bit smarter (and with much more legal muscle) than Shawn Fanning, the founder of Napster, tried to avoid the same mistake. But every day, there is a new video removed from YouTube on the request of one media company or other, which feels its rights are being infringed.

That is not the only link between YouTube and Napster, and hence between television on the web and peer-to-peer systems. While there are large similarities in the technologies used, there are also enormous differences (there have been huge leaps forward in technology as shown by the different media types they handled – Napster was primarily a music download site, YouTube is about video). They both use the same model for enabling users to find content: one central directory, which you browse and search to find what you are looking for. The difference is that YouTube hosts the content on its own servers, whereas Napster (and its successors among the file-sharing peer-to-peer sites) hosts the content of the computers of the members of the network (the peers). For the user, browsing the content and accessing it is not too different.

What has happened in the almost 10 years between Napster and YouTube is that the mental model for content has started to change. When Napster came about, only a handful of bands considered promoting themselves by uploading their music and letting others listen to it for free. Now, this is something of a grassroots movement, and even the established record companies are beginning to wake up and smell the coffee. Big, established bands such as Radiohead make their content available on the Internet – for a fee, if you want to pay it. Small, independent bands can even count on the ability to use downloads as a marketing tool, gaining much better communication with their audience than when they have to trust the record companies and their marketing departments to help them.

Even so, most of the content on Napster was copies of professional recordings, by professional artists and produced in professional recording studios. The change with YouTube is enormous: a very large part (some estimates say 60 %) of the content is produced by the users themselves (the site was even founded, or so it claims, to share the holiday videos of the founders). While the editors remove content which they judge unsuitable (such as sexually explicit content – for which there are other, unaffiliated sites), the content is not limited to babies taking their first steps, hats being blown off in the Alps, and Aunt Bertha at Blackpool. The commercial aspect, which can be anything from a band promoting their latest song to a lecture on how to use the obscure features of a company product, has become an important part of the content. And so has user-produced animations, mostly featuring computer-generated graphics (since the process of producing the cels for animations manually is mind-numbingly boring). These are often made with the graphics engines of computer games (and called "machinima"), something that was not possible 10 years ago, but which has become feasible with the combined spread of computer games for the PC, and the increased processing capacity of that device.

The fact that pornography is increasingly user-created is an interesting aside (and that is all it will be in this book): pornography requires two or more participants (someone has to hold the camera), and professional actors do not do a better job of having sex than amateurs. They may have better developed bodies (sometimes in absurdum), but according to such a venerable source as the *Economist* magazine, user-provided pornography is hurting the professional industry – since viewers find it easier to imagine having sex with someone who looks like the girl next door, not as a pumped-up supermodel (of course, there are various specializations – but even there, the user-provided content is making inroads, and has been driving the sex industry for a long time). Pornography was the driver for both the emergence of the VCR and the Internet, and so is something to watch in this space – even if it is probably losing its appeal as the major driver.

There are two ways to restrict content distribution. One way is to make sure that the content cannot be physically redistributed, which is how cable-TV networks used to be built, with one secure connection per set-top box. The other way to control the distribution is technically harder, and much more difficult from a socioeconomic perspective. Note that in the US, file sharing and trading of a work is considered commercial use (it is the opposite in most of Europe). The Creative Commons license terms, for that reason, have a specific exemption for file sharing, as well as for the use of DRM. There is no technical issue – only the fact that the way DRM is regularly used conflicts with the spirit of the Creative Commons license. In essence, DRM is a very clever way of not having to trust the end-user.

Digital Rights Management

As we have already discussed, the metadata about a document can contain information about who created a document, and who owns the rights (which, it ought to be clear by now, do not have to be the same). This is especially important in the movie world, where the rights belong to the studio, but your next job is only as good as being in the credits for the next movie. The right of recognition can, in other words, have an economic value as well.

While the creators have an interest to be included in the credits, they have already been paid. Their contract was fulfilled when the production was created (well, they may have royalty rights, but you get the idea). There is only one party that manages the rights. This can be the television broadcaster, the movie studio, the production company, an aggregator, or someone else. Irrespective of who they are, they have two interests: they want to make sure that the content they have the rights for is not distributed in ways they cannot control; and they want to get paid for the work.

In interactive IPTV, controlling distribution does not necessarily mean getting paid more. Making the program available to as many people as possible means more of them will interact with it, and if they pay for the interaction, that is where the owner will make money – not from the distribution. So, in interactive television, there are good reasons to forego the control of the distribution.

However, there are plenty of other media formats where the rights owner has a perfectly good reason for controlling distribution. If a person gives copies of a film (or music, or a television program) to his friends, the thinking is that this is a sale lost; at the very least, it is a missed rental opportunity.

Digital content can be copied without quality degradation. That simple fact has created a shambles of entire industries, from photography to music, and everything in between. The rights holders – who mostly are record companies, having bought the rights from the artists – have reacted in four typical ways:

- preventing possible infringement through automatic technologies;

- developing tools that enable the management of rights;

- lobbying for making these technologies law; and

- trying to teach people how they are allowed to use copyrighted material.

Copyright is the basis for the IPTV industry, and for the entire digital content industry. It is also the basis for the sometimes ludicrous efforts to protect copyright. Companies do have a case: if you have a copy, and make it available to millions of people, it is likely that only some of those people would have bought it.

Advertisements and copyright

If the copy of the program had advertisements in it, and those advertisements were now seen by millions who had never seen them before, that is probably a good thing for the advertiser. Right?

Not necessarily. If you put advertisements in a video, and intend to make money from the advertisements rather than video sales, there is always the risk of a runaway hit. It may not happen in ways you expected – if a Bollywood film becomes a hit in the US, the commercials for a curry maker in Chennai are probably wasted advertisement space. And it is not that clear-cut, even if the advertisements were only released in your local market. It may be that if the film was old, the product advertised is no longer available; or it may be that the manufacturer was not ready for the storm of potential customers from the Internet. So advertisements are not automatically a way to monetize rights over content, even if they work much better in that regard than charging the end-user for the content (and of course, there are different models here too).

Nevertheless, most people are ready to pay with their time instead of their money, if they can get entertainment without too large disruptions. Even though users would often like to remove the advertising, this is not something the content providers like to see, since that would lower the exposure the user has to the advertiser's promotional content. There is no easy resolution, and no simple technical solution either (those that exist make use of tricks, which relate to how the video stream is composed, but such tricks are not necessary and the video stream is not always composed in a particular way).

In the end, digital rights come down to trust: the user has to trust that the content provider gives him the content he has been promised; the content provider has to trust the user to pay, or watch the advertisements. When the trust is no longer there, for example when one party has consistently broken the trust of the other (by publishing the content on peer-to-peer file-sharing networks), then the other party will have to resort to countermeasures. And in the IPTV world, those countermeasures are called digital rights management (DRM) techniques.

DRM: Simple Philosophy, Complicated Mechanism

The philosophy behind digital rights management is very simple: only the person who has been given the right to the document receives the key to use it. The document – file or stream, depending on the delivery mechanism – is encrypted, and can only be opened if you have been given the key by the rights owner. The key may be given conditionally, depending, for example, on how many times you are allowed to use the file, or for how long; this can be solved using software in clever ways. The encrypted file can be distributed freely, since it can only be opened if you have the key, which you can only get from the rights holder.

The super-distribution based methods are elegant because they do not take the network into account – but they also present a problem if you want to protect your content at any price, since they allow the user to distribute the file freely, and the methods to limit the key use to one single client (which is a "trusted agent" in the parlance of the DRM technologies) are difficult and costly in terms of computing power and memory.

There are quite a few mechanisms to enforce the appropriate use of content (or rather, prohibit the inappropriate use of it). There are also mechanisms to verify the distribution of content (rather than enforcing it). That is a completely different matter, and much easier technically.

If you know who distributes content, you can enforce the license terms other ways than technically. A lawsuit is a very efficient weapon, but unless it is to end up in the small claims court, the misuse has to be flagrant. And as the court will probably require that the copyright owner prove the misuse, there has to be a way to verify where content came from, and how it was used. There is a technique to do exactly this, and even though it is not strictly a technique for digital rights management, we will discuss it first.

There are two aspects to this. The first is to check whether a piece of content has been used at all. The second is to check where that content came from, so the misusing distributor can be taken to task, or to court.

If distribution is done physically, for example by carrying DVDs around, it is very easy to control. You can not share the DVD unless you get it first. But if there is no physical medium, if the only copies that exist are digital, then it becomes difficult to control the distribution. In traditional cable-TV systems, the set-top box is connected directly to the cable (or satellite antenna). Adding IP means the connection is decoupled from the location – in principle, an IPTV service could be anywhere, and the set-top box somewhere else. This method to control distribution works only if there is no way to redirect the media stream from the sender to the receiver, even to storage. If media can be redirected to storage, it can be found and played again.

The control of the receiver is a big thing for the cable-TV industry, and the solution is to encrypt the data stream. The decryption can only be done by a set-top box. This creates a problem of distributing keys, and if the content can be stored on the set-top box, it also creates a problem of how you distribute the keys for opening the content – in particular what rules are associated with the keys. The sender may not want the receiver to do just about anything with the file once it has been opened; for example, the sender may only want to allow a file to be opened once; or not redistribute it; or something like that.

If your music has been sampled, you have to prove it. That may not be so difficult if the only thing that has been done is to replace the song track with somebody else's rendition. It is still possible to hear that the song is the same, and easy to check the license terms. But if the music has been sampled, or if the issue is not music, but a web page where the CSS has been reused, how do you prove that?

The easiest way is to embed something in the music which lets you check who had it last. This is tricky: there is no way for a content provider to enforce that consumers have readers, which mark content with who has used it. And it may also not be particularly meaningful, in particular if the markup is human-readable (in which case it is also human-erasable). Anyone who is clever enough to share content with others is clever enough to erase the markup which shows he had the content last.

There are, however, ways to embed markup in content so that it cannot be identified, but can be recovered later. Using ultrasound tones, which can be transposed to be heard, could be used to include the "watermark" that brands the content. There are similar ways to embed information in images, text and other media.

Embedding a watermark is a simple way to safeguard the distribution. But if the content was impossible to open without the appropriate encryption key, and the copyright owner could make sure that the key was only distributed to people who should have it – and that they could not redistribute the key – are what digital rights management is all about. Traditionally in the cable-TV industry, DRM was not about this, it was about enforcing conditional access – that only devices which come from a specific connection can consume content. This means having explicit control over the access network, and making sure that there is no way the user can change the connection (or the identification of the connection). Compared to the ease of super distribution, this is a very cumbersome, primitive and limiting way of managing content.

Anyone can send anyone else an encrypted file. Not everyone can open it. Encryption is a mathematical process that changes the content according to a secret process – the sender creates the encryption. The receiver gets the encrypted content; but cannot change it to something usable without having the key. There is much more to say about key distribution and management, which could fill several books (and which do); but this is not something which we need to go deeper into here.

Standards for DRM

There are many different ways of doing encryption and key distribution and management. When you associate this with a set of rules that relate to how the key and the decrypted content can be used, you have a mechanism for digital rights management. As a matter of fact, this is how the available DRM systems work. There are a number of them, using different formats for encryption and mechanisms for key distribution; again, the differences could fill a number of books. And while several of them are standardized, there is no one single standard for digital rights management – partly by virtue of the mechanisms being patented, and nobody wants to grant anyone a monopoly over their work.

DRM systems are already heavily in use in the publishing industry, especially in the music industry (and have a very bad reputation among users; even publishers are unsure if they work as well as they should). The few that provide the most effective control over content distribution are those which are in control of the end-user device (Apple's iPod and its Freeview system are intimately coupled, for example).

That said, there are a number of initiatives in the standards. MPEG-4 provides mechanisms for DRM, by supplementing the coded media objects with an (optional) Intellectual Property Identification (IPI) data set, carrying information about the contents, type of content and (pointers to) rights holders. The data set, if present, is part of an elementary stream descriptor that describes the streaming data associated to a media object. The number of data sets to be associated with each media object is flexible; different media objects can share the same data sets or have separate data sets. The provision of the data sets allows the implementation of mechanisms for audit trail, monitoring, billing and copy protection. Besides enabling owners of intellectual property to manage and protect their assets, MPEG-4 provides a mechanism to identify those assets via the IPI data set. This information can be used by IPMP systems as input to the management and protection process. All the DRM systems follow the same generic architecture, which is shown in Figure 7-1.

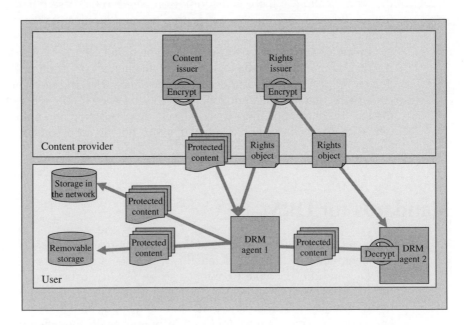

Figure 7-1. Generic DRM architecture.

Most of the DRM systems are not standardized. The metadata may be, but typically they belong to industry consortia or individual companies who are not very willing to share their secrets – because this how they retain security, and how they get paid. There are a few examples of both standardized and open source DRM, however.

While open source and security may seem a bad idea to combine at first (how do you keep anything secret if you tell everyone about how you create the secret?), it turns out to be much more useful than you might think. If the mechanisms are open to review, but the secret is shared, it means anyone who is competent can find out if there are any problems with them. While that may mean they use the problems to break the system, it turns out more often than not that they are willing to help, if they can get credit. Sadly, this is not the path chosen in IPTV standardization, and it seems the few such systems that exist are not presently used by any IPTV players.

The developer is only one of the actors in the DRM ecosystem, however, and it is an actor who goes away once the application is deployed. DRM systems map to the IPTV value chain we discussed in Chapters 2 and 3, and the actors in DRM overlap to a very large degree. Depending on the deployment scenario, different actors can play different roles in the system. A good generic description of the system has been created by the Open Mobile Alliance (despite the fact that the system is not deployed very widely), so it makes sense to look at their model.

The OMA DRM provides a very good generic model for DRM systems because it separates the actors involved in the creation and consumption of content, and the different parts of the DRM process itself. All other systems have the same functions and actors, but they may be co-located so that the creator of the encrypted content and the rights object are the same (while they are different in the OMA model).

Apart from the encrypted content, the Rights Object is the cornerstone of the OMA model. In other systems, the Rights Object is usually mixed up with the content or the key distribution, which may be efficient under certain conditions, but may break as soon as you try to generalize it. The Rights Object is a document that defines the permissions and constraints associated with the DRM content. Since a DRM Agent is needed to consume the content, the DRM Agent can refuse to open the content if there is no Rights Object. However, a Rights Object can be delivered separately; and a new Rights Object may be created when the user requests a piece of content for which the current Rights Object has expired, provided the correct preconditions (e.g., payment) are being fulfilled.

To understand a little better how the system works, there is a minimum number of actors to consider. The first is the DRM Agent, which embodies a trusted entity in a device. This trusted entity is responsible for enforcing permissions and constraints associated with DRM Content, controlling access to DRM Content, and so on. It is a piece of software, and it is possible to create a Rights Object that is valid for a group of agents (how the grouping is done depends on the implementation). The DRM Agent has to be implemented in such a way that it cannot be tampered with. Note that it is not the same as the content player; that is a different actor altogether, and in principle the responsibility of the DRM system ends when the DRM Agent hands the content to the player.

The second entity is the Content Issuer, which is the entity that actually delivers the DRM content. This is delivered to the DRM Agent. The packaging of the content may be done by the Content Issuer; or it may come pre-packaged from somewhere else (e.g., if the issuer is an aggregator).

The Rights Issuer is the entity that generates the Rights Object, which contains the rules for how content can be used. The user is the person who uses the content. The user has to use a media player to access the content, and that media player has to go through a DRM Agent, which is set up to enforce the permissions and constraints assigned through the rights object. The system will enable the user to view permitted content under the assumption that the content is received through a media stream or some other means, but since content that has been encrypted according to the mechanisms described in most DRM specifications can be assumed to be secure, it can be stored somewhere else (such as on a memory card, or in an external server). Rights Objects can also be stored outside the device, provided they are not associated with a session or state, in which case they will not work when moved outside of the current session.

Compare this with another system, the semi-proprietary Marlin DRM system (Figure 7-2), which basically works in the same way. It is not standardized, however, and it is managed by a special organization, which gives out licenses for the technology, as well as management of the encryption keys needed. This is a device-centric architecture, different from the OMA architecture.

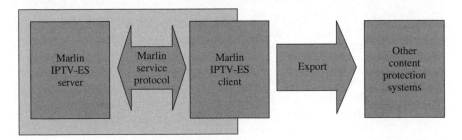

Figure 7-2. Marlin DRM specification scope.

The system has two main components: the Server and the Device. Devices are anything that can receive and play content; Servers are run by service providers and provide content. There is an agreement which you have to sign before you are able to become a Marlin developer, and that agreement sets forth exactly what the different parts of the system are supposed to do.

When a Server receives a request of a certain action (play, export or record) for the content from the Device, it checks the availability of the action for the content. Devices have to publish information about themselves, which include *SpecificationVersion*, that is, the versions of the Marlin IPTV-ES specifications the client supports; *Capabilities*, which indicate a certain functionality the client supports (e.g., secure clock function); Manufacturer, which indicates a unique identity of each manufacturer; *ManufacturerModel*, which indicates an identity of a model in the specified manufacturer; and *ManufacturerModelVersion*, which represents the versions of the specified *ManufacturerModel*.

When the request for content is granted, the Content Key and related information, such as a validity expression for the key, export information, or recording information are sent to the Device. Both Server and Device have to implement the Marlin IPTV-ES Service Protocol, which generates and analyzes messages for the IPTV-ES Service Protocol. They have to implement the Secure Authenticated Channel (SAC) that is used to communicate messages for Authentication and Encryption. Both Server and Device must have the X.509 Key Pair as the credential, which is used in establishment of the SAC for securing the Marlin IPTV-ES Service Protocol.

The rights transaction itself, that is delivery of keys used for decrypting content (Content Key) and their usage conditions from a Server to an Device, is done using the Marlin IPTV-ES and is accomplished using the Marlin IPTV-ES Service Protocol. After establishment of the SAC, a Device sends a Request message to the Server and the Server responds with a Reply message, which includes keys and usage conditions. Marlin also assumes that the IPTV end-point Service Device sends a Request message with *UsageRuleReference*, the identifier for the usage rights of the content, to the Server over the SAC. For the identification of actions that the Device is going to execute, the Device sets an appropriate *ActionID* value in the Request message, and the Server responds with a Reply message, which includes appropriate *Status* and *StatusExtension* information along with the Content Key, if the requested action is granted. The Marlin IPTV-ES defines *Extract*, *Export* and *Recording* as actions performed by Devices. Table 7-1 lists the few parameters contained in Request and Reply messages.

There are a few constraints when it comes to what can be done with the format. As the Marlin IPTV-End-point System is designed for IP-based content delivery services for televisions and other broadcast receivers, it is assumed that the content format used should be MPEG-2 Transport Stream (TS). Although two types of content format are defined in the specification, the Standalone Format is used for pi-IPTV services. The Standalone Format is compatible with MPEG-2 TS defined in MPEG-2 Systems.

However, to have something to manage, the service provider has to have a copyright policy, and a way to communicate it to the users.

What	Function	How it is done
Extract	Used for playback of streamed/downloaded content	The Server responds with a Reply message including the validity period of the Content Key and rendering obligation such as output control information as the StatusExtension parameter
Export	Used for export, i.e. copying or moving, of received content to a certain media system or interface using appropriate Content Protection technology. The Device must specify target Content Protection technologies as the ActionParameter value	The Server responds with a Reply message including copy and output control information adapted for each Content Protection technology to which the content is being exported as the StatusExtension parameter The Marlin IPTV-ES currently defines Export to the following Content Protection technologies: – DTCP: for digital interfaces including IP network (in particular, DLNA) – CPRM: for DVD-R/RW/RAM and SD Card – VCPS: for DVD+R/RW – MG-R(SVR): for Memory Stick PRO – MG-R(SAR): for Memory Stick and Memory Stick PRO
Record	Used for recording of streamed content on the Device's local storage with appropriate protection mechanisms	The Server responds with a Reply message including copy and output control information adapted for each Content Protection technology by which the content is going to be recorded as a StatusExtension parameter

Table 7-1. Parameters contained in Request and Reply messages.

Designing Copyright Policy

If a service provider has content coming in from many different sources (not just those where the rights to the content have been professionally negotiated), the way to ensure that the copyright is correctly managed presents a major problem. For the service provider, who decides to accept user-provided content, there are three things to consider. There has to be a way to:

- communicate the terms to the user (this applies regardless of the terms themselves);
- manage the copyrighted information (especially if the service provides both user-provided and professionally created content); and
- enforce the policy.

All models for providing copyrighted content have one thing in common: the user needs to sign (or approve) an agreement, which binds them to the terms of the service. This can be done when they sign up to the service, or when they start using it. Either way, the traditional method of providing the agreement (where the important things come at the end, and the user has to scroll for a long way before getting it) is not good enough if the service provider wants to have real agreement from users; to get that, the user has to both understand and accept the terms.

When providing a service for user-provided content, the service provider typically becomes a gatekeeper when it comes to copyright, but what he does and how it is enforced will vary depending on the kind of business he is trying to provide. This model applies if the service provider is trying to build up a collection of user-provided content, which is under the control of the service. Here, the service provider needs to determine how to acquire the rights to the user-provided content, and how to make sure the user-provided content is not infringing any other copyright (e.g., when users upload movies and television programs).

Gatekeeping requires the creation of appropriate licensing agreements, which include the terms of exploitation – especially if there is any exclusivity for the service provider. This can be especially important in the case of content which derives from an established work (e.g., fan-provided content), because it is not only a way to provide the content, it is also a way to attract more users. But this, in turn, requires the permission of the original copyright holder, which may be hard to get and has to be managed carefully.

It is unrealistic to expect all users to assign all copyright either to the service provider or to the public domain (although many users would probably be willing to do so if they knew it was an option). If the service provider is paying for the use of the content (either directly, or as a percentage of the income from usage), users will have very small incentives to sign away their rights. So the policy has to be communicated to the users, and it has to be set in such a way that it both encourages further use, and discourages stealing of content.

When users upload copyrighted content without permission this is normally a problem because it disrupts the commercialization plans of the content provider. A movie, for example, is typically initially released in cinemas; generally in the case of American movies in the US first, then in Japan and Europe, and later in other countries. After this, it is released on pay-TV channels, to airlines who license film libraries, and on DVD. After this, it can be sold for free-to-air viewing, first on the dominant channel, later on the secondary channels. The permission typically applies only to a limited market (e.g., a country or a region); if the service provider is licensing the film for a cable-TV or IPTV network, the usage would be limited to the subscribers of the service provider, and would probably be subject to a business model where the content provider gets a share of the fees for the usage, or is paid a fixed fee – depending on the size of the market, and the popularity of the movie, and many other factors which are negotiated between the producer and the service provider (often at movie market events).

During the last few decades, selling content to distributors has been complemented with cross-media sales. Selling to service providers at the right time, as well as starting the DVD sales when the pay-TV market is exhausted, optimizes the windows for the different formats. But one format can also be used to drive sales in another format, and that is where the Internet is particularly useful. In such a strategy, the marginal value of each channel is not optimized (as is the case in traditional marketing), but the total user experience is maximized. As is the income, since media companies increasingly understand that they can extend their business into the domain of the service provider, by for example demanding a percentage of advertising sales instead of a fixed fee.

Chapter 8: Identities, Subscriptions, User Profiles and Presence

Traditional broadcast systems are based on the notion that the receiver is not known until the transmission is started. A radio signal is by definition a point-to-multipoint signal, and everyone who has a sensitive enough antenna can receive it. The problem is decoding the received signal; this is where the mobile phone system, for example, uses encryption in order to hide the meaning from others apart from the intended receiver. However, traditional radio and television are sent to anyone who can get the signal.

The bigger picture: identities, subscriptions, and users

When asking someone to pay for using something, especially if the payment is to be received after the service is consumed, it helps to know who you are dealing with – and that it is too hard to fake that identity for the effort to be worthwhile. The identity is the root of the subscription. When signing up for a service, the user signs up not only to get the service; he also agrees to pay for receiving it.

> *(Continued)*
>
> *The same thing applies when tracking a profile. If someone can be uniquely identified, they can be tracked, and their behavior logged – in addition to the information they provide about themselves.*
>
> *In both these cases, the identity must be easy to keep secure, hard to fake, and simple to communicate in a secure way. The same identity can then be used for tracking the actions of the user. Getting updates every time someone switches channels is valuable to broadcasters; knowing exactly what demographics (if not individuals) watched an advert is valuable to advertisers.*
>
> *Identity management, in other words, is a crucial component in an IPTV system, at least if it is to offer something better and more interesting than what is available on cable-TV today. And here, IMS performs rather well.*

Introducing encryption meant that radio signals could only be decoded by the intended receivers (this came with satellite transmission), and when the radio signal was transmitted over a cable, it became possible to send both encrypted and unencrypted programs together, creating a system with premium channels. The user had to have a receiver capable of encrypting the signal, and in addition have the secret code of the transmission. While the cable industry distributed this initially on magnetic cards, later on smart cards; the computer industry has created a huge and very elaborate infrastructure around the secure, yet open, distribution of secret information. This book is not about public key cryptography, but it is an important part of the subscriber management.

If there is a need to establish a relationship between the sender and receiver in advance, for example to get the shared secret of the program, and the sender charges the receiver to get it, there is a subscription. Charging the user to access the signal is one way to monetize the IPTV service (advertising is another). However, subscriber management is much more than just tying the access rights of a user identity to a way of sending a bill to that user. It also contains the profile management, and the way to distribute this profile information – for example, selling it to advertisers so they know who has viewed their advertising.

Subscription management can be done in multiple ways, for example, through an Internet portal where the user pays with his credit card. Once the relationship between the user and the identity has been established (and the user has signed an agreement and paid for the subscription), the user becomes a subscriber. This change in the relationship determines most of what happens next in the system, since the agreement the user signs can contain a release of the private information that has been collected as part of the viewing session – this is important, since there is legislation in many countries (e.g., all of the EU and Japan) which requires that there is an express user consent before information is reused, and that information which is collected about a person can only be reused for the express purposes it was collected.

Once such a permission has been granted (and usually it would be as part of signing the contract), the user information can be collected, and the profile can be managed. How profiles are managed, we will come back to later. For now, there are two ways of collecting information: asking the user; and logging it during the session.

Managing and Federating User Profiles: XDMS and PGM

Nobody knows who watches a broadcast. There is no back channel for the television broadcaster to tune in to and find out what the ratings are like. Different metering techniques exist, but getting high enough confidence in the figures is difficult and expensive. In an IPTV system, where every user by definition has to be connected to a network, this becomes easy. It is simple to control who watches what, as well – something that can be leveraged in parental control systems.

Before looking into how the automated logging can work in an IMS-based IPTV system, it is interesting to look briefly at how information can be actively collected from the user, and under which conditions. For starters, it may be a bad idea to ask someone to fill in this information as part of the user agreement. This is normally a paper document, and information gathered on paper has to be typed into the computer system. Not only is such typing prone to errors (regardless of how good the typist is, there is a statistical likelihood of mistakes, which may be costly). It is better to reverse it, and let the user fill in all the information, print it out, and send in the signed form. Not only does this save time and resources for the IPTV provider, it also drastically decreases the risk of errors. Asking too personal questions, however, is likely to put the user off. They are there to watch TV, not be a machine for the feeding of an advertisement flow, and an IPTV provider has to respect that. Designing questionnaires is a specific discipline, and the sensibilities and formulations vary greatly between different countries (and different cultures), so we will not go into it here. It is not something that pays to skimp on, since having correct and appropriate information about the user will be a key for the IPTV provider in getting advertisements that the advertiser finds worthwhile paying for.

Presence in IMS

IMS is intended to keep track of users and manage their traffic, not handle the services they use, or user interactions with the services. Yet it does, by virtue of keeping track of users, provide a method which makes it easy for users to interact with each other through what they do – social networking, in a sense. For mobile users, this is a key function, as well as for their service providers. In this world, the ability to find the user's location is the key.

For IPTV, it is less interesting to know where the user is watching the program (unless the user is watching from a mobile). It may be more interesting to know what he is watching, since this can be used to find others who are watching the same thing (and hence may have the same interests). Knowing the same interests means being able to have discussion groups, where members are interested in the same television programs, for example (a good indicator of other common interests, and probably one driver behind the success of dating services in the Sky interactive services in the UK). It also means that advertisers can find out what the user has been watching, if the presence information is logged. This also means potential intrusions in the user's privacy, especially if this is combined with other information collected about the user.

Presence is a built-in feature of IMS, but not of IPTV, although it is very easy to use together. Since the architecture of the Open IPTV Forum is designed to set up a session between the service provider and the user, the user's presence can be registered when the session is set up (since this means that the user has started viewing the television program). Inside this session, there is a "bundle" of multicast channels, so there is no way for the service provider to know which of these channels the user is watching – unless the user's client explicitly informs the service provider, something that can be done using presence. It is different for video on demand, because the user automatically receives only the program he has requested, and the connection is unique to that user.

One problem in handling presence is that there have been multiple definitions of it, and they are not always compatible – even within SIP presence.

3GPP and 3GPP2 have defined an aligned presence service framework (actually, 3GPP2 has taken the 3GPP definition and adapted it). This framework defines a presence reference architecture, both in the "network layer" and "application layer". This means that the 3GPP and 3GPP2 specifications define an information flow for presence end to end. The "network layer" is the communications infrastructure, which is in place between the functional elements that define the presence service (e.g. the presence server and the presence agent), and includes the various network elements defined in the network architectures of 3GPP and 3GPP2. The "application layer" refers to the signaling between the elements of the presence service (e.g. presence server and presence source).

The IETF has started to define a presence and instant messaging framework (this is where the idea of presence came from). This framework is extremely generic, and it is reused in the 3GPP definition. While most of what a service provider may want to do, and user requests of a presence system, is present in the IETF system, it leaves a number of questions open – for example, how the user is authenticated, and how the presence information connects to other profile information. To solve this, the Open Mobile Alliance took the

IETF's definition of presence, and added support for mobile networks and connections to the XDMS profile management system. The other important thing about the OMA presence system is that it is very easy to extend – for example with IPTV information.

The Open IPTV Forum architecture uses OMA presence, precisely for these reasons. OMA has also done a reasonably good job of defining a "network-independent" presence system, focusing on the data structures and the ways to share them. The presence system in OMA connects to the OMA XML Document Management System (XDMS), which is used for the handling of user profiles and groups, by also being an XDMS client, and it can deliver not only presence information to other systems, but also profile information taken from XDMS.

The presence system of OMA is based on the presence system of 3GPP, which defines presence on a network level. It also takes in the IETF definitions and marries them to XML – specifically through creating document formats and management structures, and through using the XCAP language for configuration. Using these different standards as its base, OMA defines a number of components, which together form the presence management system. To understand how the presence information flows through the system, it is important to understand the different components that interact, and how the information comes in there from the start (see Figure 8-1).

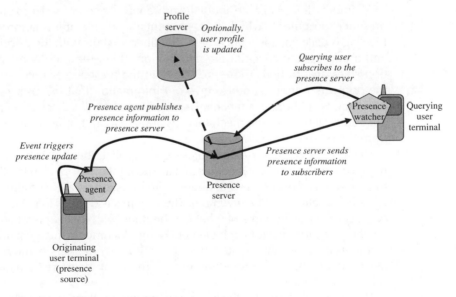

Figure 8-1. How OMA presence works.

The presence source is the part of the system that provides the presence information. It is usually implemented as presence agent, a program that gets information about changes in status from a different client, such as the

IPTV client, and communicates it to the presence server (in the IETF terminology, this is called a presentity). The communication in the OMA system is done using SIP Publish, Subscribe and Notify, the subscriber being the presence server. OMA also defines an external presence agent, which can fetch information from network elements such as calendar and email systems. There are multiple ways in which an IPTV system can provide such information as well, and the IPTV system can also use such information (the IPTV application server can access the information through the ISC interface).

The core in the OMA presence system is the presence server, which receives and stores the presence information published to it, and distributes presence information and watcher information. It is the part of the system that interacts with the XDMS system, by fetching documents from the shared XDMS system and the presence-specific XDMS. It uses SIP Subscribe and Notify to get and distribute the information from the presence source. The presence server also handles the processing needed for the publication, which includes filtering for who should receive which events (through the event notification and authorization rules filtering); and also shaping of the traffic by determining how many notifications can be sent at the same time. There is a configuration, which can be set for the shortest time between two notifications, and this can be used to pace the notifications over the network, so it does not download the system (see Figure 8-2).

All information available about the presentity are combined into one single presence document, which since it contains all the information without filtering is called a "raw" document. The filtering starts with the prioritization and a consistency check of the information. Then, the presence authorization rules are applied. These are stored in the presence XDMS server, and the purpose of them is not to give out information that the user considers personal. First, there is a mechanism to ensure that all subscription requests are authorized; then, the mechanism is applied to all notifications going out from the presence server.

The rules for content are different for each application of XDMS. In the case of presence, the base are the rules that are being standardized in the IETF. The rules essentially describe what queries can be asked on a document, and the what information can be given out. The way this is done is to set constraints on each element in terms of what can be done with it – essentially creating an access control list for each element of the document. The most important elements are those that relate to the user's device, connection and the user's personal information – including privacy. There can be more than one set of rules for each user.

The notion of the user is the key to the presence system. It has its base in the 3GPP presence system, where it is the user who is responsible for all actions – the principal. Only a user can log in, only a user can have an identity. The user logs in and becomes present on a device (which is described in the SDP). The device cannot become present in itself – it has to have a user associated with it. A user can be a program, and does not have to have any volition of

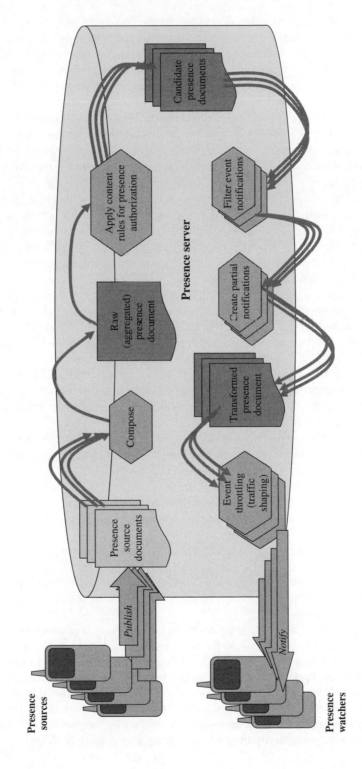

Figure 8-2. Presence information processing stages.

its own, just executes according to the logic that is written into it – which may be to log in, set up a watcher, and react when the presence changes. In this case, the watcher has to have an identity, which is registered in IMS. The IMS identity system is crucial for presence to work across applications, not just for IPTV. Since IMS is mostly applied in the mobile world, the presence is usually shown in a mobile device – but this is in no way necessarily so. The idea of a "presence enabled phone book", where you can see what your friends (the identities in your group list) are doing, comes from the telephone book in the mobile terminal, but it can equally well be displayed wherever it is possible – for example on your television set. And the presence information can be so much more than "is the terminal switched on".

There are two parts of the rules for presence authorization: the rules for who is allowed to subscribe (the subscription authorization rules); and the presence content rules, which determine what information the authorized watchers are allowed to receive. These rules can be updated at any time, and this means the presence server needs to check the rules every time a request for the information is made.

There are four types of rules for how a subscriber can be handled: block, polite-block, confirm and allow. If the rule says "allow", then the subscription is active and information will be given out freely. If the value is "confirm", then the presence server will apply the local rules – which usually means getting active authorization from the end-user. If the value is "block", then the subscription will be rejected. If the value is "polite-block", there has to be a response to the requestor, which says that the "presentity", the end-user who owns the information, is unwilling to give out the document, or that the document is unavailable.

The next step in the filtering is event notification filtering, determining how the notifications should be handled. It is possible to have partial notifications, i.e., concerning parts of the presence document, and the same rules apply to them. The events – notifications – can be filtered as well. It may not be appropriate to send a notification at any time. The filter is defined in an XML document, which can define inclusions and exlusions of the elements in the raw presence document.

The notifications of a change in the presence document, which fits the filter criteria, can be partial (only concerning those parts that have been changed, for example), or full. Whichever is true, the presence server generates a NOTIFY, which is directed at those watchers who are authorized to get the information.

The type of information relevant for presence is not limited to XML documents. It is also possible for the presence system to manage MIME objects, which means basically anything that can be included as binary information in an XML document. XML has built-in mechanisms for including documents by reference, but to get them, a server is needed, and this is what the last part of the OMA presence system, the content server, does. The information provided by the content server is retrieved by the presence server when it is needed, potentially through the aggregation proxy in XDMS.

Presence Data Format, Lists and Profiles

The raw presence document results in an XML document, which is composed according to the schema defined by the OMA, based on the data model of IETF but extended and in XML. That document, together with the lists defining the groups to which the user belongs and other relevant aspects managed by the different lists managed by the different XDMSs associated with presence, provide the presence data which a developer can access to provide services. So what happens when a user switches on his television set and selects a channel? As part of the registration, the user (if the user does not sign in individually, the family identity) gets signed in, and the presence of that user identity is registered. This is completely according to the standard; there are no specific IPTV issues in doing this. The user presence is documented in the presence XDMS.

But how do you know what the user is watching? There is no IPTV resource list XML document management server, although there probably should be. When the user selects the channel, the multicast address corresponding to the channel is selected, and the television starts decoding, and then displaying the television program in the channel. Since the user is receiving a bundle of channels on the multicast associated with the session, there is no way to know which of the multicasts the user is currently watching from anywhere else in the system than in the user terminal, and what that channel contains can be identified from the EPG.

It is different with VoD. Here, the user requests a program, which is sent exclusively to him from the media server. Since the user is making a unique request, and because the control goes all the way back to the server, the VoD server knows what the user is watching right now (including whether the user paused watching or not). The disadvantage of this is that if many users want to watch the same thing, there has to be one data stream per user, and that takes a lot of capacity in the network – which translates to latencies in the data transmission.

Regardless of whether the user gets the television service on demand or as a broadcast retransmission, the same mechanisms can be applied: there has to be a presence source (in VoD, it can be in the VoD server; in the case of real-time transmission, it has to be in the end-user's terminal). The presence source has to deliver its presence information to the presence server, and the presence server provides it to the watchers who have subscribed to it. If the information is what television program I am watching, then my friends can see that I am now looking at *Desperate Housewives* instead of doing the work I was supposed to.

The Presence Document

The presence server, just like the sources for the presence information, works with the SIP Subscribe-Notify mechanism. It enables the watchers to register for certain types of information, and to determine if it should be filtered and how. The watcher is the converse of the presence source, which gets presence information out of the system. It can get information about a presence source, or information about another watcher. There can be several types of watchers – they do not have to subscribe to notifications about changes in status of the service, although that is probably the most usual watcher; a watcher can also fetch the presence information only once (or poll for it at regular intervals). In the OMA presence system, the watcher can subscribe to a presence list in XDMS, so one subscription results in information from multiple presence sources.

The server, the source and the watcher are defined in the IETF reference model for presence (although it is not defined in those precise terms). Presence, however, is often related to two things: groups and resources. Both can be expressed as lists, and if those lists are encoded in XML, there are well-established standard methods to manage them. This is what the OMA did with the resource list server, the XML document management client, the presence XML document management server, the resource list server XML document management server (RLS XDMS), and the content server. These components are intended to make the presence information a part of the larger profile information management system defined by the OMA, and they enable the use of presence information in IPTV that makes it possible to find and manage information in an interesting way. There are four main components in the data model: the person, the service, the device and the URI. The data model uses the attributes provided in each of the elements to give a description about the respective parts of the model (see Figure 8-3).

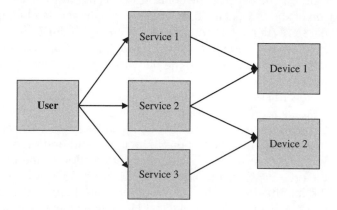

Figure 8-3. Relationship between the elements of the presence data model.

- The URI is the identifier for the entity that the presence information is about – this is normally a person, but it can equally well be a service, a computer, or anything else that can cause a state change. The URI functions mainly as an identifier for the presentity, being a SIP URI, tel URI, or a pres

URI. The pres URI is a special URI that identifies presence information regardless of the presence system used.

- The person component (the element and the attributes contained in it) does not have to be a real person, but can be anything that can have its own presence. This part of the presence document can contain, for example, attributes such as the willingness for communication, the physical appearance, the mood, activities, and so on. There can only be one "person" component per presentity, but since the presence information can relate to a group (i.e. a list in XDMS), this is not a real problem.

- The service component does not relate to any services, instead it relates to the communications services which the person (the presentity) is using, or could potentially be using. The "willingness to communicate" can be connected to different services, so that the user expresses a "willingness to communicate over PoC" or a "willingness to communicate over IM" for example. This granularity is probably far too much to expect the user to set every time he is using the service, so it is probable that it will be set by some kind of template depending on the user's situation – and this template can then be provisioned by the service provider.

The IMS standardization (in TISPAN and 3GPP) defines the different IMS communications services, and that is what was originally meant to be identified by the "services" component. However, there can be many more services than the ones defined by TISPAN, therefore there are two ways to identify a service: through the <service-description> element, or using a URI that identifies the service, included in the <contact> element. The URI is then assumed to identify the service, i.e. its canonical schema. There is, however, a third way to identify communications services, which is tantamount to composing them from building blocks. Then, a set of building blocks for media capabilities are used.

The capabilities included in the specification are heavily media-specific, as indicated in Table 8-1. These can be combined to describe a service, for example Duplex = Full, m = video, and Isfocus will describe a video conferencing service.

The communications services available to the user may not necessarily be available from all devices, or all of the time. Since the user can potentially be logged in from several devices at once, and these devices may have very different capabilities (a mobile phone can send SMS, but a television set cannot), the "device" component will have to contain components describing which devices are available during the current session. It is intended to describe only a physical piece of equipment used to execute the service, for example, a television set. There can only be one "device" component per device identifier, but there can be multiple examples published from different presence sources, which are composed and consistency checked by the presence server, and merged into one single document. To ensure that there are not several different device identifiers refering to different devices,

Capability	Attribute values	Defined in	Functional description
Duplex	Full, half, receive-only, send-only	RFC 3480	Defines whether the service can simultaneously send and receive data
Audio	Streaming (or nothing)	RFC 3480, and may be indicated as m = audio in the SDP information	Shows whether the service supports media as a streaming data type or not
Video	Streaming (or nothing)	RFC 3480, and may be indicated as m = video in the SDP information	Shows whether a service suppots video as a streaming data type or not
Message	Streaming (or nothing)	RFC 2046, and may be indicated as m = message in the SDP information	Shows whether a service supports messages (i.e. text) as a streaming data type
Type	MIME media content types	RFC 2913	Describes what data types the service can handle (MIME data types)
Isfocus	Isfocus (or nothing)	RFC 3840	Indicates whether the service is a conference server
Event-packages	List of event packages supported		The different SIP event packages which this service can handle
Methods	SIP methods list	RFC 3261, and following RFCS defining additional SIP methods	Which SIP methods can be used by the service
Extensions	SIP extension list	Extension option-tags as registered by IANA	Describes the SIP extensions which the service can use

Table 8-1. Specification capabilities.

a UUID is allocated to the device ID, and this is used throughout the lifetime of the device. If there are several different services running on the same device, they will all be able to use the unique identifier. The intention is that this device identifier then should be reused during the lifetime of the device by all services that need to idenify themselves by the device ID.

It is perfectly possible to show that the user is watching a video service in the OMA presence data format, but it is not possible to describe that the user is watching a specific IPTV program, because the semantics for this is not yet described by the OMA or any other body. Since there is an EPG entry for every program, and that entry can be identified by an XML fragment URI, there is a URI for each program, which can be used to identify it. Everything else is defined in the standards. An example of an OMA presence document which includes TV programs that the user is currently watching would look like the following.

```
<?xml version="1.0" encoding="UTF-8"?>
<presence xmlns="urn:ietf:params:xml:ns:pidf"
     xmlns:pdm="urn:ietf:params:xml:ns:pidf:data-model"
     xmlns:rpid="urn:ietf:params:xml:ns:pidf:rpid"
     xmlns:op="urn:oma:xml:prs:pidf:oma-pres"
     xmlns:sa="urn:oma:xml:prs:pidf:session-answermode"
     xmlns:xsi="http://www.w3.org/2001/XMLSchema-example"
       entity="sip:someone@example.com"
     xmlns:tv-anytime="http://www.etsi.org/tv-anytime"
     xmlns:iiptv="http://www.interactive-iptv.com/tv/schema/">

  <tuple id="a1232">
   <status>
    <basic>open</basic>
   </status>
   <op:willingness>
    <op:basic>open</op:basic>
   </op:willingness>
   <sa:session-answermode>
    <sa:automatic/>
   </sa:session-answermode>
   <op:session-participation>
    <op:basic>closed</op:basic>
   </op:session-participation>
   <op:registration-state>active</op:registration-state>
   <op:barring-state>terminated</op:barring-state>
   <op:service-description>
    <op:service-id>openiptvforum.org:IPTV-program</op:
    service-id>
    <op:version>1.0</op:version>
    <iiptv:program>http://channel1.example.com/epg/
     20080221#1625/</iiptv:program>
   </op:service-description>
       <pdm:deviceID>urn:uuid:48662e19-5fbf-43fc-a2fd-
       d23002787599</pdm:deviceID>
```

(*Continued*)

```
    <contact> sip:someone@example.com</contact>
    <timestamp>2008-02-21T16:25:56Z</timestamp>
    </tuple>

    <pdm:device id="a1233">
<pdm:deviceID>urn:uuid:48662e19-5fbf-43fc-a2fd-d23002787599
</pdm:deviceID>
    <pdm:timestamp>2005-02-21T16:25:56Z</pdm:timestamp>
    </pdm:device>
</presence>
```

In this case, the example of the television show being watched can be identified by the time it starts, the fragment ID corresponding to the time stamp of the start of the show will point to the EPG entry of the show. This has to be resolved by the logic in the presence server, because the only information available about what the user is watching in the client is the multicast address for the program; this can be translated to a pointer to the entry in the EPG by the server. The logic for this is not very complex, but it assumes that the service providers for the presence and the EPG have a contract enabling them to do this (which is necessary anyway, since the assumption would be that this access takes place over the ISC interface, the PS and IPTV AS both being application servers in IMS); alternatively, which will probably be the base assumption, they could be the same company.

Another way is to get the information from the television set, but this will require some contortions. Since modern television sets will contain the CEA-2014 HTML4CE browser, which makes it possible to run ECMAScript programs in the television set; and the content server can be used to serve up such programs (possibly through a proxy in the home IMS gateway, the HIGA), and the television set will be discovered by the HIGA when it is switched on and can be triggered to set up a NotifSocket channel, these features can be combined to make presence an integrated part of the television set. This assumes that there is a gateway in the network which can translate the user's actions into both television-related and presence-related signaling. For example, the user pressing a channel key means switching to a different multicast group, with the RTP stream being multicast at that address. This, in turn, means that the user starts watching a program in a channel; the information about this channel is available in the EPG, and the presence information can either contain the information from the EPG, or if there is a way to identify the channel, a reference to the EPG (e.g., an XML fragment identifier pointing to the program information); this can then be included by the presence server in the user's presence information – essentially the same process as if the multicast address is captured by the IPTV AS, and the information translated to EPG information by a lookup in the EPG. The difference in the latter model is that there is no need for the multicast information to be propagated to the

IPTV AS. The channel switching can, in this case, be invisible to the IPTV AS – until it is informed of the switching by the Notify from the presence server.

Lists in XDMS

The resource list server (RLS) handles the group management (in OMA terminology, a group is a list of resources). By going to an RLS, the presence watcher can get information from more than one presence source, since they are grouped through the RLS and there only has to be one single transaction. There is also a specific XDMS server for managing the resource lists (RLS XDMS), this is the system that manages the XML documents which handle the group information. It uses two of the base XDMS technologies: XCAP for configuration (an XCAP server); and it has a SIP Notifier to send notifications of changes in the lists it manages to the subscribers (a Notifier is a SIP user agent which generates SIP Notify messages to notify subscribers of state changes in a resource, it also accepts Subscribe requests to create subscriptions to future notifications).

Since the RLS – but also other parts of the system – has to be able to interact with the XML document management system (XDMS), there has to be an xml document management client (XDMC). Actually, there can be several examples of this depending on which information service it connects to, but the authorization of it is managed centrally. It is the part of the system that connects to the different other XDM systems that have relevant information (e.g., a presence system can also use the Push to Talk XDM, in which case there has to be a PoC XDMC; and there has to be others – such as a XDM for television programs). The intention is that an XDMS can be defined by any working group, inside or outside the OMA.

What the XDMS handles is an XML document. Each XDMS is a database (the standard does not describe how these are handled), which can be somewhere in the network – the flexibility of XML and SIP makes it possible to address them anywhere. The only restriction is that since the system relies on SIP messages, and all SIP messages by default have to pass through the IMS infrastructure, the repositories that are being used have to be connected to the same IMS domain as the consumer.

In XDMS, the XML document is handled using XCAP, which is a protocol to manage XML documents in a secure way – it would be a potential problem if anyone could update the presence information of others, so you have to be authorized to do this, the authorization has to be secure, and there has to be a secure method to handle it – all this is covered by XCAP. The language defines a number of operations, which the applications can reuse. Queries on the documents can be done using the XML query language (XQuery) defined by W3C.

The base specification for XDMS defines four different types of shared XDMSs, the databases shared between users:

- the shared list XDMS;

- the shared group XDMS;

- the shared policy XDMS; and

- the shared profile XDMS.

In these databases, there are four possible document types:

- the XDM list, which can contain a list of URIs, as well as a list of how documents can be used in the group (the group usage list);

- the group document, which declares what resources are available to the group;

- the access policies declared in the user access policy document; and

- the user profile, which is declared in the user profile document.

In addition, there is also a mechanism for advertising things to groups, the extended group advertisement.

The presence document list is a separate document, and OMA XDMS defines a separate presence XDMS system, which is also an XCAP server and SIP Notifier. It manages the presence-related documents, which in this case are the authorization policies relevant to the presence system; it enables single subscriptions (using SIP Subscribe) to the presence documents; and it sends out notifications about changes in the presence.

There is not yet a standardized XML resource list for IPTV presence, but the documents relating to groups will be the same – these are assumed to be available across different services, because the user membership in the group is what matters. An example of a group list looks like the following:

```
HTTP/1.1 200 OK
Etag: "et53"
...
Content-Type: application/list-service+xml

<?xml version="1.0" encoding="UTF-8"?>
<group xmlns="urn:oma:params:xml:ns:list-service"
xmlns:rl="urn:ietf:params:xml:ns:resource-lists"
xmlns:cr="urn:ietf:params:xml:ns:common-policy"
  xmlns:ocr="urn:oma:params:xml:ns:common-policy"

  xmlns:ext="urn:oma:params:xml:ns:shared-group">

 <list-service uri="sip:myconference@example.com">
  <display-name xml:lang="en-us">Friends</display-name>
  <list>
```

```
<entry uri="tel:+1-212-555-1234"/>
<entry uri="sip:hermione.blossom@example.com"/>
</list>

<max-participant-count>10</max-participant-count>
<ext:subject>My conference</ext:subject>

<cr:ruleset>
 <cr:rule id="a7c">
  <cr:conditions>
   <is-list-member/>
  </cr:conditions>
  <cr:actions>
   <join-handling>true</join-handling>
   <allow-anonymity>true</allow-anonymity>
   <allow-invite-users-dynamically>true
   </allow-invite-users-dynamically>
    <ext:allow-subconf>true</ext:allow-subconf>
            <ext:allow-private-message>true
               </ext:allow-private-message>
            <ext:allow-media>
               <ext:audio>mand</ext:audio>
               <ext:message>mand</ext:message>
               <ext:video>supp</ext:video>
            </ext:allow-media>
   </cr:actions>
  </cr:rule>
 </cr:ruleset>
 </list-service>
</group>
```

IPTV Profiles

One potential way of using the presence information is to log the user's actions. Since the presence is updated every time the user changes something, and because these changes can very easily be related both to television programs and groups to which the user belongs (including groups based on demographics, for example), the presence system provides a very easy way to build user profiles.

Apart from the metadata, which shows what the user currently is watching, and provides facts such as director, producer, duration etc about the currently watched program, there is other interesting information to collect. This includes the profile information of the user, which can reuse other existing presence documents. It also includes the buddy list of the user – something else that can be reused. The third important piece of information is where the presence slides over into interactivity: what the user thinks about the current program. There can be several ways of determining this, but the simplest would be to give the user the power to rate the program.

Logging the presence information means building a history of what television channels the user has watched, and by extension the user's preferences.

Some of these may be easy to deduct – for example, watching an action movie on Friday night, or football on Saturday afternoon. But others may be trickier, for example, deducting during summer what preferences the user will have for the new year's eve concert. The issue with logging is partly that the data must be logged during a sufficiently long time and to a sufficient detail; and that the algorithms for drawing the conclusions have to be clever enough. Deducting the information and storing it means creating a profile of the user, which can be used to manage the information to be presented to the user.

To compound the problem, there are several types of profiles that are required to provide an IPTV service (see Figure 8-4). These relate not only to the end-user profile, but also there has to be a set of service profiles. They are typically kept in the Home Subscriber Server (HSS), and contain items such as charging information, iSIM public identities, authentication vectors, services triggers etc. This is information that the system needs to provide the service (and to charge for it). In addition to the service profiles, however, there has to be a profile for the IPTV service: if the network provider and the IPTV service provider are different, these profiles will have to be different and the information stored either in the IPTV service provider's database, or in the XDMS system of the operator (which also can be used to federate the information stored in the database of the IPTV service provider).

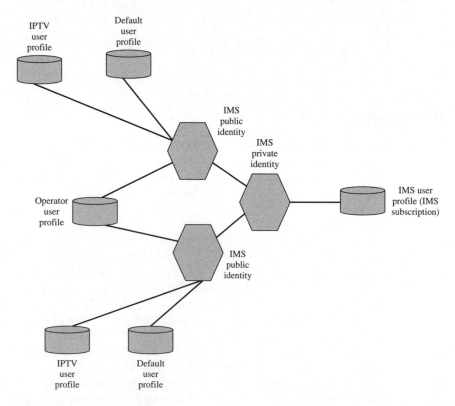

Figure 8-4. Profiles in IPTV.

If the network provider and the IPTV service provider are different entities, federation of identity information can be managed through XDMS, although this assumes that the operator is in control of the database management system – which is not unusual, since this book assumes that the identity is managed through the IMS system, and it comes from the network.

However, if the service comes from a different network than the one the user is sitting in, there will be a different kind of problem. Then the two operators, whose networks the user and the service are in, have to interact to enable the delivery of the service. See Figure 8-5.

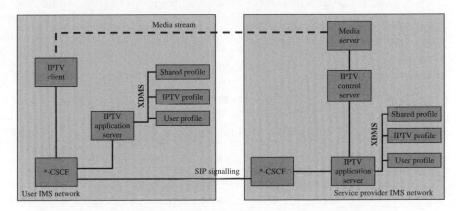

Figure 8-5. IPTV profiles in an IMS system.

There is no IPTV profile defined in XDMS yet. This does not mean there are no user profiles, however. The XDMS shared profiles define which data in the different XDMSs can be shared with other services. These include, for example, the lists in the RLS, which contain the group information. There are also user profiles defined in XDMS, which can be used by the IPTV system to provide service. They contain the generic information about a user (which the user has provided about), such as the preferred language (if this is applicable – it may be in a multilingual country such as Quebec in Canada, for example; but less likely in a unilingual country such as Germany). What information can be collected here is of course sensitive, as it is (according to the privacy laws in most countries) only permitted to collect information that is going to be used to provide the services. This means location, contact information such as address, e-mail, phone number; and possibly demographic information such as age (a note about data quality: this needs to be updated every year). The income of the household, the size of the house, the drinking habits and shoe size and a lot of other personal information about the user is something many IPTV providers would love to collect, and in some countries (like the USA) they are allowed to do so. But the risk for abuse is self-evident, so it is not very likely that the provider will be allowed to collect the information. Even if the user provides the information without being prompted and knowing what he is doing, it is unlikely that it can be used for anything (even providing advertisements) – it is simply too sensitive.

Then, there are the IPTV-specific parameters. These include the subscription parameters – what the user has paid for. The multicast addresses of the different channels, the different VoD services and other parameters relevant to distribute the service can go in here. Pay-TV can be included – it may be possible to include even more business rules in the system, which would mean applying the profile information to generate the service itself. Other subscription parameters can include parental control – if the user has the right to set parental controls – or if the parental control should be applied when the user wants to watch TV shows which may contain violence, or nudity, or other things which may not be allowed. The parental control would probably apply not only to television, but also to other services (which would probably also include the right to subscribe to new services online, something many parents will probably require).

In XDMS, a user agent lets the user manage his own profiles. This is typically a specific agent (for security reasons), which interacts with the user information in the XDMS through XCAP, an XML-based language that handles configuration management information. Here, the sequence will work as follows:

• The user enters the new settings into the XCAP client.

• Depending on the requested operation the XCAP client will instruct an XCAP server, the IPTV XDMS, to add, modify or remove information from the user profile in the IPTV and other XDMS databases.

• The XDMS will notify the IPTV client and the relevant IMS Application Servers about the changes.

• The IPTV Client is notified (through a SIP Notify) about the change, and will download the updated user profile.

When using the IMS services through an IMS Gateway, the user logs in to the IMS Gateway, which then performs the IMS registration of the user's PUID. The user is registered in the HSS, and the user profile is downloaded. The profile is managed using XCAP and XDMS. The user is also registered in the presence service, which means that she can get status information on all users that are defined in what commonly is called his "buddy list". Users may also be collected into groups. An application as an address book may include the buddy list status and mix users with presence status with users that don't have a current presence status.

Advertising and Presence

Advertisements are messages, which are not part of the main program, and are inserted into the program at a point decided by the programmer. It is often assumed that advertisements have to be paid for, but this is not really true; the way to manage a public service message is exactly the same.

Advertising in television is an economically important activity – of the estimated US$450 billion spent on advertising around the world every year, $150 billion go to television. The lion's share of this is directed at the traditional television advertising, where a television channel sells advertising directed at the audience of the programs it is broadcasting. A much smaller share is directed at locally targeted advertising, primarily in cable television networks (where it is possible to know the location of the user); a smaller (in this context disappearingly small) portion is advertising in IPTV.

Advertising in television is effective – at least, more so than the alternatives – or else advertisers would not keep coming back, investing billions of dollars in advertising every year. However, it is not particularly efficient, in that it does not provide a direct channel to the end-user, and no feedback loop to generate leads from the advertising. Measurements on audiences give a rough idea of the statistics, but they are a very blunt tool and cannot be used to influence a person who is about to purchase something (unless that is a daily good, such as shampoo, where purchases happen several times every month). And it is hard to tie local content, for example from a car dealership, to the global brand advertising. For certain types of products, such as cars, this would be highly desirable. For others, such as personal hygiene, it may not matter – the store chains normally want to handle their own advertising.

With traditional advertising, there is a big problem: How do you measure its efficiency? The method, which is traditionally used, is audience measurements, where a selection of users are asked to respond to a questionnaire; or in more modern methods use a mentometer (a device that allows users to express what they feel about the current program) or other way of interacting with the measurement house. This means that the measurement house gets feedback on how the user perceived the advertising, in some famous cases (when placing cameras on the television sets to record the user's activities) on what people actually do when they are watching television (answers include eating, ironing and having sex).

Measuring Advertising in IPTV

If, as the famous adage goes, half of all advertising is wasted, then everyone wants the other half. Advertising that does not lead to business is just money sunk into making the Internet prettier (or maybe more irritating, depending on your point of view). Advertising that cannot be measured is wasted; advertising that leads to business is an investment.

But how do you measure which advertising has led to business. Here, the identity management of IPTV comes to the rescue. Instead of competing with traditional television for the $150 billion which go to traditional television advertising, IPTV also has the potential to compete for the $200 billion which go to direct mail. The reason is simple: the back channel in IPTV allows for direct feedback; the personalization opportunities, which come from IMS, enable the advertising not only to be selected based on the user's preferences, it can also be modified – even to include the name of the user. Under certain conditions, of course.

In traditional broadcast, the feedback channel is out of band from the television program. When broadcasting traditional TV, there is no feedback channel – you do not even know if the television set has been switched on. With IPTV, the first thing that happens is that the user is registered as having switched on his television set. The same is true for web-based television, and the advertising business has created a set of guidelines for measuring the effects of web-based advertising, which partially apply to Internet television as well (however, it is possible to create a much richer set of feedback – and more precise targeting – as we will see in the next section).

The Internet Advertising Bureau (IAB) guidelines for the counting of advertisement impressions are: "A measurement of responses from an ad delivery system to an ad request from the user's browser, which is filtered from robotic activity and is recorded at a point as late as possible in the process of delivery of the creative material to the user's browser – therefore closest to actual opportunity to see by the user." These guidelines are created on the assumption of a web-based VoD system, which means that the protocols involved are HTTP and possibly RTSP (or some other protocol initiated from the web page, such as RTP). They also contain one important part: the filtering from robotic activity. On the web, there have been discussions – although very few proven cases – of "click fraud", where an advertisement has been accessed automatically by software (or, in some rumored cases, by people paid a few fractions of a cent per click). The reason is that the advertiser pays per click, that is the number of times an advertisement has been accessed, not displayed. The payment models try to take this into account, and Google for one has put in place a very sophisticated system to distinguish real user interactions from programmed ones (again something we will get back to in the next section), protecting the advertisers from fraud (since they have to pay every time someone clicks on their advertisement, whether it leads to business or not).

When an advertisement placement costs less than a dollar, advertisers do not particularly worry about measuring the effect – it is not going to provide a significant enough hit on the bottom line regardless of how many clicks there are. It is even cost efficient to use advertising on the web to replace (or extend) microtargeting, for example, using it to make surveys (writers of books now put out a number of proposals for names as advertising, and the one that receives the highest interest will be the title of the book). One reason for this is how incredibly cheap text advertising is to produce; it costs only a small investment in time of a not very skilled editor. Producing video is a completely different matter in terms of time, cost and equipment, regardless of whether the commercial is filmed in your parking lot (or showroom). So a video commercial will be a bigger investment right from the start, and the investment will require follow-up in order not to be sunk cost.

As an advertiser, you are likely to want a feedback mechanism to provide information about how many people watched your commercial. Preferably, you would also want to know who they were (this is something we will discuss in the next section).

As we talked about in the first chapters of this book, a user will have a client on his computer or television (or whatever equipment he is using to view the content). The client will be connected to a server, which delivers the content. When counting the number of times a video has been viewed, you can start either in the client, or in the server. Counting in the client is more reliable than in the server, although this is not entirely true when streaming media, since the server has to initiate the stream and the client terminating it will mean that the termination is sent to the server, and can be registered. When downloading a video, however, there is no other reliable way than measuring the number of times the user has actually looked at the video than making the client count them, and this is a huge problem because most browsers (when used as client) are not set up to give this feedback; other types of clients (such as BitTorrent, eDonkey and other peer-to-peer file-sharing clients) are even less amenable to this. Of course, it is possible to use the proxy mechanisms in HTTP to help the service provider count the number of times a file has been accessed from a cache, but they are hardly reliable (if the user is offline, there is no way to know if they are watching the program).

Hence the need to count the impressions – the number of times an advertisement has been displayed to a user – on the client. There is, however, no way of telling to whom it has been displayed using this technique (unless you give the browser a unique identity number or use cookies to set session markers, and start registering the user profiles – which means significant risks for user privacy, and significant problems in terms of user permissions). To handle this, you really need active user subscriptions and a persistent method of session management – which is what we talked about in the first part of this book.

When viewing advertisements over HTTP, the IAB limits itself to counting the advertisement impressions only when a counter (in the server) receives, and responds to, an HTTP request for tracking from a client. The count must happen not when the page is opened, or when the underlying page is retrieved, but when the actual content is opened (as it might otherwise mean that the advertisement is received but directed to /dev/null, or received but the display is blocked by an advertising blocker). This means that the requests should be initiated by an HTML element such as <IMG> (when an advertisement is downloaded as an image), <IFRAME> (when an advertisement is downloaded into a separate frame in the browser), or <SCRIPT SRC> (which means there is a small ECMAscript program in the browser which initiates the advertisement download and display).

When the counter program in the server receives the request, it responds by delivering something which can be counted (it may be the same for multiple advertisements on the page, which means that the Internet Advertising Bureau automatically assumes pages as the main display format, not videos).

This can be a "beacon", typically a 1×1 pixel image, which does not show on the screen of the user but can be used to count the user's reception of the advertising; it can be a "302 redirect" to the advertising; it can be an ECMAscript program (as html/javascript); or it can be the advertising content itself. The advertising counter should deliver the counting asset using standard HTTP, including cache-control headers.

When triggering a video from a web page, however, some different rules will apply. Since this type of video essentially is a small video embedded in the web page, but does not contain other content than the video itself (i.e. there is no television program), it can be downloaded using HTTP as part of the page, and trigger the player when it is received.

However, when you embed commercials in content received in this way, you get a problem. The commercial can be displayed before, during and after the actual content (the same problem as you get in any display of advertising in television). When the content is streamed, you have essentially the same measurement problems as you will get with IPTV.

The IAB response to this is to measure the responses from the advertising delivery system to a request for advertising (request from client, response from server). The measurement should be triggered as late as possible in the display process – otherwise you do not know whether the user was watching. This means setting a trigger in the video itself, so that it triggers a download (e.g., downloading a web page in the background). And as said before, there is a problem of programmed activity (robots). The advertising provider is recommended to make sure that each request is unique (not to enable cloning of successful requests, for example); this means having a session becomes really handy (another plus for IMS-based IPTV). Other methods of doing this include using cookies, or JavaScript programs that generate random values in beacons.

One problem for advertisers, especially those who advertise in media where the display is not well controlled, is caching. Buffering, where the client downloads the video and it is stored in the local computer to be retrieved again when the user opens the page, will be a problem since there is neither a way to count the times the user has retrieved the content, nor make sure it is fresh. If an advertisement is old, it may have curiosity value, but there will be a problem if it is advertising something that is no longer valid – for example, if the user views the advertising for a Christmas campaign for Mother's Day. The advertiser may even become liable (a big problem in many countries).

Hence, an advertiser is obliged to make sure that the advertisement is erased from the local buffer as soon as it has been displayed; he is also obliged to ensure it is removed from any intermediary caches that may display the content. This can be done by setting the appropriate HTTP (or, as the case may be, RTSP) headers; it can also be accomplished (maybe in combination with) by cookies; but by far the easiest way is to tie the display to the session, and make sure it is only possible to view the content during the session which initiated the display, which is only possible to do with a session management system such as RTSP – or SIP.

In HTTP, and protocols based on HTTP, it is possible to make sure the content is automatically refreshed when the user requests it. This is done by setting the cache expiry header in HTTP so that the content has to be downloaded every time, and that intermediary servers have to check if there is a new version of the content.

Fraud and IPTV advertising

A major problem for advertisers on the Internet is fraud. Filtering out robotic activity (that the requesting user is not a person, but a program, which is only downloading the content to create a clickthrough, ence making sure the advertisement service provider gets paid by the advertiser) is a major issue, although the countermeasures are fairly easy to discover – because this is not a condoned behavior, the investment in the technology typically comes from the "bad guys", and the countermeasures from the "white hats", who have a legitimate interest in tracking the viewing.

Fraud checking is the converse of behavioral targeting – the same techniques can be used to discover and trace behavior which is encouraged, as well as behavior which is discouraged. Filtering, according to the IAB, can be done in all the standard ways: using robot.txt (a file with instructions to robots which a webserver owner can set, but which is hardly respected by anyone using robots to generate clickthroughs); or a filter based on the originating URI or browser type (the IAB maintains a list of known search engines and other robots that may be doing automated searches). In the latter case, it is a judgment call if you want to include Google and Yahoo! in the audience. If the website is a company website, it is also an issue whether to include the associates of the company in the audience or not (and they can easily be filtered out depending on the origin address).

However, this is not sufficient to track any reasonably intelligent fraudster. It is very easy to switch between a range of IP addresses, which may not be associated other than through the company that has received them; it is also very easy to look like you are a number of users, when really the originator is the same robot. Here, activity-based filtering is required to distinguish the non-human downloaders from real users.

The filtering can be done in real time, but it is much easier to do after the fact, in the log file from the web server (which also means that the reporting of which advertisements have been viewed to the advertiser cannot be done until after the fact, and not in real time, but only when the log file can be analyzed). There are a number of data-mining techniques which can be applied to see if advertisements are being defrauded, but the basic filtering can be done with very simple programs. It is a matter of looking for multiple sequential requests from one single user (it is very unlikely that a person would click on all the links on a web page – or in a video – but it is typically something a robot might do); multiple parallel activities, for example, retrieving all linked items from many web pages at the same time; and outlier activity, where a user is in the top percentiles of website users.

Knowing who has accessed a website is not very useful if there is no way to tell anyone about it, and the IAB provides a number of parameters to make sure that the reporting is consistent (apart from the definition of "impression"). These include time periods – a "day" is from 00:00 to 24:00; a month can be reported either according to the broadcast standard – the month starts the Monday of the first full weekend of the month; or a four-week period (of which there are 13 per year); or a calendar month. It may also be relevant to know during which time the impressions were made, in which case it makes sense to keep track of the time zones.

It does not make much sense to receive a report if you cannot trust it. Anyone can make up a report; so the reporting must come from a well-known brand (which has a reputation at stake) or be verified by an auditing or certification agency – at least if the customer is to be expected to come back. This is no different from certifying any other Internet advertising, however.

One major problem with traditional advertising in IPTV is that users are taking control of the programs. They record them to view them later, in the process often tuning out the advertising. Few things create as much irritation as advertising that interrupts the viewing of the program, and attempts by telecom companies to standardize methods that make it impossible to tune out the advertising have just given an opportunity for hackers to show off their breaking ability.

Part of the reason is that the advertising that interrupts programs is not interesting to the user. If advertising could be made more interesting, the user would be less likely to tune out. This implies the selection (or modification) of advertising to accompany the program, something which is more likely to happen first in video on demand and mobile television (where the user has had to log in before the programs start). In both cases, the use of IMS will give a significant leverage, because the profile system and automatic identification of the user can be leveraged. According to innumerable market reports, the addressability of advertising is the key to the market of the future.

One of the most important possibilities is the increased targeting and measurement, which come together under the heading of "behavioral targeting". The more personal the advertisements, the theory goes, the less likely the user is to skip them. Personal is not just a function of guessing the age and sex of the user. Behavioral targeting makes it possible to measure what people actually watch, and adapt what they will see next based on it. To the user, it might mean more interesting television; to the advertiser, higher response rates; and to the programmer, higher income from advertising, since prices can be raised when advertising becomes more personal. However, personalizing advertising in this way also raises concerns about how the information about the user is reused, and if it can be sold to others than the service provider. We will come back to this in a later chapter.

One aspect of interactive advertising, which is already being deployed in VoD systems, is "telescoping", where the user clicks on an advertisement and gets a longer video, with the option to get more information sent to them. In theory, it might also include more interaction, which can be collected and used to target the user further.

Chapter 9: Beyond the EPG – Metadata in Interactive IPTV

How do you know what is on television? Look in the newspaper? The use of one medium to give information about what is going on in a different medium is yet one more aspect of the IPTV system, and it does it in the present-ation of the metadata about the program – which we usually think about as the Electronic Program Guide (EPG), but which should really be known as the Electronic Service Guide (ESG) or the Electronic Resource Guide (ERG). The guide describes the television programs, the time and transmission channel – data about the program, its metadata. The definition of metadata is simply "data about data", but this does not cover half there is to say about it.

The bigger picture: metadata mining and matching

Once you know what someone has done on the network, and how they character-ize themselves, this can be used to tailor the television shows they are receiving. Either tailoring in real time, as we discussed in Chapter 3; or tailoring in terms of selecting the video clips that will compose the show (or the advertisements which will be inserted into it).

Why IPTV? Interactivity, Technologies and Services   Johan Hjelm
© 2008 Johan Hjelm

(Continued)

This, however, requires not only knowledge about the demographics and inter-actions; it also requires a way to match the video clips to the knowledge. This information is metadata, data about data.

Metadata is regularly produced as part of any digital video production, but to make it match a user's preferences requires somewhat more effort. If I say I like "beer", there has to be a way for the system to understand that "happoshu" or "ale" belongs to that category, since these are ways in which a video clip (especially an advertisement) may be characterized. There are ways to automate this know-ledge matching – either the blunt and not very effective way of a small controlled vocabulary, which while offering serendipity also limits the selection; or by using ontologies, a system that has its roots in artificial intelligence research. The system can then draw conclusions and create recommendations, based on the matching of one user's preferences in his profile, with the metadata describing the show.

A second problem is how these metadata are collected. Making the producer of the video clip create them is time consuming and expensive; having archivists doing it no less so. There are standardized formats to characterize television pro-grams, as well as movies. And these are rapidly moving in a way where they can be converted to formats that can be turned into ontologies. However, there is still a huge gap between what metadata can be automatically collected and what can be created by a person.

Content in an IPTV system can be identified by the metadata that accom-panies it. You can see what is available, not only in your own television set, but also for download and viewing in your own IPTV service and in other IPTV services, as well as in servers in the home. But that is not all. Metadata can be used for searching and retrieving data, as well as controlling present-ation. There are a number of standardized metadata mechanisms that can be used for this.

You may think of the ability to interact with the metadata as interactivity: it is not. Changing the way you see information about a program (or even changing the information about the program) does not change the program itself, which is what it takes for interactivity; unless, of course, you regard the metadata as data itself. It may be – the trailers can be the program. But usually it is not.

When selecting a piece of IPTV content, be it a live stream from an event or content from a VoD system, the underlying technology is the same. The main difference is in the way the content is presented to the end-user; this is also where branding can happen, making it possible to distinguish between differ-ent types of content and content providers by their visual presentation. Usu-ally, the EPG is shown as a table, with the time on one axis, the channels on the other, and the program titles (and other information) in the cells. There is no particular reason why it has to look that way; it is a just a convention. With XML as the base, it is possible to present metadata in many different ways.

What the user sees when browsing through the list of available content, or looking through the television table, is not the programs themselves (although previews and small "thumbnail" pictures are becoming increasingly popular). It is data about the programs, which describes them in a way that makes it easy to choose. However, metadata is much more than what is on and when. There are several vocabularies for different aspects of metadata, from the authorship to the time of transmission and the location in the channel.

Since all vocabularies for metadata are in XML, they can easily be mixed and matched through the namespace mechanism. The same content can also be presented in different ways using style sheets, something that a television browser is likely to implement (including even more flexible models than presenting the content on a web page, such as adding SVG, Scalable Vector Graphics, to create animations). This is not an XML book, so we will assume that you have a basic knowledge about XML and how it works, as well as CSS.

XML makes one more thing possible: it enables the transformation of the data into formats that can be used for automated conclusions about the program, the user and the behavior. These automated formats are based on graph theory (the same science that underlies social software such as Facebook and MySpace), and the representation in XML is not straightforward. This is different from taking the XML data in a database and drawing conclusions from it using a program; this enables only conclusions about what is already programmed, whereas the use of ontologies – as the format that enables automated conclusions is known – makes it possible to make any conclusions that can follow logically from the data.

Metadata is not only used to provide information about what is on next to the end-user. It is also the mechanism used by producers to keep track of what they have in their libraries. This may not be such a big problem when it comes to video on demand, but it is harder when it comes to managing production assets, consisting of video clips, which may have no relation to each other, but which can be combined to create new productions. A digital asset management (DAM) system is used to manage the video clips. This consists of a database where the "media assets" – the video clips – are stored, and an efficient system to manage (store, locate, and retrieve) them. Sometimes, the hardware is also included.

In general, a digital asset is any digital media file that has value. Digital assets often include rich media such as video, audio and graphics, but this is not a requirement. Images, graphics, logos, video and sound files, web pages (HTML, XML), PDF documents, Quark and Illustrator files, MS Office files, free text files, advertisements, marketing collateral, brochures, product packaging designs, etc., all qualify as digital objects. Most companies and organizations have many kinds of digital assets such as text, photos, logos, music and video that may have value in the future, if (and only if) they can be relocated and reused.

Traditional broadcasters managed their assets on tape. Huge tape libraries had to be cataloged, the tapes kept in good condition, and made easy to retrieve. Since this was expensive, it was made possible only by the fact that production was also expensive. Today, production is so cheap that if content wants to be reused, it has to be readily available at the flick of a switch.

In a "hierarchy of media and metadata", the "media" part is the video clip or program, the media itself – which is essential for media production but of low value without any descriptive information. Think of a video tape of your summer vacation, fresh out of the camera. Combined with descriptive metadata this essence is referred to as "content" – you just wrote a label on the tape. If there is copyright or licensing metadata available it is called an "asset".

A typical DMS would take care of the ingestion of the media (getting it into the system), whether from tapes or directly from a broadcast; manage export (including transcoding and wrapping); and handle the metadata import, export and throughput (changing the format of the metadata as required). This also means managing a unique identity for each asset, as well as access rights and workflows (who should do what next with this piece of content).

As part of the workflow, the producer may want to see a preview of the finished product. Here, there is an interesting problem, which did not exist when all television sets looked alike: Do you need to see a preview for all different platforms on which the content can be viewed? Even today, the notice that "this film has been edited for content. It has been edited to fit your screen" in the airplane means a different version had to be created. If viewers can come with their mobile phones, their PCs, and anything from a 100-inch plasma screen to a 10-inch kitchen television, do you have to view all possible formats to approve the quality? And more importantly, does this mean creating a number of different versions that have to be managed in the asset management system?

For a producer, having a function to handle the rough cuts and different clips is very nice. However, extracting the data about the content – any copyright management that may be required, and any other data that can be derived – is necessary if the system is managing a very large library of content.

Metadata can have many dimensions. The video from your summer vacation is both a vacation video, a summer video, and a video. It contains beaches and parties, probably. If someone else gets hold of the video, they would most likely sort it completely differently. Translating the metadata into formats, which can be reused, is what the semantic media management is about, making reuse possible without having to sort everything into predefined categories.

Search and retrieval in multimedia databases is heavily dependent on the metadata that is provided together with the multimedia items. Metadata for this purpose is either generated manually or semi-automatically. Metadata also forms the basis for search in the system databases (most asset management systems today only allow keyword-based searches), which presents a problem because in most cases the keywords assigned to the media items is different than the terminology used when searching for the media. The use of ontologies to overcome the limitations of keyword-based searches has been one of the motivations of the Semantic Web, although this is not fully realized yet. This is also the subject of a huge number of research projects.

Most semantic search solutions today give the possibility of searching by predefined concepts (essentially, keywords), which were assigned to media in the indexing phase. Synonyms and meanings are looked up in a canonical source. Some semantic search engines allow searching using ontologies, which also allow the user to see concepts in context and explore databases around the topics.

Large digital repositories require efficient tools for the search and retrieval of digital assets. Current media asset management systems are highly tuned to index textual content of any form to make it searchable and, for this purpose, provide the possibility to either manually or automatically annotate audiovisual content with metadata. However, this manual annotation of content is time consuming and in many cases subjective.

Searching on the semantic features of the content is a common requirement in media asset management systems: this requirement can only be met by a semantic indexing process with the ability to identify the semantic features of the content.

Some digital asset management systems allow the indexing of audiovisual content by extracting keyframes from the videos, and by applying speech or other sound recognition to use the audio to generate metadata, for example by transcribing speech to text, which can be full text indexed. However, keyframes only support quick human recognition and possibly allow searching by example. On the other hand, speech recognition is still prone to errors and works only when the system is well trained (in a limited vocabulary), or with one single user. There is also no way to solve the problem that the audio track either does not reference everything in the video, or that it describes something outside the video.

Searching for content means either creating a production from it, or handing it over to the users to do what they want with it. In a live interactive television system, the user would get different video streams, which would need to be stitched together to form the video experience (it is hard to call it a program) that would result, for example, when watching a football game, seeing the match from the side of the home goal.

However, if you belong to a group (supporters of a certain football club), other group members' preferences could influence yours – if they could be compared with the video. Knowing who is a group member and what their preferences are is the function of the profiling system, which we talked about in Chapter 8. Retrieving recommendations based on metadata is different, however. Here, a recommender system has to be used.

Recommender Systems, Social Software, Presence and Personalized EPGs

Social software lets a user track the activities of friends, and manage groups of friends. Today, social software websites typically use their own group and profile management systems, and do not interact with the profile and group management systems of IMS-based IPTV. The group management system is crucial, because it allows the software to draw conclusions about which groups a user would like to belong to, based on which groups the user already belong to, and what preferences they have registered in their profile.

The preferences, and the grouping of users, is also the basis of another type of software – a recommender engine – which is usually not considered as "social software" but which can draw conclusions about what a user may want to see or do, based on preferences, history and which the group the user belong to. The most famous example of recommender engines is Amazon, where a user is met with the message saying "Hello, Johan. We have recommendations for you". In one way, you could see it as a program guide to the Amazon database of books and goods. The recommended books are not instantly added to the shopping basket; in the same way that recommended television programs are not instantly displayed when you switch on the television – they are merely displayed in the program guide.

A recommender engine and a social software system work in basically the same way, that is, looking at a list of items and drawing an (automated) conclusion about what other items from a larger list should be included. There is a lot of research in recommender systems, and there are a number of different algorithms that can be used; but generally, they are mechanisms to base the selection of what a user sees on selections made by an associated group, for example, a list of other users. If you have a system that enables your friends to see what you watch, then you can set up a system that enables your friends to choose programs based on what you see.

While TV-Anytime does a credible job of representing the broadcast stream from the point of view of a program, and MPEG-7 represents the production values, they are still XML formats. XML is excellent if you want to take data into a database management system and work on it with already created programs. But drawing new conclusions from XML data is very difficult, unless it is transformed into a suitable format.

One such format is based on graph theory, and has been the foundation of one of the major efforts in the later years of the web world: the Semantic Web. While this has often been hit by the same hype as artificial intelligence (including the same overconfidence by the creators), it does solve a different set of problems of formats which are "only" based on XML. The reason for the "only" is that these formats may not be based on XML originally, but they can be represented as XML, although many of the industry pundits prefer more obscure notations, which are easier to use mathematically.

Recommender systems is a very active research field, with a huge number of commercial applications. A number of different techniques have been developed in different research groups around the world, and the research community continues to develop. The two most widely used approaches are the collaborative-based recommender systems and content-based recommender systems. Collaborative recommender techniques are not very fast (in today's implementations), and for this reason are more often used to generate offline recommendations. Content-based recommender systems, on the other hand, can be used for archive recommender systems, consumer recommender systems and live event recommender systems. Combining the two into a hybrid is also possible.

A collaborative recommender system creates plausible recommendations for the user based upon his ratings and those made by his nearest neighbors (people who have similar taste). Ratings may be explicit, for example, giving a number of stars to a video; they may also be implicit, derived from the user's actions and interactions. And the user has to be uniquely identified.

The first stage of a collaborative recommender system is selecting nearest neighbors – in terms of interests. In order to find the user's nearest neighbors the system must run a series of algorithms that compare the main user and the compared user. When the system finds the required number of neighbors, it can calculate recommendations for the current user. This approach gives good results, but also requires a significant amount of processing time. An alternative solution is to find clusters of users, which provides the system with a less accurate list of nearest neighbors, but significantly reduces the amount of processing time required.

The system can then directly recommend items to the user by sending him the recommended item's id (CRID, see later), or it can store this information in a database for access at a later time and possible reuse by a hybrid recommender system. One of the main advantages of collaborative recommendations is that the system does not need any content descriptors – recommendations are based on the other user's preferences. Because of this, it is also possible to use the system in an environment containing different content types, for example, movies, documents and images. This makes the system much more flexible and presents the user with a wider choice of recommended items. The downside of a collaborative system is that it is based on searching for people with similar taste and does not work very well until there are enough users in the database. It is, however, possible to bias the recommendations in the same way that Facebook has with "friends" who are companies, not persons– by inserting artificial user preferences into the system. Integrating new items into the database is another problem, because only those items that have received a rating from the users can be used, so until the item has been rated it is invisible to the system.

Content-based recommenders do not interact with the profiles of other users. The principle of content-based recommendation is tracking the metadata of items that the user rates. If a user prefers a certain type of genre, this will show in positive ratings given to that genre description. This makes it independent of other users and integration of new items is very simple – provided the necessary metadata is available. However, the system can only work with the selected metadata format – different metadata types are a problem. Content-based recommenders work best with one specific type of content at a time. They have a hard time handling different content types or metadata standards.

The classification in the content-based recommender system is based on the history of content ratings given by the user and the metadata of a particular content item. To make it work, there has to be a training dataset, containing descriptions of content items and the corresponding ratings. Examples of metadata are a list of genres, list of actors, list of directors, list of screenplay writers, list of awards given to the movie, and additional metadata attributes such as the country of origin, production year, synopsis, etc. There are standard formats for all of these, and they can be collected from the different metadata items used to characterize a film.

In a recommender system, the classification of content items (videos) is based on calculations of similarity between the content description and the user model. Similarity is calculated separately for each of the description attributes (genre, keywords, actors, director awards, etc.). The calculated attribute similarities are then combined into a total similarity measure using the support vector machine (SVM) or other classification methods. The output of the system is a list of recommended content items (movies or video clips). If certain conditions are met (usage of regression methods for classification), the output can contain an ordered list of content items, including the indication of content suitability (a number ranging from 0 to 10). Other types of weighting are also possible.

Recommender results can be improved by combining both the individual and group approaches. There are many possible ways to construct a hybrid recommender system (hence the intense research in the area). For example, a parallel design, where both approaches are run at the same time and the final decision is based on both outputs; or a serial design, where both approaches are run one after another; or a more complex hybrid design such as a cascade design, where the output of one recommender represents an input for the second recommender system – this can be done in steps, or recursively, depending on the inputs.

Filtering and Personalizing IPTV Content

There are two ways (generally speaking) of blocking access to content. The first is to remove the content itself from the data stream, making it impossible to access. The other is to remove any reference to it, so it becomes impossible to find. Content filtering, which leverages part of both, is among the most requested features in IPTV systems – although it turns out that when users have access to the mechanism, they do not use it, preferring other means of filtering (e.g., agreeing within the family what is acceptable viewing is one popular option). The reason is simple: content filtering is based on the same premise as DRM – that you do not trust your users. And while this may work in the relationship between the service provider and the end-user (the distrust is frequently bidirectional – many people do not trust their telecommunications provider), a family based on mutual distrust will usually face a disruptive crisis sooner or later.

Content filtering, especially if combined with alerts, may be a sign of distrust, but the exact same mechanism may be used for personalization of the content delivered to the user. The filtering is applied to give the user the content he has requested (and only that content), instead of content where someone else has set up the filters.

While filtering in video on demand is pretty straightforward, and based on the use of a recommender system to draw conclusions and create recommendations about what the user wants, it is harder in a broadcast system – especially since the premise here is that you cannot know what the content will be until it has been broadcast. However, the important aspect of personalization (for the purpose of this book) is the ability to insert individual interaction items in the user dialogue.

What that means is putting personalized items on the screen. It can be done using the overlay browser (if items are to be inserted in the data stream, it becomes much more inefficient, since this is tantamount to providing one single stream per user – losing synchronicity and the network efficiency gained from multicast). For certain types of personal items, however, the browser overlay may not be sufficient (such as closed caption or subtitles – essentially translations of the show into text in another language). These need to be sent as part of the content stream, as they are dependent on lip synchronization. Or else, there must be a very good mechanism for the synchronization of the browser content with the media stream, something that is not in place at the moment.

For the purpose of this book, the important part is the personalization of interaction based on the program being viewed (or selected, if the user has requested a recording). In addition to the program data, the personalization is based on the user profile; therefore it is important that these two can interact. The data formats available in the IMS-based IPTV systems can also include the program currently being watched – with a reference to the TV-Anytime program guide, so it is clear what the content is about. Reasoning about the content comes into play here as well. Including the profile in the reasoning is no harder than reasoning on any other kinds of lists (social networks, which can be reduced to graphs, actually make reasoning easier).

Metadata Types and Models

Metadata for content can be generated automatically or created by a person. There are a few automated technologies that create some sense in automatically generated metadata – usually based on recognition of the sound and parts of the image. However, more often, automatically generated content (unless it relates purely to production factors such as the camera angle, light intensity and so on) has to be reviewed by a human to make sense.

This is because humans have a much deeper understanding of the meaning of images and objects. On the "subject-matter level" the information describes, at different levels of abstraction, the "meaning" of the material. Such meaning is usually created manually to ensure a further reuse of the audiovisual material: broadcasters let experienced archivists assign labels to the audiovisual material – the labels are taken from a controlled vocabulary in order to enable efficient search and retrieval of the stored material. The controlled vocabulary or classification scheme has been optimized by the archive experts to cover the most important use cases for search and retrieval within their usual workflow. It does, however, mean that the metadata is not generated until the media has been archived.

Metadata generation systems can be classified into manual annotation tools, semi-automatic annotation tools and automatic annotation tools. Manual annotation systems help the user to label the data by some useful classification schemes, which are important for finding relevant material later on in an archive. Free-text annotation is used in combination with terms stemming from classification schemes (e.g., in YouTube ranging from "automotive", "comedy", and other categories). The user selects those classification labels that are most helpful for later search and retrieval – which of course depends on subjective factors and on the retrieval organization. European broadcasters have tried to agree on this, and to some extent have created both automatic metadata generation systems and methods.

Automatic and semiautomatic tools for generating content metadata are intended to extract as much useful information from the actual media item (the video clip or other audiovisual content) as possible. A number of algorithms from computer vision, pattern recognition, speech recognition, spoken document retrieval and signal processing have been applied in various ways by different tools to solve this. However, extracting meaning from an object is not straightforward. Only a few of the classification tags that users are interested in can be defined automatically, and machine systems cannot make sense of humor and other human concepts. On the other hand, hard cuts generated by the editing are easy to detect automatically.

Algorithms that give acceptable performance already exist for temporal video segmentation, transition classification into cut and dissolves, face detection, camera estimation, tracking of objects, audio segmentation, speech–nonspeech detection, melody detection and audiovisual fingerprinting. For example, video segmentation and face detection are already built into cameras, although the amateur cameras have far more automated features than the professional equipment.

IPTC News Codes, NewsML and SportsML

Broadcast material is, as described, usually annotated manually. This is not only for archival purposes. Major news companies, such as Reuters or Sports-Ticker, perform online annotation of live sports events and produce real-time feeds to consumers, for example, via RSS. Media observation companies monitor various broadcast feeds and annotate the live material. Tools support MPEG-7 analysis (thus combining the automated analysis with the manual annotation), and can handle selection from ontologies. The main standards body for these annotation libraries is the International Press Telecommunications Council (IPTC). It is an international consortium of news agencies, editors and newspaper distributors. The IPTC has developed standards such as the Information Interchange Model (IIM), NewsCodes (formerly the Subject Reference System), the News Industry Text Format (NITF), NewsML and SportML. Almost all of them are defined by using XML technologies. There objective is to represent and manage news along their whole lifecycle, including their creation, exchange and consumption.

The IPTC NewsCodes is currently split into 28 individual sets of technical terms in controlled vocabularies (CV). Some of these sets are structured in a hierarchical fashion similar to taxonomies. The set that defines the subject codes is the most important one of the IPTC NewsCodes standard. The subject codes are designed to classify the subject of content items (mostly news items). They are defined in a hierarchy with three levels. The first level defines subjects, which provide a high-level classification of content items. Subjects include categories such as "arts, culture and entertainment", "crime, law and justice", "disaster and accident", "economy, business and finance". The second level is built by subject matter, providing a description at a more precise level. The final and most precise level is defined by subject details. Both subject matter and subject details are assigned to a category by using an eight-digit numbering system where the first two digits represent the subject, the next three the subject matter and the last three the subject detail. Other controlled vocabulary sets describe audiocodecs, confidence, genre, media type, MIME types and urgency (which is important to a news organization). All of these are used as values for metadata fields defined in other IPTC-related standards such as NewsML or SportML.

NewsML defines three levels of envelopes, which describe the news item. The first envelope is the content layer and provides a uniform interface to content irrespective of the media type. The format, MIME type, encoding etc. are examples of metadata items defined within this envelope. The base XML-complex type for this layer is called the ContentItem.

The second envelope level provides metadata about the internal structure of the news item and is called the structure layer. This layer uses NewsComponents as base XML-complex types. NewsComponents define metadata and the internal structure. A NewsComponent is therefore a container that can hold one or more sub-items (NewsComponents, ContentItems, NewsItems or links to NewsItems). The metadata of NewsComponents is separated in administrative metadata (provider, creator, source, etc.), descriptive metadata (language, genre, subject code, date line, etc.), rights metadata (copyright, usage rights) and miscellaneous metadata.

The third envelope defines a management infrastructure above the other two envelopes. This layer defines NewsItem as an identified and publishable piece of news. This is the object that a provider will create, store, manage, reuse, link to and from other NewsItems; it is the unit of interchange in a news environment.

Each NewsItem is identified by a UID (a globally unique identifier) including the ProviderID, DateID, ItemID, RevisionID and a PublicIdentifier (a URN combining all the other IDs). In addition, a NewsItem provides management properties such as information about the first creation date, status, urgency, revision history, etc. Based on these management metadata, the standard defines some usage cases (so-called management strategies) for NewsItems. NewsML uses the CV defined by the NewsCodes to restrict values of the different defined metadata fields. This design ensures interoperability between different systems and organizations.

SportsML is another similar markup language (in XML) created by the IPTC. It is a cross-sport cross-language standard for the interchange of sports data and statistics. SportsML supports the identification and description of a large number of sports characteristics. Highlights include scores (who won, how many goals were scored, how did the score change); schedules (which players/team are playing and his/their opponents, when is/was the match and where); standings (who is leading the championship or a specific tournament, who's the closest to qualifying for the championship); statistics (how do the players and/or teams measure up against one another in various categories); and news.

SportsML consists of a core XML schema/DTD containing a great amount of properties that describe a wide range of sports coverage. In addition, SportsML contains several "plug-in" specific-sport schemas/DTDs, which are only necessary when the publisher needs to go in-depth for a specific sport. There are only seven sports covered in SportsML's initial release, but users can also define additional plug-ins.

Dublin Core

One of the oldest, and most famous, metadata vocabularies is Dublin Core (see Figure 9-1). Originally based on a similar standard for libraries, it was standardized by an industry consortium based in Dublin, Ohio (hence the name). Today, it is a generic vocabulary that keeps track of and manages digital resources. It is used to describe an individual example of a resource, rather than the generic resource in itself.

The DCMI metadata terms consists of elements, element refinements, encoding schemes and vocabulary terms. The basic metadata exchange scheme is a list of 15 metadata fields (Title, Subject, Description, Creator, Publisher, Contributor, Date, Type, Format, Identifier, Source, Language, Relation, Coverage, Rights). The element refinements may detail the basic elements by adding specific qualifiers (e.g., Created, Valid, Available).

There are two levels in the Dublin Core: the Simple and the Qualified levels. The Simple level has 15 elements, the Qualified adds three elements, and also qualifiers on the Simple elements, intended for use when discovering a resource. It was never intended to describe complex relationships

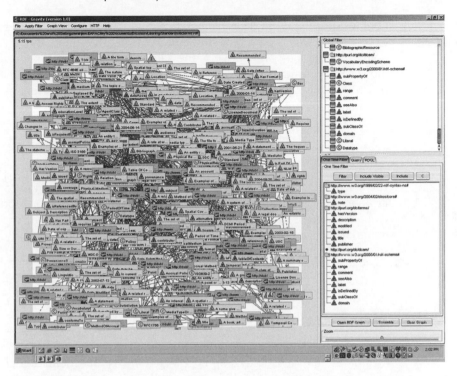

Figure 9-1. Dublin Core RDF graph visualization. Reproduced by Permission of © 2008 Salzburg Research Forschungsgesellschaft mbH.

between resources, but was created to handle a limited set of attributes with refinements. Even before RDF mapping, there were abstract data types in the language as well, defining memberships and whether something was a part of something else.

Today the DCMI metadata terms are mainly used to exchange (descriptive) metadata on the web. To some extent, the Dublin Core metadata overlaps TV-Anytime, but as both are XML namespaces, it is not unusual to see them in the same document describing the dataset, where the respective properties are more precise. Dublin Core, showing its legacy as a library data format, can show when an object is available, and when it is not – something not possible in TV-Anytime (useful, for example, for a digital video lending library).

The main advantages of Dublin Core are that it has extensive functions to describe authors and contributors, and is very good at describing publications and serials. If a video is part of a journal, for example, it can easily be described using Dublin Core. There is also a structural advantage: the mapping to RDF.

Dublin Core is one of the first complete metadata formats to have completed a mapping of the format into RDF. There are RDF schema documents for all the Dublin Core elements, and this means they can be used to draw conclusions about the datasets they describe. For example, when there are two objects, which both have the author "Evelyn Waugh", the system can conclude that they are works by the same author; if they also have dates, there can be conclusions about when they were created and if one comes before the other. The conclusions can be drawn on the different aspects of the document, but also on the object itself.

P/Meta

The European Broadcasting Union (EBU) has developed another format, P/Meta, which is intended for the exchange of information about programs between broadcasters, primarily for the purpose of exchanging archive material. It defines attributes, attribute types, sets of attributes, and sets of sets. The focus is on machine readability, and the ability to automate data exchange.

SMPTE Metadata Dictionary, MXF and UMID

Defined for a professional broadcast production environment, the Society of Motion Pictures and Television Engineers (SMPTE) has defined a collection of metadata items and definitions, which is intended to serve as a reference for all media descriptors used in broadcast production. Its aim is to overcome interoperability constraints between network devices for digital broadcast production caused by incompatible formats for audio and video. All items are uniquely identifiable by their so-called universal labels (UL) which are 16-byte strings, defined in SMPTE 298M. Some items are nodes that combine the leaves (i.e. the single metadata entities) of the dictionary to classes and subclasses. It can be managed using the Materials Exchange Format (MXF), which is a file format developed by a group of experts from the Professional MPEG Forum, and standardized by SMPTE. Part of the standard is the Descriptive Metadata Scheme Part 1, (DMS-1), which defines a set of common descriptive metadata items derived from typical scenarios within the broadcast production chain.

SMPTE also defines the UMID (unique material identifier), which identifies audiovisual material at different levels of granularity (from a single video frame to a whole program). It consists of a 32-byte string, which contains a UL of 16 bytes. The remaining 16 bytes are used to form a globally unique number. The extended-UMID makes it possible to add information about where, when and by whom the material was created. The UMID is used to uniquely identify material within media asset management systems and to associate any essence item with its external metadata. The UMID might be transported using SMPTE 291M ancillary data packets of the serial digital interface (SDI). UMID is also part of the MXF specification, where it is a built-in part of the structural metadata.

Metadata and the EPG: TV-Anytime

While the underlying technology may be the same, the navigation through the resulting document depends very much on the device used to access it. The way you navigate through a table on a television set is very different from how you navigate the same information on a mobile phone, or a PC. And this is what makes television on the web, mobile television and IPTV seem different today. The distribution mechanisms are the same, the presentation mechanisms as well – but the user navigation is different, and this makes for a different way of viewing the content, and hence how the navigational content is perceived by the user.

Streaming content, as we have explained in previous chapters, means that the content is sent as a continuous flow of packets to the receiver, and not sent as a single huge chunk at once. The protocol used to control the streaming, however, is different in the web, mobile television and IPTV – as is the program used to actually transport the media.

Different service providers will have different priorities for steering users towards selected programs, and users will have different preferences for what they want to view. In a static EPG, it is hard to do the latter option, but easy to do the former – but the selected program may not be of interest to the users, who will skip the promoted entry and go searching for what they like anyway.

In a personalized EPG, where the content is designed to fit the individual user, both these mechanisms can be combined. Personalizing content is one reason why XML is a good format to use, since it is halfway between structured information (in the database sense) and presentation. A short primer on how XML works may be required, because it will feature very heavily in this book, being one of the primary underpinnings of the EPG systems (and other metadata) of the future.

The basis for the EPG (which after all is only a table of information presenting the content of a television channel), a VoD server, or other resource, is the metadata describing the channel. Metadata, as we saw in previous chapters, can be taken from the analysis of the video itself, or be provided by the programmer. With presence, it can also be created by other users, for example, as ratings.

As the format describing the EPGs is described in the XML schema documents, it is possible to define what the EPG will contain based on the XML schemas. Since the TV-Anytime format is a well-established format, which has been modified to fit many different standards, a number of different enhancements and extensions of the TV-Anytime format have been created to fit different requirements from different standards groups, in particular those of the DVB Forum and ARIB, which means it is used as metadata for the digital television broadcasts in most of the world.

The TV-Anytime process of creating metadata is a generic process, which will also work for other distribution means than broadcast over radio – for example, the broadcast can take place over the Internet. The process will be exactly the same when creating metadata for a scheduled Internet multicast as when you do it for a radio broadcast. The only difference when it comes to IPTV distribution is for VoD, since the on-demand aspect makes it possible to handle adaptations of metadata to situations (e.g., personalization or localization). In a direct transmission, the adaptation is harder because there is a timing threshold, in some cases legally enforced (e.g., there are laws when it comes to gambling and the timing of broadcasts in many countries); in other cases, there may be restrictions in agreements (most stock exchange information is delayed by 20 minutes since this is much cheaper than getting the real-time information). For VoD, the price relationship may be reversed, however – a classic movie might be more expensive than the first time it is broadcast. On the other hand, it may attract more advertising when it is fresh.

This impacts metadata, because information about the price and other conditions for sale is a part of the data about the data – which is precisely what metadata is. Included with the program or not, it can be handled in two different ways – either extracted and forwarded to a separate billing system; or sent to an integrated charging system in the case of IPTV. We will look at these two models later in the book. The metadata can either be included with the media (essentially, embedded in the media stream); or it can be referenced and downloaded separately. In practice, these two models may be parallel, because it is easier to handle updates when data is sent separately (either pushed to the television of the end-user; or pulled by an HTTP client). Referencing the metadata also makes it possible to separate the display of the metadata from the television channel, for example, displaying the metadata on a mobile telephone as the user is watching television. This is not possible when the metadata is embedded in the video stream. The way metadata is created and used looks schematically as shown in Figure 9-2.

TV-Anytime Document Structure

XML is a framework, which creates a way of defining elements that makes them understandable to a computer. The XML framework also specifies a way of defining what the individual elements contain, how they are related, and what constraints there may be on them. This is known as an XML schema. Each XML document type is associated with a specific URI, where there is an XML schema document for the particular document type.

Since the XML schema defines what the elements are, the EPG has to have an XML schema that essentially defines what a television program guide is, and what it contains. If each company that creates an EPG does so according to its own schema, taking an EPG from another service and using it in your own will be problematic at best, impossible at worst. If all service providers could agree on one EPG format, this would make life much easier for anyone who wanted to display EPGs from many different service providers.

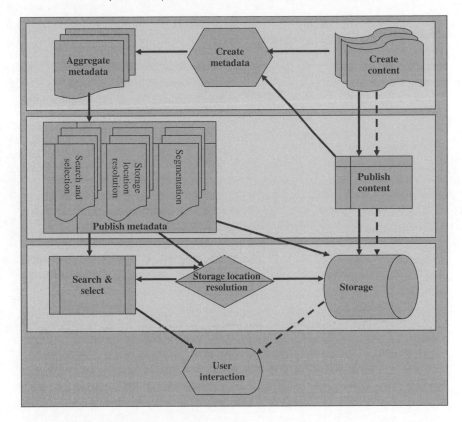

Figure 9-2. Metadata and content flow.

Actually, this has been done, and quite some time ago. The TV-Anytime group, which was created primarily by television companies to define an interoperable EPG format, is one of the few standardization groups who have successfully concluded their work and now dissolved. The results have been handed over to the European Telecommunications Standards Institute (ETSI), which, despite the name, is an international standards forum. Handover took place at the end of the 1990s, but there have been occasional updates of the standard to correct errors, and also to conclude the "phase 2" work, which was left open when the consortium closed. While the documents the TV-Anytime forum submitted have been updated by ETSI, the content is basically the same as the TV-Anytime consortium agreed. The TV-Anytime program descriptions have also become very influential, becoming the foundation for many other standards, including the program guides in the DVB Forum (used in terrestrial broadcasts in Europe).

The TV-Anytime standard consists of three main documents: the overall systems description; the metadata specification; and the content referencing and location resolution specification. There are actually nine separate standards documents (see Table 9-1), detailing different aspects of the standard (with subdocuments). And then, there are schema definitions, which belong to several of the standards, included in subparts of the different documents.

Document number	Name	Contents
ETSI TS 102 822-1 V1.3.1 (2006-01)	Part 1: Benchmark features	Essentially a set of requirements for TV-Anytime, which the standard is measured against to determine if it has fulfilled its goals
ETSI TS 102 822-2 V1.3.1 (2006-01)	Part 2: System description	The overall description of how the TV-Anytime system is intended to work
ETSI TS 102 822-3-2 V1.3.1	Part 3: Metadata	Metadata elements, schemas, vocabulary and structure
ETSI TS 102 822-4 V1.1.2	Part 4: Content referencing	How the CRID (the URI used in TV-Anytime) works
ETSI TS 102 822-5-2 V1.2.1	Part 5: Rights management and protection (RMP)	How metadata is applied to create content protection
ETSI TS 102 822-6-1 V1.3.1	Part 6: Delivery of metadata over a bidirectional network	Defines a format for SOAP/HTTP bidirectional data transport
ETSI TS 102 822-7 V1.1.1	Part 7: Bidirectional metadata delivery protection	Security and transmission protection for bidirectional data transport
ETSI TS 102 822-8 V1.1.1	Part 8: Phase 2 – Interchange data format	Defines retrieval of TV-Anytime data from "alternative sources", essentially website download
ETSI TS 102 822-9 V1.1.1	Part 9: Phase 2 – Remote programming	Defines how to program a recorder, for example a PVR, using TV-Anytime

Table 9-1. The nine standards documents of TV-Anytime.

The system model for TV-Anytime (described in Part 2 of the standard) is actually very simple. The basic model has three major elements: a service provider delivering the TV-Anytime service; a transport provider that carries the service; and a piece of equipment in the home that stores the content and plays it back at the consumer's request (which may mean immediately). There is a complication, though: if you introduce digital rights management, the system becomes much more complicated.

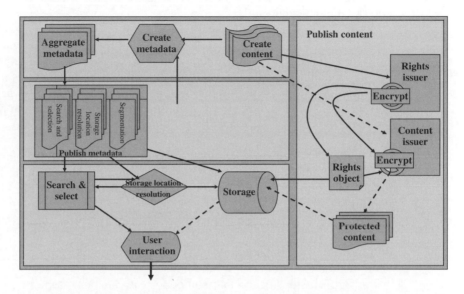

Figure 9-3. TV-Anytime with and without rights management metadata.

The model is generic enough that it can be implemented in many different ways: the boxes in Figure 9-3 are functions, not actual physical boxes (or even software functions). While the model does not (wisely enough) include the user interactions, it does assume that the metadata is created when the content is created; and that it is used for the search and navigation (something which is very probable, but there are other ways of searching multimedia content). The content itself flows from the content creation, to the storage, to the presentation, to the user.

In the TV-Anytime model, there is also a DRM mechanism included. Based on the user's location, this has not been implemented (and is part of the phase 2 work); it relies on the content being accessible only at certain locations, but the management of this is as tenuous as other mechanisms to manage content by forbidding the user access.

The model is not unidirectional: content metadata is sent to the user, and not returned to the service. However, because TV-Anytime began when set-top boxes with telephone connections started to become popular, it was designed with feedback from the user in mind. The "fully interactive" model, however, is different from the semi-interactive model where the user has an uplink to the service. The fully interactive model was designed primarily for networked systems, not those where the content is received via broadcast.

The system only has three functional elements: the content creation; the content service provision and access; and the personal digital recorder (PDR) – a kind of catch-all box for all user-related events, such as receiving, storing and displaying content, including search, navigation, and so on. It is a generalization of all the consumer electronics devices in the home, and not a real device (yet).

While the base assumption is that the content should be sent from the content provider via the service provider to the user (unidirectional), there is nothing that states how the interaction should take place in the other direction – detailing how the information about the user's interactions, i.e., the feedback, gets from the user to the content provider. The roles of the content and service providers, however, are not spelled out in detail, which helps in reducing the confusion brought by pushing a business model onto a media format.

TV-Anytime is primarily intended to be unidirectional – the metadata, as well as the content, goes from the broadcaster to the viewer. This reflects its history as a format for television metadata. But it also tries to be bidirectional, defining how to send SOAP messages from the user to the content provider, containing both interaction (primarily with the EPG) and personal data.

There are six steps associated with the use of metadata (apart from the step of transmitting the data, which in itself can be a complex process): Publish, Search, Locate, Acquire, View, and Finish.

The Publish step means that someone – it can be a content creator, it can be an aggregator – publishes a CRID (explained in detail below), which represents a show. The metadata about the show is associated with the CRID. The Search step is the reverse – a user wants to find a show. The user searches in metadata, which means the search process is no different from any other search process on the Internet. There is however, potentially a different way to search, which means looking for patterns inside the content (e.g., the video or audio stream itself); this model is more complicated both technically (because it is harder to index multimedia content) and from a user interface point of view (because it is harder to explain what you are looking for). Finding an actor called "John Wayne" is much easier than putting in a non-verbal search term – how do you define what you are looking for? And how does the system know which of all the possible polygons in the picture to look for?

Having found something, the next operation is Select. The CRIDs for the content which the user has chosen is sent to the user's PDR (i.e. his television). Following the Select is Locate, which means getting the content referenced by a CRID; it has a physical location, which is resolved by the server in the same way as the physical location of a resource underlying a URI is resolved, and the resource retrieved. When the resource is located, it has to be acquired; the Acquire operation gets the content from the location where it is stored to the location where it can be played.

When the content is acquired, it should be played. The View operation displays the content on the user's screen (the metadata is stored along with the content, and can be displayed if desired), and after that there is only one operation left: Finish, which potentially can be used to store the user's pref-erence information. The TV-Anytime standard defines a number of elements that can be used to create a history of the user's usage, as well as his personal data, in addition to the metadata describing the content.

Since the format of TV-Anytime is XML, it can easily be presented by most modern browsers, and it is relatively easy to write a dedicated client for it. It can also be transformed from XML to HTML using well-defined standardized methods, so that it can be presented in older browsers. However, a CE-2014 compatible browser, will not have any problems with the XML presentation. The issue is probably rather one of data transmission, where the TV-Anytime format may be too heavy to transmit over a constrained network (such as a mobile network). The XML adds quite a bit of overhead. There is no immediate solution to this.

There are six basic types of metadata that can be used to create any EPG, as described by TV-Anytime:

- content description

- example description

- consumer description

- segmentation description

- metadata origination information

- interstitial and targeting

The schema documents for these six basic types contain XML fragments, which make efficient distribution and extensible structure possible (see Figure 9-4).

Content description metadata

The content description metadata are descriptions of content items, such as television programs – but also segments of programs (which is used in the "interstitial" metadata). The program information can contain information about title, genre, synopsis, originator, cast members and similar information; this part also contains information about how to purchase the program, as well as reviews (or links to them). The data is organized in XML documents, but it is structured to be taken into a table.

The elements in the description list are of two types: the ones that organ-ize the metadata, for example, the type and the packaging; and those that describe it, for example, those referring to the actors, the synopsis, and so on. TV-Anytime covers almost anything you need to say about a television program, and at a very considerable level of detail – too much detail here. All the information that appears in the credits for a television show, including the production staff, can be covered by the TV-Anytime metadata.

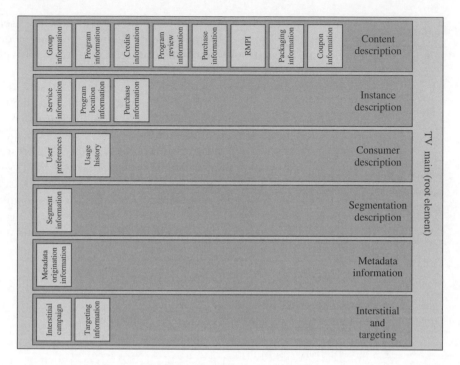

Figure 9-4. TV-Anytime metadata structure.

Example description metadata

A program may be part of a series, but it can also be broadcast at different times, on different channels, and in different countries. The locations – in time, or on a medium – are saved in a separate document, including the freshness date when the changed synopsis expires. The location of the data is linked to the CRID, essentially the URI for the program, which is also represented in the example description.

Consumer metadata

The consumer metadata are intended to contain user preferences (e.g., language preferences), as well as the usage history. It is not clear why this data should be handled by TV-Anytime; there are plenty of different models, including the IMS-based profile model described in this book.

The specific goal of the metadata system is to provide data about the content the user wants to consume (not data that the broadcaster needs to manage copyright issues, to take a tangential example). The metadata describes information relevant to the user about the program in a human-readable way (i.e., in text format). The metadata that is presented can be the title of the program, the category, the actors and the director, and so on. TV-Anytime is based on MPEG-7 DDL (description definition language), which is a scene description language (as we have discussed in Chapter 11, when you analyze the scene for MPEG-4 encoding, you can also create an automated description in MPEG-7), but it is not equivalent to MPEG-7 (as we will discuss later in this chapter).

Like most standards not defined by the consumers themselves, the TV-Anytime consortium guessed what users would want. Based on their own wants, they sometimes guessed right, but not always. The user preference data recorded by the PDR at the Finish operation is one example of the user data; this is intended to allow for sharing of data, as well as exchanging profile information. There is one additional form of unidirectional user-related data which is supported by the TV-Anytime standard: coupons.

Coupons in the TV-Anytime system are not primarily intended to be used in the supermarket (although in theory, they could). They are intended to be used with IPTV services to enable a service provider to provide value to the end-user, for example, as a discount on video on demand over the weekend. The metadata described by TV-Anytime provides ways to signal the existence of a coupon, explain it (what value it has, what the method of using it is, what is the subject of the discount, a textual explanation, etc.). However, there are no specifics, other than the framework, that is left to the individual operators.

Segmentation or packaging description

Originally, the function of the segmentation was the same as that taken over by the interstitial metadata: to describe how a program can be composed from different parts, but when the TV-Anytime standards were moved to ETSI, the data that could be handled extended far beyond the television programs. Irrespective of technical platform, the data which the user could receive – and which could be used together with the program – included games, text document, even additional video and audio. It then became necessary to create a "package" to hold this, and it is what the packaging description is about.

The package consists of a collection of different content items intended to be consumed together, or in some combination, to provide different consumer experiences. In the TV-Anytime world, these are add-ons to the television shows (e.g., carrying subtitle texts in a separate document, which is packaged with the television show). The user preferences can be used to determine the content of the package (if I want to have my subtitles in Swedish, then this should be possible). Packaging can also contain multi-camera angle information, and rights management information for the package and its components.

Metadata origination information

This contains information about when (and potentially, but that is not defined, where and by whom) the metadata was created. In the current TV-Anytime system, there is only one element for the origination information.

Interstitial and targeting

"Interstitial" means the intervals between something that is stitched together, and just like an XML document, it is possible to compose a television program from program segments. The references for these, and the way they are handled, is done using the TV-Anytime interstitial control data. This also contains copyright management information, although it is only a representation of the content.

Targeting is the process of automatically matching and delivering relevant content to customers, whose profiles are stored in the system. There are two ways to do this: push (where the broadcaster sends the metadata along with the data to be included in the substitution); and pull, where the user's terminal retrieves the program content based on the data it has about the user. In the push scenario, this data is kept with the service provider, and the service provider creates the substitution data.

Identifying the Data: the CRID

The first of the cornerstone documents in the TV-Anytime set of standards is the system description. The second cornerstone document is the metadata definitions. And the third cornerstone is the content referencing specification. A keystone for the content descriptions is the CRID (content reference identifier), which is unique for every program (or group of programs, such as a television series). It consists of an authority and the data. An authority can be a television channel, but the concept becomes somewhat dodgy in the new IPTV world, where traditional television stations start being one of many different types of content managers. A CRID can be any type of organization; it is quite possible for a church, for example, to create a TV-Anytime program guide, and include its own references. The intent was that the CRID should point to the program as a resource, but as URIs can be references, that is not necessarily an issue.

The CRID is a URI, and consists of a multilayered set of descriptions of the content as a resource. The same CRID can point to a whole series, instead of a single program example. It does not describe the file format and location, but the title of the program and the location it is coming from. A CRID has the same syntax as a URI, with the prefix CRID, and contains the authority and the data that describe the document. Like an URI, it points to a resource, not a fixed location (in time, like a broadcast time); it could be used to point to a server on the Internet, as equally easy as pointing to a television channel. Figure 9-5 shows an example of a CRID.

The CRID, being a resource description, can reference other resources (the "Star Trek" CRID referencing both the "Original Series", the "Next Generation", "Deep Space Nine", and "Voyager", in additon to a series of films). The referencing can go further, down to the individual programs – or even fragments of programs – just like any URI referencing an XML fragment. Since

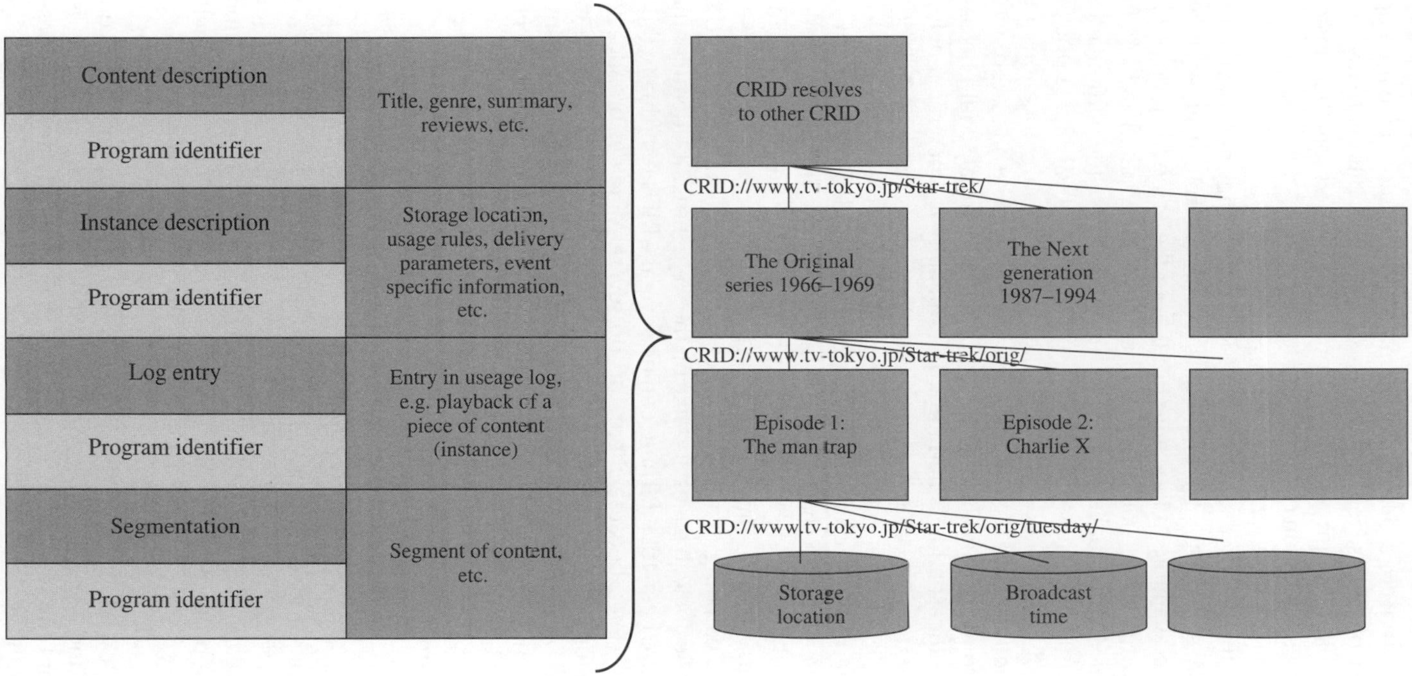

Content description	Title, genre, summary, reviews, etc.
Program identifier	
Instance description	Storage location, usage rules, delivery parameters, event specific information, etc.
Program identifier	
Log entry	Entry in useage log, e.g. playback of a piece of content (instance)
Program identifier	
Segmentation	Segment of content, etc.
Program identifier	

CRID://www.tv-tokyo.jp/Star-trek/

CRID resolves to other CRID

The Original series 1966–1969

The Next generation 1987–1994

CRID://www.tv-tokyo.jp/Star-trek/orig/

Episode 1: The man trap

Episode 2: Charlie X

CRID://www.tv-tokyo.jp/Star-trek/orig/Tuesday/

Storage location

Broadcast time

Figure 9-5. Example of a CRID.

the CRID is not absolute, but relates the authority to the data and can be referenced, a different content provider could create a completely different set of CRIDs, referring to the content in a different way (e.g., all episodes containing furry creatures could be grouped together, instead of grouping the episodes in order of broadcast).

Since the CRID refers to the program, there can be more than one example of the resource to which the CRID refers. They may have different characteristics, for example, different aspect ratios (16:9 versus 4:3), even though they are available at the same time, from the same location, etc. To identify the particular version, there can be an example metadata identifier, which identifies individual examples within the scope of the particular CRID to which it has been assigned. The IMI is of the form dvb://123.5ac.3be;3e45@2001-12-07T12:00:00.00+01P02:10, i.e., it contains the location, an identifier, and the time and date of the transmission. This resolves very well to traditional broadcast, and instead of the broadcast format, could be used to provide a multicast address and time. The difficulty is when referencing stored media, used for video on demand. Then there is no need to include the timestamp, unless the reference is to a recording of a live broadcasts, for example, created by the user's networked personal video recorder. The address has to reference the channel where the media was broadcast, but using the address and protocol to show that this is stored. The timestamp has to be given in the same form as timestamps in HTTP.

Searching in a metadata set such as an EPG is similar to searching in an index of web pages. Just like on the web, the CRID identifies the resources that have been found. There can be several CRIDs referring to the same resource, however, because there can be several different broadcasters sending the same program at different times – it can also be available in a VoD collection. As on the web, additional metadata has to be used to distinguish between them. This can be the IMI, or a higher level CRID. which groups all CRIDs with the same characterstics.

While most major standards bodies have adopted a TV-Anytime-based format for their program guides, they have not necessarily done so without extending it. One of the major features of XML is its extensibility, and while this makes it very easy to add features to an XML format, it makes it difficult for anyone creating a system that cannot be extended easily – such as most consumer electronics manufacturers. If a browser has to download new versions of the rendering software every time it gets a new television channel, creating content, which is universally accessible across equipment types, will be problematic. While this is not unsolvable, it is not something which has been solved today, so there is a risk that IPTV service providers may find it a good idea to add features that all receivers cannot render. Luckily, you are allowed to ignore those aspects of a presentation which you do not understand if you are a program, and while this means rendering is not a problem for browsers, it may present a problem for software that does something else with the data – such as parental control.

Metadata for Production: MPEG-7 and MPEG-4

Where does the metadata in the EPG come from? One answer is: from the program metadata. That, in turn, can be enormously richer than the bare facts of the title and the author. A television program is a complex production, and the content can be described along multiple dimensions. One of them is the production: who the actors are and who directs it, who the producer is, who the cameramen were, what type of camera was used and what its settings were, and so on. Some of these can be created automatically, for example, by the camera itself. As yet, there are no standards for the creation of such metadata, but it can easily be included in the XML metadata describing the overall production. This is also true when it comes to the individual objects in an MPEG-4 scene.

MPEG-7 is a metadata standard developed by the Motion Pictures Expert Group. Unlike other MPEG standards (MPEG-2 and MPEG-4), MPEG-7 does not say anything about the coding of the audiovisual data, it deals only with the description. Since 1996, the MPEG family of standards has developed three content models starting with the metadata standard MPEG-7, the definition of digital items in MPEG-21, and MPEG-A, which is an attempt to bundle the growing universe of metadata requirements so that they become manageable in real applications. The more MPEG diverges from its roots in the audiovisual coding, however, the less useful the standards have become – although MPEG-7 has become the de facto standard for production metadata.

MPEG-7 has a set of standardized tools to describe different aspects of multimedia at different levels of abstraction. They mainly serve the purpose of annotating what can be seen on the image or video, but do not offer means for adding highly structured background information. Furthermore, the possibilities of MPEG-7 are not formal enough to allow reasoning in order to deduce new knowledge from known facts. This lack of semantic formality should not come as a surprise: the model was conceived of in the mid-1990s as a metadata scheme for broadcasting. Early on, its proponents realized the need to extend the modeling capabilities and this was done by allowing the definition of add-on descriptor schemes. However, any descriptor scheme is only as useful as the operational semantics defined for it. This makes it difficult for new descriptor schemes to become accepted by the MPEG user communities. However, the MPEG standards do offer a platform for convergence of knowledge-based and metadata-based content annotation.

Apart from the rather experimental MPEG-A, the latest work from the MPEG consortium is MPEG-21. This aims at defining a framework for multimedia delivery and consumption, which supports a variety of businesses engaged in the trading of digital objects. MPEG-21 builds on the previous MPEG standards to characterize multimedia in terms of objects.

Structuring by the objects, which constitute the content, is another way of creating metadata structures. An MPEG-4 scene is structured in objects, ordered into the foreground and background of the view. There is a language, which describes how the different objects are organized in the scene, as well as a number of other parameters relevant for decoding, encoding and transforming. While there is some rights information embedded with each object, it is not possible to embed a complete description, much less link to external descriptions and provide translations between the schemas used to describe the content in the scene description language and a metadata language, which describes other features.

The MPEG working groups have not only defined the encoding and distribution formats for video and audio content, they have also created a metadata format, MPEG-7 (the numbering in MPEG is obscure; it goes from 1 to 2 to 4, and then to 7 and 21). Like the binary format for scene description (BIFS) in MPEG-4, which makes it possible to create very rich descriptions of how a scene is produced and embed it in the media stream, MPEG-7 descriptions are scene descriptors; they can be used to analyze how the content is structured, and also as the raw material for EPGs.

For each entity to be described in an MPEG-7 description, a descriptor is defined. Descriptors are combined to description schemes, which themselves can be part of more complex description schemes. These are formed by the description definition language (DDL), which is based on XML schema (and which is used as the basis for TV-Anytime). The MPEG-7 description schemes are divided into a visual part, an audio part and the semantically rich multimedia description scheme (MDS). MPEG-7 descriptions can be coded using XML or in binary format called BiM (Binary format for MPEG-7). The latter may be integrated in the transport streams of other MPEG formats (i.e. MPEG-2, MPEG-4). The MPEG-7 standard consists of 11 parts. In 2002, parts 1 to 6 were published and since then several amendments to these parts have been made. The newest parts are 9: Profiles, 10: Schema Definitions, and 11: Profile Schemas, which were published in 2005.

MPEG-7 is formally named the "Multimedia Content Description Interface", and it is intended to describe multimedia content data in a way that enables the meaning of the content data to be translated to other computers. MPEG-7 is not bound to MPEG-4, but was intended to be reused where possible (in practice, more specialized standards are used instead). MPEG-7 is intended to be both human-readable and computer-usable. MPEG-7 consists of a tool set, which is used to create a description that is based on a schema and a set of individual descriptors. An application can then apply these descriptor examples to use the system, for example to filter out, browse and search for content (searching in a content description, which is textual, is much easier than searching in a multimedia description, which may be structured but where the structure cannot easily be understood by a computer). Filtering and querying XML content is well known, and can be applied to MPEG-7 descriptions.

The work on MPEG-7 was started in 1996 to enable the creation of a metadata model, which could cover both still pictures, video, graphics, 3D models, audio, speech, and basically anything else that can be recorded and replayed using audiovisual equipment. The final amendments were published in 2004, and there is no further work to be done on MPEG-7. The format is flexible enough to cover everything audiovisual, as well as purely visual formats. It is possible to create an MPEG-7 description of a silent movie, or a printed newspaper.

There can be different granularities in the MPEG-7 description; the easiest way is to think about the content representation in terms of MPEG-4 encoding, which takes the elementary streams (representing each object, e.g., the newscaster being one object, and the background being another) and creates compound objects (e.g., newscaster in studio with audio). MPEG-7 descriptions can apply to both compound objects and objects within a scene. The highest level of representation is the semantic description, which would be something like "newscaster in studio". Lower levels can then be constructed ("person in suit behind desk talking in front of synthesized map"; "moving face attached to still body with wool and cotton texture and red tie behind wooden plane with papers on it, with synthetic video representing color fields and arrows in the background"). The descriptive features vary between different domains. The picture of a face represents different things in, for example, a video teaching cosmetologists and a video teaching plastic surgeons. However, when broken down, there is a level where commonalities occur, for example skin color, muscle, position in the picture, and so on. The lower the level of features, the more they are machine-readable; the higher levels require human interaction to understand them.

MPEG-7 provides a number of description tools that consist of descriptors, which define the syntax and semantics of each of the features; and description schemes, which specify the structure and semantics of the relationships between the components, which may be both descriptors and description schemes. The descriptors and description schemes are encapsulated in a description definition language, which makes the creation of new schemes possible, as well as modifications of existing schemes. This was based on the XML schema language, with some MPEG-7 specific extensions. There are also systems tools defined, which help in creating binary-encoded representations for storage and transmission (which XML in itself does not), as well as the multiplexing of the description, synchronization of descriptions and content, etc.

A MPEG-7 description is an instantiation of the description scheme and the descriptors. The descriptions of content can include information about the content (and its production), such as a list of actors, director, title, etc.; there can be information about the usage (copyright information, but also broadcast schedule); storage features of the content (such as format and encoding); and structural information about the spatial, temporal or spatio-temporal components of the content (i.e. cuts between scenes, segmentation in regions of the scene, motion tracking in the scene).

Low-level features such as color, context, and sound equivalents such as timbre and melody description are also standardized by MPEG (at least in the basics), and so are audio descriptors; and conceptual information related to the object relations (objects, events, interaction among objects); as well as information about object collections, and grouping about objects to enable the efficient browsing and searching. Each MPEG-7 description has a tight association with the media objects themselves (i.e. the objects in the content). Normally the description is stored together with the content, but if not, it is addressed using XML URIs. There are a few generic features common in all media, for example, vectors, time information, textual descriptions, etc. When these are not sufficient (and that is pretty often), there is a way to create descriptions of more than one medium. These include the representations of the perceivable information (content description); the management of the content (media features and usage); organization (the analysis and classification of the content); and navigation, access and user interaction information for the content.

There is also a way to describe production capabilities in terms of camera motion. The camera operations are well defined (see Figure 9-6), and the different frames and scenes can be broken down in terms of the movement of the camera, and translated into camera motion descriptors. Each descriptor set contains the start time, duration, speed of the motion of the image, and where the image is panned or zoomed (focus of expansion or focus of contraction). Mixed camera movements can also be expressed.

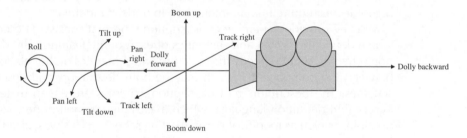

Figure 9-6. Camera metadata in MPEG-4.

MPEG-7 descriptions can be written manually, but that is not the intention. A scene in MPEG-4 makes it easy to extract the objects, and the features can then be mapped to the objects automatically, keeping track of which object represents which actress, for example, and mapping the appropriate features to it, and also to use the scene descriptions as metadata, as shown in Figure 9-7.

As it is an XML document, the MPEG-7 description begins with a root element, and this describes whether this is a complete or partial description. Different descriptions can be embedded into each other, so an object description can be complete, but only consist of a partial scene description. Top-level elements below the root then describe the different elements in terms of their purpose (e.g., particular types of descriptions). The fundamental elements, which relate to the object organization (time, vectors, etc.), come next. The linking to media files and objects is also done here.

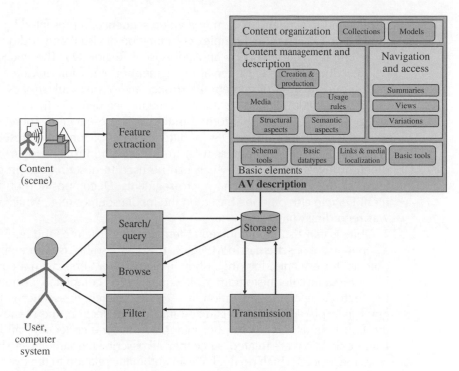

Figure 9-7. Abstract representation of possible applications using MPEG-7.

Description schemas can be organized as graphs or trees (the difference between a graph and a tree is that in the tree the structure is a way of expressing the relation of the elements to the top level, but in the graph the connections themselves have meaning, as well as the individual elements). Spatio-temporal relationships are typically described using tree structures. Graphs are more useful when searching among and within objects, as it typically is the analysis of their relation that enables you to retrieve and browse them.

The tree structures in MPEG-7 description schemas are represented in terms of segments, which represent a section of an audio-visual content item, and there are nine different subclasses that represent the individual types of metadata (Multimedia Segment DS, AudioVisual Region DS, AudioVisual Segment DS, Audio Segment DS, Still Region DS, Still Region 3D DS, Moving Region DS, Video Segment DS, and Ink Segment DS). These may have properties which represent both time and space. They can be combined to create higher level descriptions (e.g., several Still Region DSs may be combined to form a Still Region 3D DS, and this in turn could be combined with an Audio Segment DS to add music). A spatial segment may be a region in an image or a frame in a video sequence, represented by a Still Region DS for 2D regions and a Still Region 3D DS for 3D regions. A spatio-temporal segment may correspond to a moving region in a video sequence, represented by a Moving Region DS or a more complex combination of visual and audio content, for example, represented by an AudioVisual Region DS. The Ink Segment DS describes a temporal interval or segment of electronic ink data, which corresponds to a set of content ink strokes and/or meta ink strokes. Finally, the most generic segment is the Multimedia Segment DS that describes a composite of segments that form a multimedia presentation. There are also several types of specialized DSs for different types of content representation.

As the Segment DS represents a tree, it can be broken down into the different levels of the tree, which can be used to describe the program in terms of the scenes, shot, and micro-segments. Decomposition of the Segment DS can also be done based on the media sources (e.g., which camera was recording what).

Objects can also be organized in terms of the structure, and here there is a separate set of schemas in MPEG-7. This can also be used to analyze the objects, for example, looking at how many different objects in a collection have the same color histogram. Collections can be completely orthogonal to the spatio-temporal organization. In a football match, the scenes with more grass than players can be organized into one collection, the scenes where the ball is in the center can be organized into one collection, and so on. These collections are related, since they all describe the same event, but they describe aspects which do not have an automatic relation to the event. There is also a description schema for analyzing probabilities and how collections can provide models for other collections.

Figure 9-8 demonstrates the breakdown of an object: in this case, it is a Still Region object decomposed into its different components and the textual annotation, as well as the media information and usage information. The segmentation here is done in the different regions, each with its own associated color histogram and other specifics; the objects in the tree inherit the usage, creation and other such overall information from their parents higher up in the branches (i.e., these values are only set at the root level, unless they change – e.g., when combining two objects).

The graph description is where it becomes really interesting, especially when extending the description in time (see Figure 9-9). There are a number of objects involved, structured as Still and Moving regions: these then reuse different elements between them; and the relations of the different regions over time.

MPEG-7 contains means to provide summaries of descriptions; these describe the organization of the main features, and the highlights of these main features (as key frames, key audio clips, and textual annotations). One way of looking at this is as a tree structure, with the actual video and audio data at the bottom, and the subsequent highlights and their summaries building up to the summary of the entire movie. This is then organized in a document schema describing the media object(s). Summaries not only can be done based on content, but it is also possible to select a slice of the video at a certain time; or a frequency band within an audio.

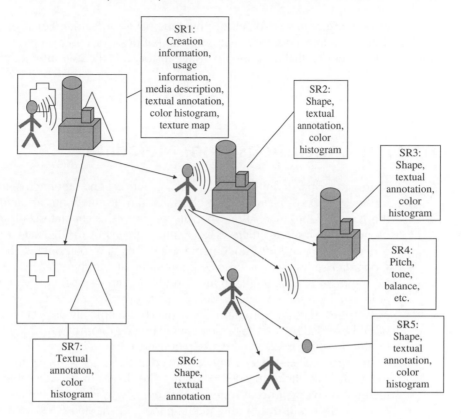

Figure 9-8. Examples of image description with Still Regions.

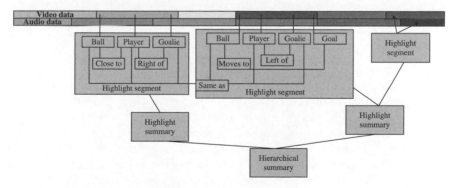

Figure 9-9. Example of a segment relationship graph.

A video can have several different sound tracks, Japanese and French, for example (with or without subtitles). These two are variations of the same content, and there are many other variations possible (where a variant ceases to be a variant and becomes a new object is something very hard for computers to determine; this is something humans have to decide). For user navigation and interaction, there is a schema that describes user preferences, and other schemas that describe the usage history and the user interaction of the content.

Drawing Conclusions from Metadata

There are formal definitions for authors, producers and other actors involved in the creation of a program. Nobody would dispute what an actor is, or the role of a producer. But when it comes to the characterization of the programs themselves, there is very little agreement on the categorization used. Is *American Idol* a documentary, a drama or a soap opera? Should the *Godfather* series of films be classified under "Italian heritage" or "Crime"? Content can be characterized along multiple dimensions, however, there is at least within some groups a consensus about what characterization(s) are most appropriate for a program. These may be completely different from what would be used by a professional characterizing the program, but they may be the ones that matter to the group of users.

The first website that really leveraged the user's creativity in terms of metadata was Flickr, the photo-sharing site. Here, users were able to create their own categories. While this was not new, combining it with a simple user interface and machine-readable encoding enabled Flickr to appear to be able to fetch pictures easily, and also to become a social network even before such networks became popular.

This is where user creation of metadata comes in. Metadata has traditionally been created by professionals (or automatically based on programs written by professionals), such as librarians and catalogers. In the library sciences, there is a set of very elaborate rules and schemas for how cataloging, categorization and classification of content should be done. The most famous of these formats is the Dewey Decimal System, which is a schema that enables books in libraries to be classified in categories such as "history", "biography", and so on. Schemas in the traditional library sciences use a controlled vocabulary, for example, the Library of Congress Subject Headings.

Professionally created metadata takes time and effort to produce, and while the categorizations may satisfy professionals, end-users frequently see things differently. The professionally created metadata also does not scale well – it assumes that there is a professional available to classify the content, or that the software used for the classification can assess the categorizations automatically – which may not always be easy. When people have the power to create their own metadata, they create completely different categories, and assess content in different ways.

In the "Web 2.0" debate, the term "folksonomy" has been spread around; this is a popular description without formal definition (defined, as it were, by the object of its expression). This makes it very appropriate to describe a program or scene, but it becomes very hard to draw machine conclusions from it. This is not the case with ontologies, which is a formalized way of structuring descriptions. Ontologies, which are structured vocabularies with logically formal definitions, are used in many different areas because they can be used to characterize objects and then use software to draw automatic conclusions based on the descriptions (the conclusion mechanism is based on the constraints that define the objects formally). Ontologies are usually represented in languages based on the RDF format, which expresses the relations between objects as graphs; this means they can reasonably easily be transformed to MPEG-7 descriptors – and the other way around.

Folksonomies emerged in photo-sharing sites, first in Flickr (now owned by Yahoo!), and then in Picasa (owned by Google). The images in Flickr, or the content of an audio program, can be characterized as a self-contained narrative world, inhabited by agents, objects, events, places and times. Flickr is actually more of a community where the users generate the content. Picasa supports free-text annotation as well as metadata according to the IPTC standard. It helps the user to produce slide shows, photo books and web pages by offering additional tools for image manipulation, image enhancement and layout as well as upload functionality to photo finishers and online communities in order to publish the results. There are also more advanced tools available, enabling contextual annotation when taking photos from mobile phones, including the location and other relevant parameters from the user – including whatever folksonomies they choose to add.

This description can be encapsulated in an MPEG-7 Description Schema (DS), the SemanticBaseDS. It relates the objects in a scene to semantic descriptors (a moving object can variously be characterized as "running man", "player offside", or "hero of Inter Milan in full stride"). The object can be abstracted ("Zlatan Ibrahimovic" being generalized to "football player"). These objects can be described in terms of the events that they experience, for example, "ball passing and missing kick". These events can be generalized into terms of properties ("forward"). Even abstractions can be characterized as MPEG-7 DSs, which can be generalizations of events, or specific types of an instantiation of a sequence of events ("football match").

YouTube categories such as "Arts and Animation", "Sports", etc. must be manually entered before the upload of audiovisual content. The portal provides search and retrieval on a clip basis, in which user preferences, for example frequency of viewing, are directly used to generate recommendations. These categories are really folksonomies, because they have been haphazardly created (although, true to its roots as a company managing unstructured databases, Google also enables the categorization to work on the content of comments and the video clips themselves). However large YouTube has become, it is still a fairly unordered collection of video clips, and these are characterized with just a few types of metadata: the author, the length and the category.

The categorizations in YouTube give an ad-hoc feel, and since there is no tagging function in YouTube, the characterizations can be somewhat haphazard. On the other hand, that makes for surprises for the user, and also makes them stay longer on the site, which may be an end in itself for a company like Google, which makes its living from advertisements (even if YouTube is so far advertising free). The categories in YouTube are:

- Autos & Vehicles
- Comedy
- Entertainment
- Film & Animation
- Howto & Style
- Music
- News & Politics
- People & Blogs
- Pets & Animals
- Sports
- Travel & Events

For anyone who wants to create a site for user-created content of their own (and there is a big difference between user-provided and user-created, as we will discuss later), the more appropriate for the target group, the better. In the case of YouTube, the target group is too wide to be easily characterized. And there are several other problems with the tagging in YouTube. For example, it is scoped only to describe the video as a whole, not the individual scenes in it. This is fine if you do not expect any reuse of individual segments of a video, but if individual video segments are to be reused, the informal tagging approach does not work, unless the tags are normalized to a formal vocabulary. Furthermore, the tagging is discreet in time, attached to the sequence only, but a television show works because there are rises and falls in tension between the individuals in the segments. If the segments can be marked up with finer granularity, you could use the tagging system to put in tags that show the tension; then the system would have to keep track of the change in the information, rather than the individual tags.

Many television stations are already experimenting with user submissions of content and are creating programs from them; automating this mechanism (similar to the "channels" on YouTube) is rather easy – if you have the right metadata. However, making users fill in forms and provide data is tedious, and a sure way of putting people off (unless the precision is part of the user experience. It is possible to draw conclusions about the user from the data provided by the user himself, or by the service provider about the user. However, to make sure the conclusions aree correct, the user should be able to sign off on the data. The more this can be automated, the better.

The informal type of "folksonomy" tagging used on YouTube and Flickr works when the content is not intended for real-time consumption – then speed becomes of the essence. A restricted vocabulary, which is selected from a menu, can be used to solve this (and can in itself be an interactive application for end-users); but the restricted vocabulary must be able to compare with other data types, as mentioned previously. Either all the different data formats are standardized to work together (something which is possible, but likely only in a very large organization); or they are created in a format that documents not only what the data is, but also what it means.

On the web, metadata means adding tags to the HTML. These can be generated automatically (the author and other information often is), and they can be used by other applications (like RSS). There are a large number of ways by which this data can be created and used: some systems try to derive annotations from the low-level features detected in the raw media data; other systems try to analyze the different modalities of a video, the usage context of the media or rely on human annotation and interpretation. One way to do this is to combine low-level features with the background knowledge available on the web or other sources. Some approaches allow several interpretations described in the metadata to be attached to the content (such as music service LastFM), others attach descriptions with concepts using ontologies. One application in particular, Rich News, provides automatic annotation and extraction of semantics from news videos: in a first step, the system extracts text from speech and then it tries to extract the most important topics from that. With these extracted topics, the system starts a Google search to find news stories on the web that cover the same topic(s). Others, including Google, have done the same. All these projects go back to the PLUM project at MIT in 1996, which tried to provide background for news stories (e.g., if there is a flood in Kenya, what does it mean in terms that would be comprehensible for people in Massachusetts?).

There can be many different sources for information. Knowledge about the content comes from archive information, from automated extraction (archival information may come from automated extraction, and then be annotated); users can put in content; there can be content created during the process of viewing the video and of managing it. Since there are many different models by which content can be represented, the system has to be able to manage them.

When there is markup on a document describing what it is (automatically generated or created by a person), it is possible to compare it to the input from the profile system (and the information about what the user is doing), and create a program that is adapted to the individual. This works, but there are a few hurdles. One is getting the presence and profile information (see Chapter 8). Another is making the comparison with that data (discussed briefly in Chapter 3). These are big hurdles, but there are standards for solving them.

The major hurdle, when it comes to using the metadata annotations of a television show, is to get all the descriptions into a usable format.

TV-Anytime was originally developed as a set of tables to go into a database management system, but later was changed to XML. MPEG-7 was created using XML, which is also a very database-friendly system. The profile information in IMS is also documented in XML. So it should be possible to lift all the element names into the columns, and the element names into the rows, and start using the data as input to a program, right?

Not quite. Even if you merge all the data from these different XML document types, you still have a major problem (apart from the minor problems of datatypes – sometimes, the same things are described using numbers in one format, text in another): interpreting the meaning of the tag. The first part of this paragraph is about making content machine-readable; the interpretation of meaning is about making it machine-understandable, so that it can be reused by other programs. In the process, the metadata is turned from "data" into "knowledge".

The LIVE project (see Chapter 1) addressed a number of these questions. Here, the major problem was to associate the different sources of information with an event, and then to align the different types of information with each other. This they turned into eight requirements, which will be true for any system that tries to build programs automatically.

1. Knowledge about the program must be encoded using a formal language. If knowledge is to be machine-understandable, then it has to be formally encoded – tags and folksonomies are not enough.

2. Separation of the knowledge from the application. This is true for all reuse of knowledge, because the tight bond might make it more usable in certain contexts, however, there will be a major problem in the case of information that is independent of the application (such as the video editing system), yet has to be reused. In particular, this is true for copyright information and information about the media coding; while the latter is easily encoded in a formal way, this is certainly not true of the former. Standardizing on an application does not even work in the corporate environment (Microsoft is no longer dominant), and it will certainly not work in the multimedia domain (neither Adobe, Apple nor any of the major players can provide a complete suite of tools).

3. Separation of the knowledge from the content. This may be less intuitive; but the same knowledge can refer to many different media objects (e.g., there are many examples of scenes that can be described with the term "car chase"). There is a many-to-many relationship between knowledge and media objects, and this has to be supported by the system.

4. Different perspectives create different knowledge representations, which have to be coordinated. Production-related aspects (e.g., camera angles, lighting) are different from distribution aspects (e.g., compression), and

different again from genre, content, and so on. These may be different in other dimensions as well: the content may be about a test for banned substances in sports, but is it then about banned substances or sports? This depends on the perspective of the viewer – and the categorizer. They have to be reconciled.

5. Content description systems are not coherent, and mergers of knowledge models have to be managed. Different models (depending on perspective, to some degree) support drawing different conclusions about the content, but when merged, they may lead to an entirely different set of conclusions. So it must be possible to merge different content models, which includes managing the relations between content models.

6. The easiest way to manage the fourth and fifth requirement is to normalize the descriptions to the standardized formats of the Semantic Web.

7. There will be a specific need for managing user interests and preferences, to create a system that can adapt to what the users want.

8. The content delivery has to be fast, so the content management has to be fast, too. Otherwise, the system will not be accepted by users.

If you look up "ontology" in Wikipedia you get a number of hits from philosophy, and that is where the idea comes from. It has been used in computer science since the 1980s, first in artificial intelligence research, then applied to database management. The field of knowledge engineering is more or less founded on defining and managing ontologies. A simple ontology (an example is a glossary) is just a controlled vocabulary. Add some more structure to the definition, and it can be used to translate meaning between systems, as a kind of lingua franca. Put in additional structure and formal definitions again, and you can do more than translate meaning – you can draw conclusions about your representations. It is important to remember that computers are stupid. You have to tell them everything, including what background information you share; people require much less formal definitions of the knowledge space to understand it, and to reason from it.

There is a hierarchy of ontologies based on the level of structure in the definition. The simplest is the controlled vocabulary, where the structure is just a list (anything that is not on the list is not part of the ontology). A thesaurus connects the terms through relations ("reading" in English means "läsning" in Swedish). The hierarchy can be modeled – one to one, or one to many ("läsning" can also mean "interpretation"). An informal taxonomy adds a hierarchy to the concepts. The hierarchy is explicit (an oak is a kind of tree, and trees are a kind of plant), but there are no rules for how properties are inherited between the levels. That is added in a formal taxonomy, where it becomes clear that all trees have branches, and the branches have leaves or needles, and so on. Framing is the next level: here a concept similar to classes is added, the frames have properties, and they can be instantiated (the oak tree in my back yard is of the tree family, has 39 branches, and is 6 meters tall). Frames can be embedded in frames, so the concept of framing is easy to add to the concept of taxonomies. However, this is where the human understanding of concepts starts to diverge from the conceptual frameworks. Frames can be mapped to human understanding of concepts, but there is no one-to-one mapping.

Adding value restrictions makes it possible not just to say what something is, but also what it is not. Logic constraints add a further level of restrictions, based on other properties. First-order logic constraint is based on formal logic, allowing more detailed relationships between terms such as disjoint-ness, inverse properties/relationships or part-whole relationship expressed first-order logic constraints.

There are three levels of ontologies:

- Top-level ontologies, which describe very general concepts such as space, time, matter, object, event, action, etc., which are independent of a particular problem or domain. Unified top-level ontologies have been the aim of the artificial ontology community, but this turns out to be a harder problem than many researchers believed. This has held up the development for a long time, since researchers have tried to find mathematically provable concepts for things which may appear the same, but are expressed differently in different languages.

- Domain ontologies and task ontologies describe the vocabulary related to a generic domain, and it is actually here that the concepts often grouped into the top-level ontologies belong. These vocabulary are related to a generic domain (such as medicine, or automobiles) or a generic task or activity (such as diagnosing or selling), by specializing the terms introduced in the top-level ontology.

- Application ontologies, where most of the ontologies needed to describe television programs belong, describe concepts that depend both on a particular domain and task, which are often specializations of the related ontologies. These concepts often correspond to roles played by domain entities while performing a certain activity.

While there has been a huge amount of research on upper-level ontologies, the practical applications have come in the application domain. In some ways, both the domain of television and the web are applications, so it may be suitable that it is here that most development has taken place; the web has become a hotspot for ontology development with the discovery that Web 2.0 was not enough, and something extra was needed. That extra is often called "semantic web" – that is, adding meaning to web pages.

The combination comes with mapping existing metadata models onto graph systems (such as mapping Dublin Core into RDF). The cornerstone of the ability to draw conclusions from data is ontologies, and the ontology formats used on the web are based on the Resource Description Framework (RDF). This, in turn, is a way of representing graphs in a text format.

These particular graphs are easy to understand – they are very similar to a family tree, with a network of branches connected at points where they can be combined, and the branches can intersect. A branch of mathematics called graph theory defines what it is possible to do with the network of branches – the graph – and how the operations can be performed. By mapping such graphs together, it is possible to draw conclusions about users and their behaviours, and combine this with other information in ways that the creators of the metadata probably did not foresee – and hence, could not program into the database system generating the packaging. Mapping metadata documents to graphs makes it possible to map the many different metadata formats for IPTV together.

An ontology is a graph applied to an area of knowledge. Just like XML, the ontology requires two things: a schema and a vocabulary. There has been an unfortunate tendency to equate schemas with vocabularies, but as in XML, this is not true. The schema is generic, the ontology vocabulary specific. The only difference is that in the ontology schema, you specify not only how the elements are constructed, but also how they relate to other elements – which is what enables conclusions about them. Graph structures is one such relationship. Web pages are actually full of relationships, both between the elements in them and outside, so this is not so far-fetched.

When adding multimedia, the result is somewhat more than just text; how to map the content onto the ontology needs to be spelled out (this is also true for automated web page analysis). Using a concept similar to framing, a content model is a structural definition for a "type" of object (e.g. a learning object, image, book). It defines what components an object (document) or a certain part of an object (a section) can contain. It may also contain a set of constraints on digital objects and a set of rules for creating these objects. In a narrower sense, a content model mainly describes the kind of content that a specific type of document may contain. Content containing information with explicit semantic descriptions of its properties is, in other words, a notation for semantically annotated and structured data – also called "intelligent content".

"Intelligent content models" can be seen as an integration architecture for metadata about content, allowing the combination of knowledge from different (and diverse) sources, and covering different properties of the content. There are a number of such models that have been developed in the past 10 years; most of them have not yet made it out of research, but they can easily be applied to help the system draw conclusions about the content. How easy it is to manipulate and re-purpose the content using this model is called the "smartness", and a "smart content package" consists of three main parts:

- The content itself (which may or may not be embedded in the same document as the metadata).

- The structural component (the "smart content"), which is described by the metadata. Here, the metadata is separated by the knowledge about the media objects, the knowledge about how the object is interpreted in the domain at hand; and the knowledge about how the content is normally used.

- The functional component, which contains interfaces to the environment in which external actors can act on the smart content objects. One way is to see these interfaces as query interfaces or interfaces for the system to act on the data (e.g., display it). Here is where the content can be instantiated, and changed (if required) based on contextual parameters such as terminal display size and so on.

The definitions diverge somewhat – this is still research. Whether the trust information (DRM metadata) is part of the first or second layer is a discussion that is not resolved (although copyright, being a universal concept across the media domain, ought to make a nice top-level ontology).

Once you know how to characterize the content, and how to draw conclusions about it, you need a way to get the user's input to the service, where the conclusions are drawn. One way is presence, but there are others. The important thing is that the conclusions can be sent to the service and that they do not get there too late – otherwise, only video on demand will be interesting, and new applications based on user interactions will be harder to monetize. One way is designing in appropriate time slots for interaction into the program; another is to make sure that the protocol used for communicating the metadata and the user interactions does not get in the way.

Chapter 10: Protocols for Interaction

Three protocols used for the signaling in IPTV systems are now being stand-ardized. One of them, HTTP, is also used to some extent for data transport. Contrary to what you may think, this is a bad idea. It makes it harder to optimize the part that is important – either the signaling (telling the other party what should happen); or the transport (making sure the data can be packaged in the best possible way, and the delivery assured and optimized).

The bigger picture: how protocols are used for signaling

In Chapter 1, we talked about how the communication between clients and servers (or peers) works on a high level. In any computer system, the format that computers send messages to each other is standardized, so that the formats (i.e., protocols) can be developed independently (having different implementations that interwork is one of the requirements for standards to be accepted). This standardized method is called a protocol.

There are three protocols which are particularly important for IPTV: SIP, HTTP and RTSP. The Session Initiation Protocol (SIP) is used in IMS – and also handles the crucial presence information, and enables the interactivity communication.

> **(Continued)**
>
> *The Real Time Streaming Protocol (RTSP) is a control protocol for "remote video players" – in other words, video on demand. The HyperText Transfer Protocol (HTTP) is a way to send documents, such as metadata, around the network.*
>
> *These protocols, especially SIP and HTTP, suffer from "feature creep", however. They have been extended, and the methods, which were defined into the protocols, misused (at least compared to the original intentions) to create new services and ways of doing old services. In this chapter, we will look into how these three protocols are designed, and what they contain that can be used with IPTV.*

In Chapter 11, we will look at the transport encoding and protocols, since these go hand in hand. In this chapter, we will look at the signaling aspects of the protocols. The issue with signaling is how the connection is set up, how authentication and authorization are handled, how actions are handled (e.g., switching channels), and how disconnection is handled. This means the protocols either have to contain ways to do all of these, or they have to have ways of hooking them in.

All IPTV systems work in roughly the same way, irrespective of whether they are standardized or not, and they all use these protocols – although in different ways. We looked at this in Chapters 2 and 3, but the following is a summary from a signaling perspective. In the signaling view, the media is an opaque channel, and what happens inside the media session (that there are reports about how the media stream is transmitted, etc.) is essentially invisible to the signaling. Once the media channel is opened, it remains open until the user or server closes it. However, there is an exception: interaction.

In the traditional media view, the media channel is a black box from the signaling protocol point of view. If the user can do something during the media stream, signaling then becomes important. As we have defined interaction in this book – an action by the user which can change the media stream – this can be done in two different ways: the first way is that the media stream can be interrupted by the signaling, and a new media stream can be started. This is done when the user switches channels, but it can also be done when advertising is inserted. In principle, this can be done in two places: in the client, or in the server.

If the media stream is terminated, and the new media stream is inserted in the client, this can mean that the media stream was paused while an advertisement was displayed, for example. This is not very complicated when it comes to advertising, but it means there must be enough memory and processing capacity in the client to handle this. It must also be able to either buffer the media stream while the advertisement is displayed, or it must be able to handle the signaling to the back end, making sure the media stream is paused there until it can be displayed again in the client. In all of these cases, there is no real signaling over the network. The remote control (or, in case of a PC, keyboard and mouse) sends its actions to the client device, and the client performs the interaction. This normally takes some skill to program, but once the application is loaded, nothing happens on the network (unless the application goes back to a server and fetches data). How the application is loaded is usually determined in the interaction standard, for example OCAP or MDP.

The second way is that there is a signal which goes to the back end, something happens to the media stream there, and it is propagated to the user – or users, as in the example of voting. Here, the media stream is still decoupled from the signaling (once it has started, the media stream runs without interruption, probably as a multicast stream, and the signaling happens in parallel). The signaling has to be synchronized with the media stream, even if there is no direct interaction with the media. There is a window in time during which the interaction can be effective (if users are able to vote out a participant in a show, it is no use sending the vote after the participant has disappeared). This means there has to be a mechanism to trigger the interaction, and a timer – invisible or visible on the screen – during which the user action is possible. As we went through earlier, the most effective triggering is to include it in the program, for example, have the announcer request interaction.

The HyperText Transfer Protocol

The HyperText Transfer Protocol (HTTP) was defined as a format to transport files. It is a generic format, and has been extremely influential (it was used as the template for both RTSP and SIP). It is tempting to use it for signaling in IPTV as well, but in practice it does not work well for that. There are two reasons HTTP in IPTV should be constrained to manage the EPG (in itself a very important function): it is not a good streaming format; and it is problematic when used to send back messages to the server.

HTTP is based on messaging, and it is a client–server protocol. The user (or client) triggers the message; the message contains the address of the resource to be addressed, and the instruction for what is to be done – the method. There were eight methods defined in the original HTTP protocol: OPTIONS, GET, HEAD, POST, PUT, DELETE, TRACE, and CONNECT. If these seem unfamiliar even to a programmer who works with HTTP, it is because there are two methods that are used to the exclusion of any other: GET and POST. GET contains the identifier of the resource as an argument, and was designed to retrieve content from that resource in a return message. POST, on the other hand, sends the message to an application, identified by the resource. The return message from the application is a confirmation – in both cases, a message containing the "200 OK" success code, and the applicable content of the resource. In the case of POST, this may not be anything beyond the success code. If it does not work, however, there is an error code that describes what kind of problem occurred.

POST assumes that the receiver is a server and that the resource addressed will take some action based on the content of the POST message, for example, process a form, which is delivered into the server, and return a result. When HTTP was created, the servers were huge computers and the clients weak, with small processing power. However, it is important to stress that the server as well as the client are just programs, and the actual server program does not have to be particularly big or complicated. There are servers that literally fit on the head of a pin, so the server does not have to be either complicated or large. It is perfectly feasible to implement an HTTP server in a limited computer such as a set-top box, but while the server does all this, it is cannot do anything alone; it depends on having programs to do the actual processing, databases to hold the data, and so on. So while the actual server can fit on the head of a pin, the complete server environment looks more like a laptop at least.

That is the extent of the signaling in "native" HTTP. However, there are two ways the protocol can be extended: with additional headers; and with additional commands in the body of the message. The latter is probably familiar to many developers, as it has been popularized using additional information in the message body in XML, under the name of SOAP (Simple Object Access Protocol). And, of course, this has become the foundation of REST, AJAX, and Web 2.0.

Another thing the enormous popularity of HTTP has meant is that developers are familiar with the client–server paradigm, where the client is used to receive and present the data, and the interaction with it takes place in the server application. This "thin client" model is very similar to the emerging model of interactive television, much more so than the "thick client" model of MHP and OCAP. The HTML4CE specification also contains a method to send messages to servers on the Internet directly.

While the HTTP model has been tremendously influential, not only through the hype and reality around Web 2.0, but also much earlier, as the foundation for the other protocols used for signaling for interactive television, RTSP and SIP. This paternity, by the way, is explicitly recognized by the developers of these protocols. HTTP could not do the job, in their view, because it has a number of shortcomings in terms of real-time interaction. The reason for the shortcoming of HTTP in this regard is actually its greatest strengths: the use of TCP, the caching mechanism, its statelessness, and the fact that it cannot handle real-time signaling.

The web did not get very old before developers started tweaking it. Tweaking the web means adding things to it, and in HTTP (as in HTML) this meant adding elements, or headers in the case of HTTP. As with HTML, added headers, which are not understood, are ignored. Something that quickly was added to HTTP was a way to handle state – which, unfortunately meant breaking the idea behind the protocol, in the guise of enhancing it.

The underlying problem is the idea behind HTTP, as expressed in the name: HyperText Transfer Protocol. The main difference between hypertext and television is that hypertext is a cloud of information objects, but a television program is based on a timeline. The statelessness of HTTP, that a request–response pair is history-less, comes from the fact that a hypertext information set is not related in time, only in terms of the declared relations between the objects. A request for one object does not mean that the user will request another, unless there is a link to the other object that has to be retrieved and used as part of the presentation (this can be an image, but there is no limit on fetching an object from other servers in HTTP – an XML document can consist entirely of documents retrieved from other servers). Once all of the related objects have been retrieved, the dialogue between the requestor and responder(s) is finished, and if the user wants another document, a new dialogue has to be started. There is no "memory" of the user's interactions. Nor is there, in the original HTTP, any security relation – each dialogue has to be authenticated again.

This was one of the first things where HTTP was tweaked. The mechanism to maintain state is called cookies, and is equally familiar to any web developer as the main part of HTTP. Cookies is an added header in the HTTP protocol, and it maintains the state in the server by setting a marker, which the client sends back to the server, and the server looks up the history of the client so it can be used in adapting the service delivered to the user.

There is a second way to maintain state in HTTP, which most developers do not think about as state: the TCP connection. The Transmission Control Protocol (TCP) is one of the foundations of the Internet. Often, it is mentioned together with the Internet Protocol (IP), which is used for addressing, in a way that could make you believe they cannot be separated. However, there is an alternative to TCP (actually, several), which is much more suited for streaming of data: the Universal Datagram Protocol (UDP). The addressing is the same, that is IP. In IPv6, which is used in some networks (e.g., the Japanese "New Generation Network", the new generation of the fiber optic network of the Japanese operator NTT), there are much better ways of handling streaming and flow control of data from the start; however, outside Japan and China, IPv6 is not very much used.

The major difference between TCP and UDP is that in TCP, the packets of the data stream are strung together, and the flow control is managed by TCP, not the application protocol. In UDP, the application protocol is responsible for ensuring that all the data packets are delivered, and that they are ordered in the right way (datagrams are actually bigger than packets). TCP sets up a connection between the endpoints, and makes sure that the data stream between the endpoints is handled in the fastest possible way. The problem is that when the stream is interrupted, TCP tries to establish the best possible connection by starting at the lowest speed, and increasing the transmission speed. This takes time and lost packets will lead to restarts. Latency, that is packets are delivered later than expected, will also result in delays since the packets have to be resent or reconstructed.

This is a big problem in media transport, where a lot of data is sent with the assumption that there is little data lost. However, for signaling, there are also problems which stem partly from the use of TCP, partly from the use of SOAP.

HTTP for IPTV Signaling

HTTP is used together with the other protocols we discuss in this chapter (RTSP and SIP) for what HTTP is good for: the retrieval of documents asynchronously. The signaling in these protocols is done in real time with quality of service guarantees, setting up and controlling the data stream; HTTP is used to retrieve documents (e.g., EPGs) where there is no expectation of an interaction happening instantly (or within a strictly given time window). There are three reasons why HTTP has problems handling real-time signaling. The first is the use of TCP; the second is the way messages are structured; and the third is the way HTTP is designed – minimizing traffic over the network by enabling retrieval of data from the nearest server where the data is stored. Intuitively, that would be expected to lead to less delay, but the problem is that when sending an interaction message, the message has to be sent back to the origin server. When HTTP is used for the signaling, the network has to be optimized to minimize the topological distance and latencies between the clients and the origin server, and that means the network operator has to be directly in charge of the infrastructure – which goes directly against the idea that connecting to the Internet and providing services can be done by anyone.

There are two other major problems in using HTTP for signaling. The first is that HTTP has become used for so many other things; the second is that the messages in HTTP tend to generate a lot of overhead – the proportion of signaling content to packaging tends to be low, because of the amount of header information and XML structural information required. Even if this does not matter in traditional HTTP signaling, where the dialog basically contains two messages (request and response), and the proportion of the argument data in the request is low in relation to the response, this matters more the greater the number of messages sent. While IPTV systems can be designed as a HTTP request–response dialogue, interaction with the stream is more probably going to be a continuous stream of status messages, much as in voice over IP (VoIP) systems. These messages will be sent, for example, when the user has viewed an advertisement; when the user changes channel; and when the user interacts with the service in other ways, such as voting.

Session management in HTTP is a problem in this regard, because it assumes that the state information is maintained only by the origin server, and that it is the origin server which initiates the state management (whereas in SIP and RTSP, state is initiated by the request from the originating peer). The cookie mechanism was standardized in RFC 2109 in 1997, and has not changed since, even though the requirements on state management have changed.

The reason is that HTTP state management assumes that the origin server sets the cookie in the client through the SET-COOKIE header in HTTP, and the client has to return a cookie header. This means an extra roundtrip to initiate the session, which may not be a problem if this is done only at the start of the session, and the session is durable. And this is exactly what is intended.

Sessions, in the context of HTTP sessions, as well as sessions in SIP, RTSP and IMS, are logical sessions, not persistent network connections. The logical sessions may, however, overlay one or more network connections.

Sessions, from the design context, have a beginning and an end; although these may stretch over several days, if not weeks, this is unlikely in the context of IPTV viewing, where it is reasonable to assume that the user starts viewing at some time during the day, and finishes when the day is over (i.e. viewing starts in the afternoon, when the children come home, and ends around midnight, when the parents go to bed – if the current behavior of users is reflected in IPTV). In HTTP, the session is assumed to be relatively short-lived (where "short" is not defined), and it can be terminated by either the client or the server. The state information, for example the user's history, is not explicitly exchanged between client and server – it is stored in the server and reused. This is another weakness that the IMS-based systems address: state information such as user history can be federated between different application servers using XDMS.

While it is possible both for intermediaries to insert additional cookies to maintain their own state (something which adds to the overhead by adding an extra header) and for cookies to be reused by intermediaries, it requires some major contortions to use the state to handle network issues, such as quality of service prioritization. Here, the popularity of HTTP is its damnation: it is impossible to know what a given HTTP message is used for only by looking at the address – you have to look at the content. And this is costly in two ways: it takes processing capacity and slows down the traffic – introducing latency, which is the biggest problem for IPTV applications.

An HTTP message consists of two parts: the head and the body. The body is where the HTML is carried, as well as XML. In SOAP, the body is used like an additional protocol to carry the semantics of the protocol in XML. There is no need to add header fields, which may be discarded by proxies and firewalls if they do not understand them; the XML body contains everything that is needed for the application to act on the carried data, including the calls for the API.

The original way to extend HTTP was with header fields, and that is done as additional standards outside the standard. There are several which have been standardized in different standards documents, such as the cookie headers. But this has become a problem because of firewalls and other intermediaries which interfere with the signaling. Since the HTTP specification expressly says that everything which is not understood should be ignored, it is doubtful if this is correct; but another problem is the very flexibility of SOAP: anything can be added, and anything has.

That problem might be overcome, for example by using a different port number. However, there is no different port number for IPTV applications using HTTP. They all use the standard IP port 80, or a completely unregulated non-standard port (which can be selected at will from the address space above 2316). The only way to know what is in the message, and whether it is addressed to an application that is allowed, is to open the message and analyze the XML inside. This is not as simple as it sounds, since all XML documents (except XHTML, which can be treated like HTML) have to be parsed completely before they can be analyzed. There is as yet no agreement in any standards body about what the standard XML for IPTV signaling should be like either, so anyone who introduces an XML-based IPTV system will have to be sure they have complete control over both clients and servers, to ensure they are able to introduce the right semantics.

A firewall, which tries to understand the content of an XML document, introduces latency. So does caching.

Caching in HTTP

While users have a legitimate interest to cache content so that there can be lower latencies in the access (and a better user experience), the advertisers have the opposite interest: they want to make sure the content is not viewed when they cannot control it. In HTTP, and protocols based on HTTP (such as RTSP), this is done by managing the caching system (which is a built-in feature of HTTP), and protocols derived from HTTP.

A response from an origin server can be cached if the request method (in practice, HTTP GET is used almost exclusively, although HTTP does allow for other methods), the header fields of the request, and the response status message all indicate that the response can be cached.

The expiration time of the document and the validation criteria for dates are not explicit commands from the origin server to the cache (which, again, might cause a problem for advertisers). However, HTTP allows for a method to control caches directly, the Cache-Control header. Both client and server can use it to transmit directives, either in requests or responses. These directives override the default caching algorithms. Cache control header information can be applied only to some of the fields in an HTTP request or response, by including a field-name parameter.

All caches in the chain of requests and responses from the originating client to the origin server must obey the cache control directives. These directives can change the behavior of caches that might interfere with the request or response, and override the default algorithm for caching. The directives are unidirectional, so a directive sent in the request does not automatically mean that the same directive has to be given in the response. Any intermediary caches must send them on to the server or client, however, regardless if they themselves act on them.

The cache control directives come in five general categories:

- Restrictions on what can be cached. These criteria can only be set by the origin server.

- Restrictions on what can be stored by a cache. These can be set either by the client, or by the server.

- Revalidation and reload controls. These can be set only by the client.

- Transformation control (i.e., whether it is allowed to transcode – change the format – of the content).

- Extensions to the caching system (anyone can extend HTTP, but in practice it is very rare that these extensions actually get implemented by more than the vendor who produced them).

If the cache should not be able to transform the content (e.g., adapt it to a mobile presentation), there is a no-transform directive which can be set in the cache-control header of HTTP. There may be various reasons for doing so, from artistic to charging, and coupled to the potential presentation formats. It may, for example, be more important to make a cache go back to the origin server if the file to be presented has been personalized in some way, and the personalization in the case of the mobile could be based on the user location. If the response contains a Vary field, it may be the result of a content selection (e.g., the language may be set to SE, which would give a Swedish response); in this case, it is possible to cache the response.

In HTTP, a content provider can set a duration on the time a file can be cached, and whether it should be renewed from the cache. How to manage these is set in the HTTP headers, and it is a built-in feature. Normally, caches are invisible and the caching is set automatically; but when the origin server (the first server from where the content comes) or the client requests it, the cache functions can be shut off or reporting switched on.

One important base for this is the requirement that the response from the cache should always be the most recent version of the file. The cache is required to serve the latest version, but there is also a way in HTTP to check if there is a later version (and this is what is used to shut off the caching). Checking that something is fresh implys having a method of verifying the date and time when the file was cached. That, in turn, means going back to the origin server – so if a cache is not able to communicate with the origin server, it is supposed to send back an error message. The same goes if there is an error on the side of the origin server, and it returns an old version of the file, or a message saying that the file is OK but which itself is old. This message should be forwarded to the client, since it is not actually the problem of the cache if the origin server has a problem; rather, the user should get an error message and can decide to watch it regardless of it being old (although this is not how most browsers are configured).

A cache should return a response to a client that is either the result of getting the content from another cache, or that is not fresh enough (meeting the least restrictive requirement on freshness from either the client, the origin server, or the cache; however, the origin server can specify that it should meet the freshness requirement of the origin server alone). If it cannot do this, it should return a warning message. There is a "warning" header in HTTP that is intended to be used for this purpose. Table 10-1 lists six different types of warning that a cache can send in this case.

Warning message number	Text	Conditions
110	Response is stale	Must be included whenever the response is not fresh
111	Revalidation failed	Must be included if a cache response is stale, and this is because it has not been able to reach the origin server so it could revalidate it
112	Disconnected operation	Should be included in case the cache has been disconnected from the network intentionally for some period of time
113	Heuristic expiration	Must be included if the cache has selected a freshness lifetime longer than 24 hours, and the age of response is more than 24 hours
199	Miscellaneous warning	Can contain free text, which can either be presented to a user or which is intended to be logged
214	Transformation applied	Any changes to the content encoding must be reported this way
299	Miscellaneous persistent warning	Same as 199

Table 10-1. Types of cache warning.

The 1xx warnings indicate the freshness or revalidation status; they should not be sent if there has been a successful revalidation. The 2xx warnings describe that there was some problem with the revalidation, or that some part of the document could not be revalidated. There can be multiple warnings attached to the same response, and they can be chained. It is also possible to provide warning messages in local languages (instead of English); although this is rarely done, it would probably be helpful to get "Vi ber om ursäkt för att programinformationen inte finns tillgänglig för ögonblicket (Sorry, the program guide is not available right now)" instead of a "404 File not found" if you are Swedish. The configuration of HTTP messages is done in the web server, and is something any reasonably competent webmaster can do.

There ought to be very few HTTP 1.0 caches left, since that standard was overridden by HTTP 1.1 more than 10 years ago; but if there are, they do not implement the Cache-Control header. It is unlikely that they are used for video distribution, however.

A cache can be commanded to handle the response (i.e. the message and the contents) in different ways by setting parameters in the Cache-Control header. There are a rather large number of these, some of which are used to control how long a message is allowed to be stored. Header fields in HTTP requests and responses are either end-to-end or hop-to-hop; the end-to-end headers are always transmitted from the client to the origin server (note that these can be caches which use this model to refresh themselves). Hop-by-hop headers are valid only for one single connection between two entities, and are not stored in proxies and caches. Normally, the headers are end-to-end, unless they are intended for connection and proxy management.

Two messages are especially important for VoD and for the EPG associated with IPTV (since it is likely to be distributed over HTTP): no-cache and private.

Setting the parameter "private" in the cache control header means that the response message is intended for one single user only, and must not be cached in a way that can be shared by other users. The origin server can in this way specify that a personal EPG, which has been generated as the result of a specific user's preferences and the current metadata parameters, cannot be cached – it will not be meaningful for other users. The opposite is "public", and it makes the content viewable by any user (or rather, possible to cache so that it can be viewed by other users).

If the content will need to be refreshed when the user wants to watch the content again (e.g., to change the content of the television guide), however, setting the "private" in the cache control header does not help. It only indicates the degree to which content can be shared. In that case, the parameter must be set to "no-cache". Here, you can specify a field name (so if you set "no-cache = body" in the cache-control header, it will not cache the body of the message, i.e. the personal television guide). No-store, on the other hand, applies to the entire response (or request) message, both header fields and body, and implies that a cache is not allowed to store any part of the request or the response. This directive was originally intended to make sure that information was not backed up and then inadvertently released. In cases where an advertiser, for example, is concerned that old advertisements may appear, this directive should be used, as it forces the cache to go to the origin server every time.

The content of a cache should not be used when it has gone stale. But how do you tell if that has happened – you can hardly open the lid of the server and smell it? When a HTTP response is delivered, the origin server sets an expiration date in the protocol headers, so that the cache knows if the response can be cached or not. The "cache-control" header can also be used for the control.

The originators of HTTP did not intend for the protocol to always have to go back to the origin server and validate content, even if this is what happens in practice. They intended the protocol to work the other way: to go to the cache as often as possible. To that end, they set an explicit expiration time on cached data, so that one response could be used to fulfill a number of future requests. Normally, the headers are used to set the expiration time, but HTTP caches can calculate such times themselves. If the expiration time has not expired, and if content is marked "public", then the cache will not have to go back to the origin server, and there will be less traffic in the network, especially if the document that is to be delivered does not change. If the expiration time on the cached data is set in the past, however, the content will be validated every time (since it will be expired per definition). Another way to achieve the same thing is to use the "cache-control" and set the parameter to "must-revalidate". This also means that the cache has to go back to the origin server. To make the cache go back and get the content after some time, the "cache-control" parameter max-age can be used. If the max-age is short, say five minutes, the cache will download the content every five minutes, instead of at longer intervals. Two other parameters can be used to the same effect: "only-if-cached", which means that the process will be applied only if the content has been stored in a cache in the path from the server to the client (if the content has not been cached before, there is no risk that it has become old on the way); the second way is to set the header to "must-revalidate", in which case the cache will automatically check if the file has been changed since it was fetched from the origin server ("proxy-revalidate" has the same meaning for any resource which can be shared with more users).

To make the expiration date calculation possible, the origin server has to send a Date header, and a time if required, with the response. The cache and the origin server are assumed to have their clocks automatically synchronized. To make sure that the delay from the request to the response is not too long, there is also an "age" value that can be set in the HTTP headers, which shows how old the response is in terms of when the response was sent from the origin server. The "age" header is the time that the response has taken from the reply, and this could potentially be used to compensate for latencies in the streaming (if it was consistent).

The cache does not have to download the entire file from the origin server again, if it has not changed. It is sufficient that it checks that the file has not changed, something it can for example do by requesting a check sum of the file (it is also possible to check byte ranges in the file, to see if the parts of it, which have been stored, have been changed). Validating the cache, instead of refreshing it, is faster since the entire file does not have to be downloaded (if it has not changed) – an advantage if the file is large, or if the connection is slow. This can be done using e.g. the HEAD method of HTTP, instead of a full GET. The response is 304 Not Modified, and it is then OK for the cache to go ahead and serve up the file to the client (as it is if the response is a 206 Partial Content). The validation request must include the reference to the file and other pertinent information (such as the size), plus the date and time at which it was downloaded (and the date and time of the request); a response can be the Last-Modified entry. Another way is to use the ETag response header, the entity tag, which can be used in place of actual dates and times (an entity tag for a resource has to be unique globally, and can be used for many other purposes than cache validation).

Video on Demand: RTSP

Video on demand was originally created to give users the same functions as video cassette recorders (VCRs) on the network. The protocol used to control the video on demand services, defined by the IETF, is called the Real-Time Streaming Protocol (RTSP). This protocol contains functions for starting, stopping, fast-forwarding, and all the other functions that are implemented in a VCR (except the blinking clock). The streaming is not done using RTSP, however, it is done using a separate protocol, the Realtime Transfer Protocol (RTP), which is intended to make the transport of data smooth.

A special case of video on demand is personal video recorders (PVRs), and especially networked personal video recorders (nPVRs). Here, the similarity with VCRs is even more pronounced. In a PVR, the user records content on the hard disc of the system, instead of a video tape. The content becomes available in the same way, but retrieval is much easier, and the storage much larger than in the case of the VCR. For the nPVR, the functions are moved into the network, and the recording takes place at a service provider's site. This can be done in various ways, but the important thing is that the users can call up the content they have recorded.

Peer-to-peer technologies, which is a portmanteau term for a number of similar technologies that are similar, aim at doing away with the server altogether. While both HTTP and RTP/RTSP are designed as client–server protocols, which require that there is one computer that does the reception of the stream and the presentation, and another which handles the storage and sending of the stream, peer-to-peer technologies do away with that. In P2P systems, there are no client or server computers. And while it is possible (today) for computers to be both clients and servers, they still have vastly different roles. Not so in peer-to-peer systems, which demand that all computers in the system have the same capabilities.

Another important feature of peer-to-peer systems is that the streaming and data management takes place in the background. It is not visible and exposed to the user; many people who use peer-to-peer software are not even aware that they are sharing their hard disc with other users. This is not strictly a feature of the peer-to-peer technology, but of how it is used by the file-sharing programs.

To view the media from peer-to-peer transmission, however, you either have to wait (in the case of download) until the file is received, or you can receive it as it streams to your computer. Streaming is faster, since it can be viewed when the first part of the media has been received; but you cannot use several different servers at the same time, as you can with download, downloading different segments at once and stopping the download when the segment is fully received. In streaming, you still have to have a one-to-one relationship with the server.

The Real Time Streaming Protocol (RTSP) is standardized in RFC 2326. While there are three authors, as is usually the case in the IETF (see the last chapter for more analysis about how standardization bodies work), the document was the result of a group effort.

RTSP is a control protocol (more about the separation of signaling and media in following chapters), and the RTSP packets do not contain the films, news, or other programs themselves – they are sent using a separate transport protocol. Like HTTP, RTSP is a client–server protocol, but both the client and server can issue requests, although their behaviors are assumed to be different. And there is no way to manage directories and other information about the media, only a way to handle the actual management and playout of the media – recording, viewing, and so on.

Like HTTP, RTSP also works with URIs (something that is true for SIP as well, as we will discuss in the next chapter). A Universal Resource Identifier (URI) has emerged as the most pervasive way of addressing any resource – a resource being a file, a stream, a program, or even a telephone number. Anything which can return a response can be a resource. A Universal Resource Locator (URL) references a definite location in a file system; a URI can be mapped to an abstract location, or it can be an absolute URI (which URLs are). By way of example, http://www.example.com is a URI, but http://www.example.com/home.html is a URL. They may point to the same resource (the page home.html), but in the first example, you can easily change the object referenced, it does not affect the link. However, if you replace home.html with homepage.html, anyone who has put the URL in its bookmarks will not find it. This matters, because the assumption behind RTSP is that you should use an XML page (or something similar) to find the object; the reference is done by way of an RTSP URI.

Another loan from HTTP, which is rarely used, but to very good effect, is the use of accept headers in RTSP. An accept header is a header field in the protocol (HTTP messages have two parts, the header and the body; all header fields are written in clear text, which makes analysis of how an implementation should work easy).

The separation between and the control and media, something which facilitates when there is lots of traffic, is not a part of other popular protocols – such as HTTP, which is used for data transfer in the DLNA (and in some IPTV solutions). The advantage of keeping them separate is that the signaling can be handled by different nodes than the media, and with a different quality of service. The separation in terms of port mapping also makes it easier to keep traffic apart in firewalls (a frequent complaint is that so much can be done with HTTP nowadays that it is impossible to keep out malicious data).

Another major difference is that HTTP is stateless, while RTSP (and SIP) are stateful. What that means is that when you click on a link to get a different HTML document in a web browser, the browser will go out and get that document – and then it forgets the server, and the server forgets the browser. History can be maintained in log files, but there are no long-lived connections between server and client (something, though, that was bolted on later on top of HTTP). When designing RTSP, it was made similar to HTTP, because similarity makes it easy for programmers to learn and use.

RTSP does not work by setting up connections, it works by setting up sessions (just like SIP). The difference between a connection and a session is that the transmission in a connection is continuous, while a session can be inactive for a long time, but still maintain a state in both the sender and receiver. Each session has an identifier, and the session can be moved from one IP address to another without being terminated – as in SIP, there is no one-to-one mapping between the RTSP session and the interface. Potentially, sessions could (also in the same way as in SIP) be forked, split and otherwise manipulated – for example, to enable several users to interact with the same event through RTSP (although while technically possible it may be somewhat hard to understand why anyone would want to do it).

The state in RTSP is defined per object, an object being the endpoint of the stream URI and the RTSP session identifier (which means it is not actually an object, but a particular example of an object). The RTSP operations result in the states which clients and servers can be in; the SETUP and TEARDOWN operations result in the creation and destruction of a session, respectively.

RTSP supports a number of operations, which are relevant for interacting with media services, especially video. It can be used to set up conferences between users, although this is rarely used nowadays, and was intended for video conferencing (which, while it exists on the Internet, never became as popular as its proponents hoped). It also supports splicing of other streams into an existing stream. However, it is its functions to support video playback that are of interest here.

The methods (i.e., the commands that clients and servers can issue) in RTSP will probably feel familiar, as described in Table 10-2 The response to the methods is a status code, and when appropriate also the resource, which is addressed with the URI. Not all of these are mandatory, and not all of them are used in IPTV.

RTSP can be extended, and it has also given rise to proprietary protocols, the most prominent of which is used in Flash streaming. However, those proprietary protocols have diverged from RTSP, and they require proprietary clients to be interoperable. The reason why standards will always win is discussed in more detail in Chapter 13; it is sufficient to notice that however good and efficient they are, there is always a bigger world outside – even if you are in 98 % of the computers on the web, there are more television sets than personal computers.

Since the commands that RTSP feature are easily mapped to familiar video operations, it is very easy to map them to the remote control, and hence to the interactions of the user. What is more difficult is to map RTSP to an authorization and authentication framework, because this is outside the protocol. Nor does QoS automatically lend itself to RTSP, because the media handling is done in other protocols –the Realtime Transport Protocol (RTP) is pointed out as the favored transport.

The most important RTSP methods in an IPTV context are SETUP, DESCRIBE, PLAY, PAUSE, RECORD and TEARDOWN. Only PLAY, PAUSE and RECORD are visible in user interactions, however.

Method	Direction	Object
DESCRIBE	Client to server	Presentation, stream
ANNOUNCE	Client to server, server to client	Presentation, stream
GET_PARAMETER	Client to server, server to client	Presentation, stream
OPTIONS	Client to server, server to client	Presentation, stream
PAUSE	Client to server	Presentation, stream
PLAY	Client to server	Presentation, stream
RECORD	Client to server	Presentation, stream
REDIRECT	Server to client	Presentation, stream
SETUP	Client to server	Stream
SET_PARAMETER	Client to server, server to client	Presentation, stream
TEARDOWN	Client to server	Presentation, stream

Table 10-2. The commands of RTSP.

SETUP is used to specify the transport mechanism that will be used for the streaming. A client can issue a SETUP request for a stream that is already playing in order to change transport parameters. Since SETUP includes all transport initialization information, it can be used by firewalls, proxies and other intermediate network devices which need this information. The Transport header specifies the transport parameters, which the client can use for the data transmission. The response to a SETUP command will contain the transport parameters selected by the server. The server also creates the session identifier when it gets a SETUP request. If there already is a session identifier in the SETUP request, the server will bundle the SETUP request into the existing session or return an error message.

DESCRIBE gets the description of a presentation, i.e., a media object, as identified by the request URI, from a server. It may use Accept headers to specify the description formats that the client understands. The server responds with a description of the requested resource. When the DESCRIBE response is received, the RTSP media initialization is completed. DESCRIBE responses must contain all the media initialization information for the resource it describes. But it is not the media initialization itself. Media initialization can also be done via other protocols, or through the command line or other standard input (i.e. through a helper application, probably the media client itself, launched through the use of RTSP).

The PLAY method is used to start sending data to the client using the mechanisms specified in the SETUP response. PLAY cannot be sent until the SETUP is complete.

When PLAY is sent, the point where to begin playing and where to end is specified in header fields as a byte range (which means that you can use PLAY to go backwards and forwards into the stream – but not to go backwards). If the media has been paused, sending a PLAY command starts the playout again. There can also be a time set in the PLAY command. There is no fast-forward or replay command in RTSP. Instead, this is done by setting p arameters on the PLAY command. The same parameters apply to RECORD. There are two parameters which can be used together for this purpose: scale and speed.

Scale is used to indicate the viewing (or recording) rate. 1 is the normal rate of viewing a file going forward at the rate of 60 seconds per minute. To speed up the viewing, the value is increased – 2 means a viewing rate of two seconds of content per second of real time. This can in theory be increased to infinity, but there is a question about what will be supported by the client and server. Both fast-forward and back can be implemented by displaying only the key frames (or only some of the key frames). A server does not have to implement any possible range of values; it can support only the most usual ones (double, quadruple speed), so the speed will be part of the response – as will the part of the video (the time range) to which the command is applied. Scaling up means going forward faster, and a negative scale, of course, means going backward in exactly the same way (–2 is going backwards at double speed).

Confusingly, there is a speed parameter in RTSP, which also applies to the PLAY command, but it does not affect the viewing rate. The speed parameter sets the delivery speed of the stream. This is typically the bit rate of the stream, but can be any value, for example, when the stream is to be delivered for previewing at lower resolution, or into a cache, not for viewing. Bandwidth set in this way may be constrained by the bandwidth set by the network – either implicitly or through the quality of service – so setting the speed may conflict with the actual possible bandwidth.

RECORD is used to start a recording – but not in the client, even if it has recording capabilities. The recording is done in the server (and can be done by setting a time stamp in a user configuration file, which can be applied to the file at any time – unless the file is streamed in real time, in which case it has to be recorded, of course). The recording can be set in advance to happen at a time during the session (e.g., in the case of a video conference), or it can be given during the session, which means the recording starts immediately.

PAUSE halts the streaming from the server temporarily (it is up to the client whether this means that the video will be displayed where it was stopped, and for what duration of time). If there is a specific URI included in the request, the specific video (or audio) will be stopped – otherwise, the entire session will be stopped. Pausing the audio track only means the same as muting the audio, as only the audio will be halted. Synchronization will be maintained when the playout starts again. A specific range value can also be used to stop the stream or presentation at a certain point (a "pause point").

When the viewing stops, the session has to be taken away – either implicitly, through the client disappearing and being timed out, or explicitly, through switching off. This is done with the TEARDOWN command, which stops any playout of the session, and closes all resources associated with it.

RTSP does not only interact with RTP, but also interacts with HTTP. The assumption is that the RTSP URI, which is used to start the streaming (from a browser or similar application), should have been delivered with HTTP. All media streams, which can be controlled by RTSP, have their own RTSP URI; there may, however, be several media streams behind the same URI (and they may be selected depending on parameters in the request, just as HTTP header fields can – at least in theory – be used to select variants of a web page). It is also possible – just as in HTTP – to separate the storage of the media between different servers.

RTSP can handle unicast as well as multicast. In the unicast case, the media is transmitted directly to the IP address where the RTSP request originated, and to the same port number that the request came from. When data in an RTSP session are multicast, it can be done in two ways: the server selects the address; and the client does. The first is the more typical case, and is used for live or near-on-demand transmissions. When a stream is set up with RTSP, it can continue even if the control connection goes away. The state of the stream is maintained in the server.

RTSP is a request–response protocol, just like HTTP, and the requests result in status messages which can returns a confirmation, an error message, or some other information (e.g., a condition for the transmission). The same status codes are used to a large extent, but there are several status codes that are unique to the streaming of data that RTSP does. Another inheritance from HTTP is its ability to handle caching, which means that intermediary servers, which hold a copy of the stream, can be used to deliver data instead of the origin server, saving bandwidth and speeding up the delivery, since the client only has to go to the closest server. Caches, however, are rather less useful when the data transmission is real time, such as in a video conference or in an event, like a football game.

When setting up an RTSP session, the client (who is normally the party requesting a transmission, although a server can also make requests) requests a file, and gets back a description. This description is a mix of formats: the request itself is done using HTTP, returning an SDP file (normally used to describe SIP sessions), which give a media description containing descriptions of the presentation and all its streams, including the available codecs, dynamic RTP payload types, the protocol stack, and content information such as language or copyright restrictions. It may also give an indication about the timeline of the movie.

RTSP URIs can reference more than one resource; in effect, the audio and video track can be different files (and the subtitles a third). The client can synchronize them using the methods in RTP. Presentations can also be packaged in a container file, which sets a common context for a group of files (or streams). Container files can also be used to set conditions, for example, that you are not allowed to play one file before another. The container is described in SDP.

RTSP is a request–response protocol. The client asks to get a video (using the commands we have just discussed), the server delivers the video as a stream. But if for some reason it cannot deliver, the server has to respond. Response messages will contain status codes, which tell the client what the problem is (the no problem status code, just as in HTTP, is 200 OK).

The status code consists of three digits, and a message, which is optional, but helps the receiver to understand what the response message is about. The first digit defines the class of the response, the last two what the message is about. Borrowed from HTTP into RTSP as well as SIP, they are the same for all the different protocols used in IPTV. There are five types of status messages possible:

- 1xx: Informational – Request received, continuing process.

- 2xx: Success – The request has been accepted and understood by the server.

- 3xx: Redirection – Further action must be taken in order to complete the request

- 4xx: Client Error – The request contains bad syntax or cannot be fulfilled

- 5xx: Server Error – There was a problem at the server end, even though the request was valid.

RTSP reuses the status codes from HTTP, and adds RTSP-specific status codes starting at x50. This is to avoid conflicts with status codes in HTTP. The status codes are shown in Table 10-3

Several status codes – for example 304 Not Modified and 407 Proxy Authentication Required – are borrowed from HTTP, and were originally intended for file transfer, not for streaming. They are status codes that reflect the status of the cache. In HTTP, a file can be stored by an intermediary server, for example, inside the home network instead of halfway around the world, and this server has to be able to verify that the file has not been changed since it was downloaded to the intermediary. If it is not changed, it should not be downloaded every time, because the this would take time and waste bandwidth for other users.

Caching is a simple and efficient way to speed up the use of the network for users, and as long as disc space is cheaper than network bandwidth, it is a better option for the user. "Cheaper" is usually counted in terms of money, but this is not necessarily the relevant measure; an equally, if not more important, measure in the IPTV world is latency – that there are no delays in the streaming. And since the more network nodes there are between the origin server (where the streaming or download started from) and the client, the more likely there will be latencies and hence delays.

Standardized almost 10 years ago, RTSP has stood the test of time. It does indeed manage to fulfill its goal: a remote control for the network, but it does not enable more than the simplest options. For more advanced operations, as for example found in the BSkyB set-top boxes, you need a combination of protocols, and a way to program or script interactions.

Status-Code	Function
100	Continue
200	OK
201	Created
250	Low on storage space
300	Multiple choices
301	Moved permanently
302	Moved temporarily
303	See other
304	Not modified
305	Use proxy
400	Bad request
401	Unauthorized
402	Payment required
403	Forbidden
404	Not found
405	Method not allowed
406	Not acceptable
407	Proxy authentication required
408	Request time-out
410	Gone
411	Length required
412	Precondition failed
413	Request entity too large
414	Request-URI too large
415	Unsupported media type
451	Parameter not understood

Table 10-3. RTSP status codes.

Status-Code	Function
452	Conference not found
453	Not enough bandwidth
454	Session not found
455	Method not valid in this state
456	Header field not valid for resource
457	Invalid range
458	Parameter is read-only
459	Aggregate operation not allowed
460	Only aggregate operation allowed
461	Unsupported transport
462	Destination unreachable
500	Internal server error
501	Not implemented
502	Bad bateway
503	Service unavailable
504	Gateway time-out
505	RTSP version not supported
551	Option not supported

Table 10-3. (Continued).

SIP for IPTV Signaling

The Session Initiation Protocol (SIP) was created specifically to address the issues that were out of scope for HTTP, and to be more generic than RTSP. Two major issues, state management and connection management, were built into the protocol from the start. However, SIP has shared the same fate of many other successful standards: feature creep. In this case, the major items added are event management, device capabilities, and a number of other mechanisms, like file transfer and short messages, which do not necessarily conform to the original intent of the protocol. Like all other IETF standards, the extensions are made by additional RFCs, which create additional features within the framework of the protocol. The combinations have been formalized by

3GPP, the 3rd Generation Partnership Project, which created the standard for how applications work in the network. Therefore, SIP is not only about session setup and management, even though these are the reasons why the protocol was created.

HTTP is a jack of all trades, but an ace at none. When you use HTTP for signaling, the control information is not embedded in the headers of the protocol, like SIP, instead, it is included in the body of the protocol, where web pages and other documents normally are carried. When interpreting a message and deciding what actions need to be taken, it is more efficient for the receiver to read fewer instructions (and for the server to create in terms of encapsulation of the actual message). It does not matter so much when sending a few messages and receiving a great deal of content (and it is how HTTP was envisaged to work), but when you want to send many messages, which potentially affect the content stream, especially if many individual users are sending messages at the same time, the overhead of the messages becomes a concern. While there are protocols created especially for messaging today, the direction of the industry is towards SIP signaling for IPTV. There are two reasons for this: it is rather simple to send messages inside a session once it has been established; and it is possible to prioritize messages using quality of service mechanisms – one effect of using IMS, that we will look at later.

SIP was created with several goals in mind, and one of the most important drivers was to create a way to handle voice over IP (VoIP) where telephone calls are set up over the Internet. While there are several methods to do this (e.g., Skype, which uses a proprietary protocol), most of the deployed systems are based on SIP. IMS is based on SIP, but it greatly enhances it compared to the original standard documenting the protocol, for example, with addressing and identity management, security functions, filters and other functions for subscribing to information and notifications about changes, and so on.

SIP was originally created to enable two computers to establish a session: one computer sends an invitation, the other a response, and then they can use the session they have set up between them to communicate – the intent being that the session setup should contain information that the two parties need in order to communicate with each other. The session setup contains information about the capabilities of the two computers relevant to the session, such as the installed codecs (and hence, what media types they can handle).

The original idea was to connect two computers directly, but there are also functions to help set up the most efficient route, and to let them find each other. SIP is a peer-to-peer protocol, with the idea of proxies and the addition (in IMS) of "back-to-back user agents", which can insert data into the headers. SIP is not intended for the media transport, however, only for the signaling. Once a session invitation is sent, and the acknowledgment of the invited party has been received, the actual media transfer can start. This is done using one of a different set of protocols, however – which we will talk about in Chapter 11.

There are three functions in SIP that are used in IPTV: subscribe-notify; session setup and maintenance; and messaging. In addition, presence is an important aspect of IMS-based IPTV systems, which also uses the messaging and subscribe-notify functions of SIP. The session setup and maintenance is used to establish a communication path between the IPTV server and the client; this session is what makes it possible to handle the charging, security and other aspects needed for IPTV (discussed in Chapter 11). The reason IMS is needed is that a SIP setup is very simple: to set up a session, send an INVITE message to the receiver; read the header fields and take appropriate actions; and reply with an OK message.

A SIP message is as equally simple as the protocol itself:

```
INVITE sip:bob@biloxi.com SIP/2.0
Via: SIP/2.0/UDP pc33.atlanta.com;branch=z9hG4bK776asdhds
Max-Forwards: 70
To: Bob <sip:bob@biloxi.com>
From: Alice <sip:alice@atlanta.com>;tag=1928301774
Call-ID: a84b4c76e66710@pc33.atlanta.com
CSeq: 314159 INVITE
Contact: <sip:alice@pc33.atlanta.com>
Content-Type: application/sdp
Content-Length: 142
```

The first part of the message is the INVITE command; the second is the address, in the form of a SIP URI (which looks like an e-mail address). The message also contains the address for the sender, and the sequence number for the message. There is a minimum set of headers that have to be included in any message. See Table 10-4

The transactions going back and forth between the sender and receiver are called a dialogue. Within this dialogue, the request–response model indicates that the setup is a simple invocation (the model is based on HTTP). However, it also means that the session setup is essentially a one-to-one affair. Sessions are established between pairs of terminals; when communication is one-to-many (as in IPTV), each of the clients has to establish a session with the server. The server has to handle all the sessions with all the clients – and here again, IMS becomes useful in setting up the communications path for the session (a particular problem if there are many sessions at the same time). During the session setup, there can be different messages confirming that a session setup is ongoing (RINGING and TRYING), and when the session is terminated, the terminating party sends a BYE message, which the other party confirms with an OK.

The details of the session, such as the type of media, codec or sampling rate, are not described using SIP. Instead, they are carried in the body of a SIP message, encoded in some other protocol format; the one most frequently used is the Session Description Protocol (SDP), defined in RFC 2327. It is carried in the body of the SIP message (in the same way that a web page is carried in an HTTP message).

Header name	Function
Via	Address where responses will be received, and branch parameters used for the identification of the transaction
To	The display name of the receiver, and the SIP URI where the request is directed
From	The display name of the sender, and the SIP URI where the request was originated. Also contains a random string for identification purposes
Call-ID	An identifier for this particular session, which is a random string together with the host name or IP address of the sending device
CSeq	Command Sequence, which contains an integer number and a method name, which is incremented to create a sequence number for each new request within the dialogue
Contact	SIP URI, which represents a direct route to contact the sender of the invitation
Max-Forwards	Limits the number of hops a request can make to get to the destination. The number is decremented by each hop
Content-Type	A description of the message body
Content-Length	The number of bytes in the message body

Table 10-4. SIP header information.

When SIP is used to set up data transport through a proxy, the proxy sends back messages to the originator (TRYING and RINGING), until the session is set up (OK). The response message is also very simple:

```
SIP/2.0 200 OK
   Via: SIP/2.0/UDP server10.biloxi.com
     ;branch=z9hG4bKnashds8;received=192.0.2.3
   Via: SIP/2.0/UDP bigbox3.site3.atlanta.com
     ;branch=z9hG4bK77ef4c2312983.1;received=192.0.2.2
   Via: SIP/2.0/UDP pc33.atlanta.com
     ;branch=z9hG4bK776asdhds ;received=192.0.2.1
   To: Bob <sip:bob@biloxi.com>;tag=a6c85cf
   From: Alice <sip:alice@atlanta.com>;tag=1928301774
   Call-ID: a84b4c76e66710@pc33.atlanta.com
   CSeq: 314159 INVITE
   Contact: <sip:bob@192.0.2.4>
   Content-Type: application/sdp
   Content-Length: 131
```

Here, the response contains the response code (just like HTTP, the responses have a number and a slogan), and the reason phrase, in this case OK. The other fields are copied from the INVITE message, but the To field has been updated with a tag from the receiver, which will be included in all future requests and responses during the session. The contact header is changed to the receiver's contact URI.

That is really all there is to setting up a session: INVITE, perhaps TRYING and RINGING, and then OK. Now the dialogue is established, and the sender and receiver can use it to exchange messages with the same context. INVITE and OK can be used as a "heartbeat" message, keeping track of the session still being active (because when the content is sent in a multicast session, there are not necessarily any status messages which report back on the quality of the transmission, as there are in RTP).

If either sender or receiver wants to change the characteristics of the session, they can send a re-INVITE. This is just a new INVITE with the same characteristics as the ongoing dialogue. The receiver sends a 200 OK message, and the requesting party responds with an ACK message. If the new INVITE is not accepted by the receiver, it can send an error message and receive an ACK – however, the ongoing session is not interrupted. This is particularly useful when watching ITPV, as the server can send requests for interaction using the INVITE, and receive a message from all session participants through the OK.

A user does not have to sit at the computer that sends the INVITE. Users can register with a SIP registrar server, which sets up an association between the URI of the user and the address of the machine where the user is currently logged in (the SIP URI in the contact field). The binding between the user and the machine is registered, and can then be used for locating the user; but it is possible for a user to be registered at many different devices, and you then need to use presence to keep track of the different devices a user has registered – otherwise, all the devices associated with the user will ring when an invitation is sent.

The original RFC describing SIP, RFC 3261, only described six methods for setting up and managing sessions: REGISTER, INVITE, ACK, CANCEL (used for cancelling a session setup), BYE, and OPTIONS (which is used to query a server about its capabilities). There are two other important functions in SIP that interactive IPTV can use. First, there is SIP MESSAGE, which was created to make instant messaging possible using SIP (for a while, there was a turf war in the IETF over who would control the instant messaging, with the result that two parallel protocols were created). Second, there is SUBSCRIBE and NOTIFY.

SIP MESSAGE

The MESSAGE method in SIP has part of its roots as an instant messaging protocol, and was part of the turf wars about instant messaging standardization in the early 1990s. Since exchanging instant messages means the sender and receiver have a session (and are part of a group, and have presence), the inclusion of instant messaging in SIP was pretty natural, but the method to do it was not.

SIP MESSAGE is a very simple messaging mechanism, intended to send short messages without prior notice (within a session). This is exactly how instant messaging (the IMS name is immediate messaging) works. When messaging is included in sessions, operators can charge for the messages, as well as set policies on them, and control the messaging. This means the messages have to be inspected, and the inspection has to be translated into charging records. This is expensive in terms of resources, however.

SIP MESSAGE is defined in a separate standards document (RFC 3428. It was created to provide a way of sending short text messages (although the content of the message is not limited to text). Even though there is a session from a user perspective, the messages carried over the session are media, not signaling; and they should not necessarily pass through the same intermediaries as INVITE, SUBSCRIBE and other SIP signaling. In the case of interaction in an IPTV system, however, they will most likely do so – since they are interactions within the session. This means it is harder to have a separate endpoint for the interaction and message setup; the IPTV AS will have to both receive and collate the messages, which means that in part it will have to be an instant messaging server (or redirect the message to one).

A SIP MESSAGE not only has a message body containing the text, it also has the same header fields as other SIP messages. In particular, it has a header with an expiry value in it, which can be used to determine when the message is no longer valid. This means a SIP MESSAGE is rather simple:

```
MESSAGE sip:user2@domain.com SIP/2.0
Via: SIP/2.0/TCP user1pc.domain.com;branch=z9hG4bK776sgdkse
Max-Forwards: 70
From: sip:user1@domain.com;tag=49583
To: sip:user2@domain.com
Call-ID: asd88asd77a@1.2.3.4
CSeq: 1 MESSAGE
Content-Type: text/plain
Content-Length: 30

Watson, come here. I need you.
```

The "Watson, come here. I need you." is both the content and the body of the message (30 characters). It can be any format that can be encoded as text – for example, XML – but it should not be longer than 1300 bytes, according to the standard. Since an SMS cannot be longer than 56 characters, this means SIP MESSAGE is a perfect replacement for SMS in interaction with television programs, provided the service provider can find a way to charge for it.

Like all SIP traffic, the SIP MESSAGE passes through the CSCFs in the network, and can be filtered (or forked) according to the logic downloaded from the HSS. It can also be rejected, especially if it is too large. IMS also provides a way to send messages to multiple recipients. This is done by creating a public service identifier (PSI) for the group of users. The group can be created using XDMS, but also through other means; in either case, the PSI is bound to an application server, which will receive the message, and distribute it to all the members of the group. This can be used to trigger a renewal of multicast addresses for a transmission, or to create synchronizations with real-time events (or overlay them on top of existing transmissions, for example, when an emergency occurs).

SIP SUBSCRIBE and NOTIFY

SUBSCRIBE and NOTIFY is a pair of SIP methods, which enable the subscription to a resource. SIP is a stateful protocol, and messages can refer back to the session setup. Since the setup has already been done, there is no need to provide additional session information; if the resource is available within the session, the user can subscribe to information about changes in it. The specification of the SUBSCRIBE method discusses the subscription in terms of "event packages", which is a format organizing what is needed to set up a subscription, and communicate information about it. It is also possible to send a SUBSCRIBE to a resource with which there is no session, in which case the SUBSCRIBE can be used to set up the session (instead of an OK, there will be an empty NOTIFY coming back – since no event has happened yet). Normally, a SUBSCRIBE is sent inside an existing dialogue, and the response a 200 OK, which shows that there is now a subscription associated with the dialogue. There can be any number of SUBSCRIBE requests within a session. The only limit is practical – how many requests the sender and receiver can handle. As we saw in Chapter 7, a subscription can be made to a group, and it is changes in the group that will be sent out in the notification.

When a session is established, and there is an agent which can receive a SUBSCRIBE message and take care of it, the user can send a SUBSCRIBE message containing the URI for the resource that it wants to subscribe to, and the event type. The request URI is typically a SIP URI. Several types of events may happen on the same resource. Like an INVITE, there has to be duration in the request headers. When the duration expires, the session ends; unless there is a new SUBSCRIBE (or INVITE).

The NOTIFY body can contain free-text information, which the designer has to put in. The SUBSCRIBE message can also have a body, which can contain instructions for how to throttle, modify, expand or filter the event class, including setting threshold values. This includes authorization, which can have three states: Rejected, Successful, or Pending. The authentication will be done against the IMS system.

A typical SUBSCRIBE-NOTIFY dialogue will look like this:

```
SUBSCRIBE sip:resource@example.com SIP/2.0
Via: SIP/2.0/TCP watcherhost.example.com;branch=z9hG4bKnashds7
To: <sip:resource@example.com>
From: <sip:user@example.com>;tag=xfg9
Call-ID: 2010@watcherhost.example.com
CSeq: 17766 SUBSCRIBE
Max-Forwards: 70
Event: presence
Accept: application/pidf+xml
Contact: <sip:user@watcherhost.example.com>
Expires: 600
Content-Length: 0
```

Then the subscription is acknowledged, as a 200 OK:

```
SIP/2.0 200 OK
   Via: SIP/2.0/TCP watcherhost.example.com;branch=z9hG4bKnashds7
    ;received=192.0.2.1
   To: <sip:resource@example.com>;tag=ffd2
   From: <sip:user@example.com>;tag=xfg9
   Call-ID: 2010@watcherhost.example.com
   CSeq: 17766 SUBSCRIBE
   Expires: 600
   Contact: sip:server.example.com
   Content-Length: 0
```

When the event occurs, there is a NOTIFY from the server to the subscriber.

```
NOTIFY sip:user@watcherhost.example.com SIP/2.0
   Via: SIP/2.0/TCP server.example.com;branch=z9hG4bKna998sk
   From: <sip:resource@example.com>;tag=ffd2
   To: <sip:user@example.com>;tag=xfg9
   Call-ID: 2010@watcherhost.example.com
   Event: presence
   Subscription-State: active;expires=599
   Max-Forwards: 70
   CSeq: 8775 NOTIFY
   Contact: sip:server.example.com
   Content-Type: application/pidf+xml
   Content-Length: ...
```

The NOTIFY also includes additional information in the body; in this case, since it is a presence message, the body will be a document containing the presence information.

Finally, the notification is acknowledged through a 200 OK.

```
SIP/2.0 200 OK
  Via: SIP/2.0/TCP server.example.com;branch=z9hG4bKna998sk
   ;received=192.0.2.2
  From: <sip:resource@example.com>;tag=ffd2
  To: <sip:user@example.com>;tag=xfg9
  Call-ID: 2010@watcherhost.example.com
  CSeq: 8775 NOTIFY
  Content-Length: 0
```

SDP in SIP and RTSP

A SIP message can carry information in the message body, but having it transport the documents themselves would reverse the advantages of the protocol. There is a set of information types which can be used to enhance the session – for example, by selecting a variant that is more appropriate to the end-user or his device, and provide him with the appropriate multicast address for the data that will work best on his device (alternatively, the user can be provided with a version from the server using unicast). To do this, the server needs to know what would be suitable for the user's terminal. This is exactly the type of information that can be sent using the Session Description Protocol (SDP).

SDP is a text format for describing the capabilities of media streams, and it is carried in the body of SIP messages (and also used in other protocols, e.g., RTSP). It can be used the other way around (and most often is) – the media rendering capabilities is the converse of the capabilities of the media stream. SDP contains information about the session, including the time it is active, what media it contains, and the information that should be used to receive the media. It is also possible to send bandwidth information, as well as contact information for the person responsible for the session.

The bandwidth information is a requirement for the media stream; the other media information contained in the SDP is the media type (using MIME data types), the transport protocol (RTP/UDP/IP, H.320, etc.), and the format of the media itself (H.261, MPEG-2, etc.). If the session is multicast (which an IPTV session is likely to be), the multicast address and the transport port are sent using the SDP (if it is not included in the EPG). In unicast sessions, the remote address of the media and the transport port for the contact address are included instead. The different types of session description information included in SDP are shown in Table 10-5

SDP field	Content	Explanation	Mandatory
v=	Protocol version	Version of SDP	Yes
o=	Owner or creator of the session, and the session identifier	Information relevant to the session identification and management	Yes
s=	Session name	Human-readable session name and identification information, such as a URI for the session	Yes
i=	Information	Text description about the session	No
u=	Description URI	Where to find the description of the session	No
e=	E-mail address	Contact information for the session organizer	No
p=	Phone number	Contact information for the session organizer	No
c=	Connection information	What connection is required; not included if the media type contains it	No
b=	Bandwidth information	The required bandwidth for the session	No
t=	Time of the session	When the session is active	Yes
r=	Repeat time	How often the session is repeated (zero or more times)	No
m=	Media name or transport address	The media type, including codec information. Can be repeated as required	Yes
k=	Encryption key	An encryption key for the media used during the session	No
a=	Attributes	A number of attributes for the media, which can be used for additional configuration	No

Table 10-5. Session description.

An example of a session description looks like the following:

```
v=0
o=mhandley 2890844526 2890842807 IN IP4 126.16.64.4
s=SDP Seminar
i=A Seminar on the session description protocol
u=http://www.cs.ucl.ac.uk/staff/M.Handley/sdp.03.ps
e=mjh@isi.edu (Mark Handley)
c=IN IP4 224.2.17.12/127
t=2873397496 2873404696
a=recvonly
m=audio 49170 RTP/AVP 0
m=video 51372 RTP/AVP 31
m=application 32416 udp wb
a=orient:portrait
```

The m= field is probably the most important from the perspective of IPTV, because it is here that the media is enumerated and the type described. The media types can be extended, but the parser of the SDP will ignore any type that it does not understand.

The c= field gives the destination address for the media stream. If this is specified using the media transport, then this should be null. For a normal session, it should be either the IP address of the receiver (for unicast), or the multicast address (for multicast).

The SDP is normally sent with the INVITE, but it can also be sent with the OK. In the IPTV case, the INVITE comes from the client to the server (since the server does not know about the client beforehand); the server then responds with an OK. Both these messages can contain SDP descriptions, and since the user is identified by the system, the initial SDP from the user can be used to set the range of possible media formats; the response can be used to select a specific version of the media that is best adapted to the user's terminal. This is part of how it is used in IMS, which we will describe in Chapter 12; but the media description can also be used by the transport protocols, which are the subject of Chapter 11.

Chapter 11: Next-Generation IPTV Encoding – MPEG-2, MPEG-4 and beyond

IPTV builds on a number of technologies now being deployed worldwide to turn television digital. This has been heralded as the great revolution for television, but in reality the revolution has already happened, or it has yet to happen, depending on whether you believe the digitization of television will have an impact. Looking at the use of these technologies for distribution, however, it is clear that we are in the middle of it, and it is a step from 2 to 4 – MPEG-2 to MPEG-4, to be precise.

The bigger picture: encoding and transporting the media

Once the signaling is done, the media should be delivered. This is no longer just a matter of pushing a button. In digital formats, media has to be encoded, and the encoding then meshes with the way the media can be distributed. Knowing a bit about how this works is important in understanding how a service can be set up, especially a service which combines video clips from different sources – because there is no "one size fits all" when it comes to encoding and media transport.

This is not a book about media encoding, but since it is important to know how a media stream is encoded and transported to understand how IPTV works, and how you can interact with it, we will go through the two dominant encodings (MPEG-2 and MPEG-4), and the main streaming formats (MPEG-2 TS, RTP, and H-264).

If one television broadcaster used one way to represent pictures, and another one a completely different way, it would be very hard to be a manufacturer of television sets. In reality, the problem is constrained at a country level, with three different formats in the world for standard definition (NTSC in the US and Japan, PAL in Europe and many other countries, and SECAM in France and some other places). This is how the television signal is broadcast (the main difference is in the resolution and the color encoding), so we will leave it out for now, because most modern television sets can handle all three. The important thing to understand is how the signal is encoded for transmission, and how it is transported into the television set over IPTV, because that is where you have to interact with it – and the interaction has to make sense.

There are, luckily, far fewer standards for the encoding of video (something vastly more complex than transmission, actually) than there are for sending and receiving it. The two dominant standards were developed by the same standardization group, the Motion Pictures Expert Group (MPEG), and they were developed to do the same thing: encode television digitally. The philosophy behind them is very different, and the efficiency of MPEG-4 is several times higher than MPEG-2.

In a traditional television set, the pictures are made up of lines. The lines are not sent out all at the same time time, but as a sequence of dots. As traditional television sets were made with a cathode-ray tube, which sent a ray of electrons to the phosphor on the inside of the tube, they would draw one line at a time, represented as dots of different darkness (color was added later). First, the electron ray would draw every other line in one sweep across the screen, and then it would draw the next set of lines. This technique, called interlacing, leveraged the afterglow in early electron tubes. The process is too fast for the human eye to see, so we believe that there is a complete picture in front of us, even though there are only lines (our eyes see the separate lines being drawn, but our minds compose them into pictures).

The lines would draw a picture, but this picture comes from an even older technology: film. Just like an old celluloid film consists of a long series of snapshots, which are exposed one after the other (also too fast for the human brain to notice, so it stitches them together into a perception of movement), the old television would draw first one set of lines in the frame, then the next set, and then move on to the next frame. This method is becoming a big problem when creating today's high-definition television, as we will see later. To point out the difference, the computer is progressive: it draws all the lines in the picture at once. This is also easier to translate to systems which are not dependent on drawing lines, such as LCD or plasma screens.

MPEG-2 is a way of digitizing the traditional television picture. It leverages the fact that within each line, there is a lot of redundant information. You do not need to draw 20 dots in a row, when you can represent them by one dot and the operation "x20". Four characters can represent 20. In reality, the compression rate can be even higher.

But MPEG-2 does not stop there. Not only does it compress the image by making the repetition of the dots unnecessary, it also makes it possible to remove many of the "images" in the "film". Since any television show is made up of a number of frames, which do not change much (how much does the background change when the news reader talks?), the frames that do not change can be removed, and mathematically calculated from the difference between the frames that do change. All of a sudden, a lot of the bandwidth needed to send the television program disappears.

This can be put to good use in many ways. One of them is what governments will do when television becomes digital, at least in many countries: sell the bandwidth to more television channels (or mobile telephone operators, who have no problems in selling the bandwidth). There is an alternative, which is also happening (sometimes in the same countries, at the same time): the bandwidth for each television channel can be increased. This means higher resolution, since the number of dots in each frame, and the number of frames, can be higher. This is where high definition television, which is now being promoted in many countries, comes in.

MPEG-2, however, bases its encoding on the way the traditional television signal is structured (a complete stream, including everything from video image to sound and subtitles are all included in one single stream, and encoded). This is not the case with MPEG-4.

MPEG-4 separates the scenes into media objects. For starters, there is a separation between audio and video (MPEG-3 is actually the MPEG-4 audio encoding, without the video). Video objects (i.e., moving objects) constitute a media object of their own. Still images in the video (e.g., a fixed background) also become media objects of their own. These are primitive media objects, and can be two or three dimensional. MPEG-4 defines how to encode sound, video and still images; and also text and graphics, synthetic (generated) sounds, and talking synthetic heads (a technology rarely used today, but which could potentially be used for interactive television – there is a programming language for setting up the animations, and very advanced algorithms for defining the lip movements, and so on).

There is a good reason MPEG-3 has become so popular: great care has been placed in getting the audio encoding right in MPEG-4. A scene where the voice is not synchronized with the images can be extremely irritating (think early Italian movies), but human cognition is designed to handle auditory and visual input differently – in particular, we have different quality metrics and different ways of drawing conclusions from the auditory and visual input (which is what we as humans spend the first few years of our lives learning). A common misperception is that the visual signal is more important than the audio in television; but it is the other way around: if the audio does not work, the perception of the high-quality visual signal is degraded. Unfortunately, standards bodies can only provide methods to encode the media in a good way, not define how producers (in particular, users providing content) should work. There is a reason, as any television producer knows, that there is a sound engineer working in parallel with the cameraman to make sure the sound is right. Microphone booms are not for decoration.

MPEG-4 has a number of ways of handling sound, making sure everything from speech to the Vienna Philharmonic New Year's concert is represented appropriately. The generic audio encoder can handle everything from very low bit rates (6 kb/s below 4 kHz) up to very high quality, with transform coding techniques built in. There are specific coding mechanisms defined for speech encoding, which can go down to 1.2 kb/s, which is very low; and also a specific encoder for text to speech, which can include different parameters to make the speech sound natural (not, as yet, defined for all languages, however). Specific coding is also defined for synthetic audio, defining "instruments" and "scores", allowing the "instruments" to play the "scores" in the same way as MDI, which is based on the idea behind real instruments and scores. These can be mapped in the same way as the keys of a synthesizer, making sampling easy.

MPEG-4 not only defines how data should be encoded and transported, it also defines a generic model for the decoder, which allows for the precise definition of how a terminal should work (but not how it should be implemented). For anyone used to implementing multimedia systems this seems like a good idea, but if the implementor's primary experience is browser implementations this seems like overkill. However, having a definition of the functions required of the device enables anyone building such devices to use it as an interface specification – and for the sender, it means you are sure of what will happen at the other end, which is not less important. When handling multimedia, a very high rate of precision is required, if the user experience is not to be compromised (or if you do not have a very large buffer, which excludes real-time transmissions). It also means you can broadcast MPEG-4 files over the traditional broadcast network without having a feedback mechanism, specifying the buffers and timing model.

Just like the compression of the MPEG-2 signal is composed of the object and an operation; the MPEG-4 object, when encoded, also consists of a number of descriptive elements. Each media object is independent of the others, and the encoding can be made as efficient as possible for the object type, including error correction (which has some importance for how media is transported). When objects are independent, it is also easier to make them scaleable. They are then grouped into compound media objects: the compound object is a tree, the primitive objects the branches. Objects can be placed in a coordinate system within the scene to relate to each other; and they can be transformed (geometrically or acoustically), and moving the point of view. The scene descriptions in MPEG is a separate language, MPEG-7, but in MPEG-4 the placement of objects reuses concepts from the Virtual Reality Modeling Language (VRML). The description is actually encoded in a format called Binary Format for Scenes (BIFS), which describes how the objects are arranged in the scene. The arrangement is hierarchical, the first level being the scene description, the second the different types of objects, the third subsets of these (e.g., a sprite which animates the movement of a person; the facial animation characteristics of that person; and a description of the voice). The object descriptions can change as the scene changes, node parameters then being added, replaced or removed. Objects have extensions both in space and time, and carry their own local coordinate system (including spatial coordinates, which means you have to think in four dimensions when working with MPEG-4; it is easier than it sounds). The local coordinate systems are transformed to the global coordinate system, describing the scene. In future versions, it is also possible to do the same things to audio, creating surround sound in MPEG-4. See Figure 11-1.

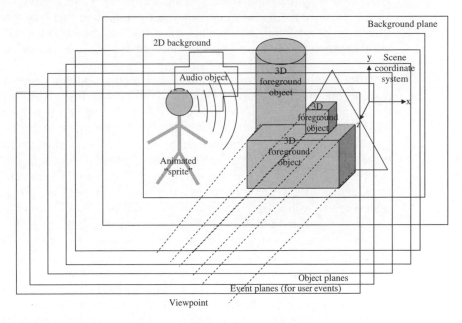

Figure 11-1. An example of an MPEG-4 scene.

The scene description is separate from the description of the media objects that compose the scene (and the media objects themselves). The objects are described through Object Descriptors (OD), which define the relationship between the different Elementary Streams (ES) which compose the object. Elementary streams can be for example the audio, the video and subtitles. The OD provides the information required to access the objects, such as the URI needed to access the stream, and the characteristics of the decoder.

MPEG also has a textual format, the Extensible MPEG-4 Textual Format (XMT), which is used to represent MPEG-4 scenes in text. This is not the same as MPEG-7, but it can easily be transformed into (or from) it, since it is an XML format. BIFS is binary, so in order to be human-readable, a text format has to exist (not that you would edit the scene descriptions by hand, any more than you would edit PostScript documents, but it is very useful for developers). There are two levels, A and Ω, which represent different levels of the spectrum; the ? being equivalent to a representation in SMIL, the Synchronized Multimedia Interaction Language, an XML language for coordinating different media streams. Another advantage of having a text format, based on XML, is that it can easily be used by other XML players, such as SMIL and VRML (Virtual Reality Markup Language – an XML language used to draw 3D scenes).

Where MPEG-2 is fixed to a static viewpoint and a static image stream, MPEG-4 allows for interaction with the individual media objects – if this has been permitted by the creator of the stream. For the user, it is (theoretically) possible to rearrange objects in the scene, or change the point of view. However, since the transport encoding is usually done in a way that results in a fixed presentation, this means going back to the encoder and getting another representation, which most television studios have not even thought about.

Various attempts at creating interactive media systems, such as MHP, assume that the interaction should be done at the client. MPEG-4 allows for both client-side and server-side interaction. The intent is that the MPEG-4 terminal should process the commands from the end-user in the same way as if they came from the original content source. Even richer interaction is possible if it is supported in the scene description. Client-side processing can be done standalone (but there is no way to program it, to limit the user's interactions, which probably makes it unpalatable to television program producers). Server-side content manipulation requires a back channel, which of course is available in IPTV by default.

These interactions include changing viewpoints (the user can pan through the scene), moving objects around in the scene, triggering a cascade of events in a scene by clicking on a specific object (this can result in starting or stopping the media stream; but it should also be possible to tie it to more advanced interactions); selecting the desired language (when there are multiple sound tracks). Interactions on the client side can involve content manipulation, for example, changing the descriptions of a node in a scene description, making objects invisible, moving objects around, and relating to user-related events. The specifications even describe a connection between the scene and reality, such as the phone in the video ringing, and the viewer watching the television can pick it up – and a communications link can be established. But while this may be theoretically possible, it is very rarely used; although it is of course easy to see how it can be leveraged in the interactive IPTV world.

Television program producers, broadcasters and consumer electronics companies (important, since a television set which cannot decode an MPEG-4 encoded stream is quite worthless if that is what it is getting) are slowly shifting to MPEG-4. Using only the capabilities of MPEG-4 for interaction would make for very boring interactive programs, and what is on the market today is really workarounds (in particular, interaction through SMS). There is an advantage, though, to the common encoding: it makes it easy to insert advertisements.

Transporting the MPEG Stream

A media object can be composed of data streams. In this case, the object description identifies all the streams associated into the media object. The data can be associated with meta-information, including copyright and DRM information. The copyright information is restricted to unique identifiers (issued by the international numbering authorities) for each video stream, enabling anyone to look up the registered copyright holder for that stream. It is also possible to embed a property and a value (author: Johan Hjelm). The system layer of MPEG-4 also has a published interface, which can be used to connect a DRM system to the encoding and decoding of content.

MPEG-4 has a language to describe the bitstreams that carry the media objects. This language is object-oriented, based on C++, and used to create a syntactic object representation, which can be used for testing. It is not something that a user would read, but it can be used to manipulate the bitstream in clients and intermediary proxies. Above all, it can be used to verify that the MPEG-4 stream does not violate the MPEG-4 syntax (and hence, could not be decoded).

The MPEG-4 system defines a toolbox and the tools works together to produce the audiovisual information. This results in a data stream (called the elementary stream, ES). The ES is composed so that it creates the actual presentation at the receiver side (in the same way as the MPEG-2 encoding does); it can also be stored and transmitted separately. The ESs have descriptors for the configuration they need to be decoded, as well as information about timing for synchronization. Each stream is time stamped and the MPEG synchronization layer takes care of the identification of the streams and the time stamps. This makes it possible to reach into the streams and synchronize the individual media objects within them. The configuration descriptors may also contain QoS hints (but the translation of these into actual QoS, in particular in IMS, is not clear).

MPEG-4 encodes streams in two layers: one groups the elementary streams to create low multiplex overhead; another packages these streams for transportation. MPEG-4 does not really define the transport streams, however. That has been left to others, in particular by reusing (and slightly modifying) the MPEG-2 transport stream format; and to the IETF, which has defined a transport format for real-time information, called the Real-Time Transport Protocol (RTP).

Since both of these protocols work with MPEG-2 as well as MPEG-4, it is possible to use them to transmit just about any stream which should be displayed in the IPTV program. The encoding can be MPEG-4 or MPEG-2, as long as the user cannot interact with the stream.

There are two transport formats commonly used for MPEG data (both MPEG-4 and MPEG-2): the MPEG-2 transport stream (TS) format; and the RTP format from the IETF. In the DLNA, HTTP is used for media transport, but RTP is optional. It may be made mandatory in later versions of the standard, therefore it makes sense to understand how it is used.

RTP

RTP is intended to be used with both multicast and unicast services, and is independent of the transport layer – and hence, does not have a QoS function built in. This is provided by systems such as IMS (see Chapter 12). Since RTP contains functions to handle payload type identification, the protocol can be used with IMS to enable QoS depending not only on the application protocol itself, but also on whether the protocol is used to handle a real-time transmission or video on demand (in which case it is less likely that the user will be sensitive to buffering).

RTP uses UDP to multiplex data, and uses the checksum function in UDP to compute that packets are correctly delivered. The underlying network is supposed to handle the delivery service, including delivery of packets in sequence and without latency. There is a sequence number function in RTP, which can be applied to packets (this is what is used in the FEC).

RTP was, like RTSP, originally developed for videoconferences with multiple participants. It turns out that the Internet itself is not really good enough for that, so the primary application of RTP is to send video and audio data.

RTP is really two protocols: the transport encapsulation for the data; and a transport control protocol (RTCP), which is used to monitor the quality of service, and convey information about the participants in the current session (like RTSP, it has a notion of sessions, but these can be layered into IMS sessions). RTCP, however, only contains control functions for the real-time transport.

RTP itself is just a framework, and has to be profiled for the application it is to transport. This is done by modifying the headers of RTP, so extending the protocol automatically. This has to be documented, presumably in an RFC (the RFC for video and audio profile is RFC 3551, and there has to be an additional document for the payload). The profile contains information about the headers of RTP, which can include a payload type. The payload types can be different media encodings, for example. The profile also defines the reporting interval for RTCP, congestion control mechanisms, and mapping to transport layer. The payload description is different from the profile, since it can be reused in many different profiles. Since an audio and video profile and payload description already exist, it is unlikely that a specific IPTV profile needs to be defined.

When transporting IPTV over the Internet, with or without IMS, it makes sense to use multicast. This is a way to send the same data to more than one user at the same time (by dividing the streams, and having the receivers listen to the same address). It has big advantages in transmitting data through the network, as the bandwidth is automatically shared between users instead of them having to hustle for it one by one (as is the case with unicast, where everyone receives different streams). One problem, however, with multicast is that there is no good way to provide all the receivers with the adaptation they require. In multicast, the stream is sent from the source and if the individual users all went back to the source and provided their requirements it would not make any sense to use multicast, since the individual requirements would essentially create different streams (and you would not share the bandwidth to get better service). This usually results in the smallest bandwidth and highest latency determining the bandwidth for all, unless there is a termination point in the network (such as the DSLAM) where the multicast data is turned into a unicast to each user.

There is a smarter way to do it: place the adaptation in the receiver, and send a layered transmission so unnecessary information can be gradually stripped out. MPEG-4 lends itself excellently to such encoding (MPEG-2 as well, although it is more monolithic in the encoding of the data). RTP can help by dividing the signal over several different multicast groups: the user can then join the appropriate group(s). This needs to be resolved automatically in the receiver, and the receiver needs to be informed about which multicast group(s) it is supposed to join; this can be done through the master session (which is assumed in this book to be set up using IMS).

If you want to use QoS control for end-to-end multicast sessions, IMS requires that you first set up a session to the IPTV service provider, and this is used to allocate the QoS resources; when the session is set up, the client sends an IGMP Join message to the multicast source, and gets the requested stream with appropriate multicast data. If a session already exists, and the user wants to join a different multicast group (which, in the OIF system, means joining a different channel), the same session can be reused, and the only message that needs to be sent is a SIP UPDATE. The intention is that the channel switching should not be slowed down by an additional session setup, since it is already slower than analog channel switching, which is what users are used to measuring against.

RTP has a companion protocol for control information, RTCP. This protocol is based on the sender/receiver periodically sending information about the reception status (over the same transmission path as the media), multiplexing the data with the media. RTCP provides feedback information about the transmission quality from the receivers, which is used to discover errors in the transmission and dynamically adapt the transmission and encoding. It also carries various identification information, as well as time stamps. The RTCP packets can be sent to all participants in a session, which means that the information can be used to optimize a video conference (the original, but for our purposes, irrelevant purpose of RTCP).

An important aspect of any control protocol is to make sure it is not "chatty", eating up the bandwidth with unnecessary information about the session. There should only be one RTCP packet per receiver and report interval (which is set by the sender), and the total RTCP traffic should be only 5 % of the RTP bandwidth. To enable multicast to work better, RTP also includes functions that can mix and translate RTP data, for example, from several separate streams into one single stream (again, very useful in video conferencing). In an IMS system, this function could be co-located with the CSCF, a SIP proxy which handles the sessions and sets the QoS (if the QoS was set, the CSCF could be used to compute the appropriate QoS for the entire multicast group). This is something that is set in the profile of RTP, and RFC 3550 (which describes the protocol) also provides algorithms for computing this.

MPEG-2 Transport Stream and the MPEG-4 File Format

There is hardly a television station which does not use MPEG-2, and most have invested millions in equipment to handle it. While MPEG-4 is a much more efficient format, the MPEG-2 equipment is not written off (much less the investment by television stations in learning to use it), so it is very unlikely that there will be a sudden surge of change over to MPEG-4. At least, not as long as most television stations are broadcasting in standard definition; at high definition, there is certainly a need for a more efficient encoding.

There is, however, a backwards compatibility between MPEG-4 and MPEG-2. This is especially true when it comes to the MPEG-2 transport mechanism, MPEG-2 transport stream (TS). It is standardized by the ISO, but the IETF AVT working group has also collaborated with the MPEG working group to create the RTP payload specification. The committee has defined several payload formats, both generic MPEG-4 and a format for the MPEG FlexMux mechanism, which enables several data streams to be combined into one.

MPEG-2 TS consists of a transport protocol in its own right, which enables the streaming over UDP. The MPEG committee defines a framework for transport; it is an umbrella document containing the various ways for for carrying MPEG-4 information over IP-based protocols such as RTP, RTSP, and HTTP, and provides the payload information for these protocols for carrying MPEG-4 data. The payload formats have normative functions, which enable the reconstruction of MPEG-4 packets, and define the MIME types used with MPEG-4.

There is also a file format defined for MPEG-4, which can be used to store the data (see Figure 11-2). As mentioned before, the streams making up an MPEG-4 presentation can be combined from a file. This design is based on the Apple QuickTime format, and is object oriented so that the file consists of "atoms", objects that are independent and can be stored outside the file and referenced using a URI or other pointer (the media itself, for example, can be stored in a different file and called using URIs).

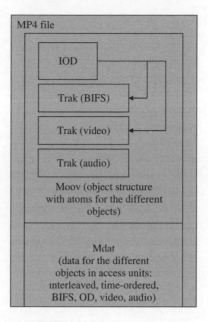

Figure 11-2. The MP4 file format.

The file format is not intended for streaming, it requires a streaming protocol such as MPEG-2 TS or RTP. When streaming a .mp4 file, the content is combined (since it can be referenced) and sent out through the transport stream. It can also be used for local playback in the same way – the presentation is combined from the referenced objects.

Forward Error Correction

Forward Error Correction (FEC) is a way of minimizing artifacts and other problems in the media stream. UDP, which is used as the underlying transport format of RTP, does not have a flow control mechanism – that is delegated to the application protocol. The issue here is that if you use TCP for the flow control over the network, it slows down the transmission if the network is constrained in any way (such as if there is a bottleneck, either because the links are slow or there is a lot of data). It makes sure the data gets through, but it slows it down.

The alternative, if the number of packets lost is not very large and they occur randomly, is to build in an error correction mechanism and ignore the packet loss. What you choose is a balance, and the tradeoff is more processing power used versus a secure transmission. However, since processing power is becoming cheaper and data is becoming larger (more than four times larger with HDTV), it makes sense to try to calculate what was supposed to be in the gap, and fill it with computed data. This is what FEC is about.

Forward error correction is a feature of RTP, and is transmitted out of band from the RTP stream. The method, described in RFC 2377, is based on providing hints on what should have been in the packet. The RTP protocol stack has to identify which packets are missing, look at the hints, and calculate the missing packets and insert them into the stream, where it can be picked up from the buffer by the receiver and used to compute the correct display for the user. The transport protocol best suited for use with FEC is RTP, which not only has functions to transmit real-time data end to end, but also contains a reporting function, which can be used to switch on the FEC. FEC information, however, has to be transported out of band from the media itself.

Chapter 12: Next-Generation IPTV Networking and Streaming with IMS

The Internet Protocol (IP) is used to deliver IPTV. This protocol is the cornerstone of the established Internet infrastructure, and it is used in routing and addressing the media streams from sender to receiver. The resilience and ubiquity of the IP infrastructure is what makes it so suitable for IPTV delivery, but is also the reason for two of its major drawbacks.

The bigger picture: making the Internet flow

Managing the identities, making sure the sessions get established, and ensuring that media flows through the network is quite a set of tasks, and it is the purpose of the Internet Multimedia Subsystem (IMS), a clumsy name for the mechanism which reaches from the session to the link layer in the traditional OSI model to establish communications paths between users and services.

IMS is crucial to the way IPTV is developed, according to this book, and we have already touched upon many aspects of it. However, there are some which we need to look into further.

IP is only the addressing structure. Each computer has a unique address (or several), and each protocol has a port on that address. The data directed to an address (encapsulated in the physical transport protocol) can be transported either reliably or unreliably. The reliable mechanism, the Transmission Control Protocol (TCP), is used in HTTP and many other protocols. It makes sure that there is a connection between client and server, and that it is reliable and persistent. TCP can be thought of as a train, which couples cars of data: when a car is derailed, the entire train has to stop, a new car put into place, and all the remaining cars brought up to speed again. The issue here is retransmission of lost packets, the acknowledgments for the packets, and the speed at which the connection is re-established. All these conspire to make TCP less useful in extreme conditions, although it is indeed better for regular file transfer in normal conditions. When TCP fails, it is hard to restart, and this makes for latencies when the bandwidth is low – or the data rate high.

There is also an issue with TCP adding information to the media. This does not matter if what is being transmitted is a file containing a web page, but it will matter if what is being transmitted is a stream. For one, TCP will add significant overhead when used with the much higher number of packets transported in a stream, compared to a file transfer.

Hence the popularity of the alternative method for streaming, the Unified Datagram Protocol (UDP), which does not have a notion of connection, and is a process of embedding data in such a way that it can be transmitted over the connection. If a UDP packet is lost, there is no automatic retransmission; it is just lost, and the application protocol has to request it again, or create a retransmission, if applicable.

There is no need to go into how it works in detail here, but there are a couple of features we need to highlight to understand how routers handle data streams. Contrary to what you might expect in a system used to distribute television (and radio, the principle is the same – it is just a media stream), there is no broadcast. Sending to "all" would flood the network and make other transmissions impossible. Or nearly impossible – this depends on the infrastructure underlying the infrastructure. For local area networks, the network used is most often Ethernet, a protocol which has its roots in systems that manage radio transmissions. To the end-user, the connection is mostly realized as an ADSL line, which may carry Ethernet signals or something else. More and more, optical fiber networks are used.

This is all incidental to the delivery of IPTV, which relies on an important principle of the Internet to become robust: the separation of layers. There should be no impact from the use of the network on the protocol used for the signaling – or the data transmission. It would be nice if this was true, but it is not completely true. There is an impact from the network on the transmission of the television data stream in three ways:

- in the addressing;

- in the way data is packaged to be sent over the network; and

- in the data quality.

The addressing and the data quality has the biggest impact on IPTV from using IMS, although there is also an impact on the data packaging and transmission. To make sure the data transmission is working smoothly, IMS is used to set up quality of service, which ensures that IPTV is given higher priority than, say, web browsing. The transmission is also affected by the ability of the transmission protocols to do forward error correction, a method to send correction information along with the data packets, so retransmissions are not required. Quality of service relies, however, on the control layer (where signaling takes place) reaching down into the network layer. In traditional systems, the communication between the layers is unidirectional, from the bottom up, apart from acknowledgments. In IMS, it is bidirectional, with the quality of service mechanisms working with both the network and the application layers.

Multicasting on the Internet

The addressing on the Internet is hierarchical, built as trees, which have their crowns in the computers at the end and the roots in the routers at national level. The naming system is built the same way, since it maps back to the addressing system. When data is streamed from a server to a client, it is streamed along the branches in the tree, following the path that leads to the requestor, to create a stream of packets from the server to the client. Such a one-to-one transmission is called a unicast. But what if it was possible to send the data stream in both directions when the branches fork?

A transmission which is intended for more than one user on the Internet is known as a multicast, and it has been around for a long time – almost as long as the Internet itself, in fact. It has not been put to work for video distribution up until now, mainly because the capacity of the transmission links was not sufficient, and also because of the assumption that video consumption would be a one-to-one experience. While it is that in some cases, any real-time broadcast will want to reach as many users as possible, which means including all of them in the group of receivers for the same stream. Multicast is, in fact, not more complicated than that.

What multicast means, however, is that instead of everyone getting their own separate data stream – with their own content and their own timing – the viewers all get the same content, at the same time, because that is how multicast works: it distributes the stream equally to all receivers. This is much more similar to broadcast than video on demand. As a matter of fact, it is so similar to traditional broadcast that to an average user, it is indistinguishable from it.

The reverse of this is that if you have a network which enables multicast, there is no reason to do broadcast. When broadcasting, you want to send to everyone,

> *(Continued)*
>
> *or at least those who are listening. Since the Internet is built so that a terminal is automatically registered when it is connected (even in the mobile network), distributing a transmission to everyone means sending it to all registered receivers. So multicast is the same as broadcast, for real-time purposes at least.*

The Internet presents another issue, which complicates things. It is a "best-effort" network. In this case, it means that data packets are let through any bottlenecks in a first-come, first-serve model. Since it is very important for packets in a media stream to arrive at very short (or at least predictable) distances from each other, a bottleneck in the network might mean that some packets get delayed. This will result in choppiness of the video, and in the worst case no service at all. There is a way around this, called forward error correction, which we will look at later. It lets the receiver compute the missing gaps, but the tradeoff is that it is resource intensive.

The reason there is a bottleneck in the network can be that the number of data packets trying to get through is larger than the capacity of the point they are trying to get through. The well-worn analogy is that it is like a highway, and one of the lanes has been shut off. Immediately there are queues in both directions.

If an ambulance comes down the highway, cars try to get to the side so it can drive through. If the police are there, they will temporarily stop other cars to let the ambulance through the bottleneck. And this is exactly how quality of service works: the packets, which must go through, will be allowed to go ahead of other packets.

The current Internet technologies won the day by being cheap (at least, relatively), and extensible. But even the originators of the Internet, the folks who gather in the Internet Engineering Task Force, recognize that the current system is flawed. Interestingly, one of its greatest (recognized) flaws is how it handles streaming media. It is simply not very good at this. They have developed a successor technology, called IPv6, which not only solves the streaming problem but also many other problems, such as the lack of address space in the current Internet (the addresses for computers will run out, according to some estimates, in 5 or 10 years' time).

However, it is the technology that delivers the data into the customers' homes, so we are going to live with it for a long time. To make sure IPv4 can stream television into people's living rooms, a mechanism to handle different types of data packets in various ways has been created. It is called quality of service, and it is a method to structure the queues in the routers in the Internet (the computers that send data packets to the correct receivers). The way routers do this is based on which protocol is used to encapsulate them, which is one of the problems with many of the web-based video systems that we will come back to. However, it is not enough, there needs to be a way to prioritize even within the protocols. And this is where IMS comes in.

The most frequently used protocol on the Internet is the HyperText Transfer Protocol (HTTP) used on the World Wide Web to transmit messages. HTTP is based on the Transmission Control Protocol (TCP). This has many advantages, but also a couple of disadvantages, especially when handling IPTV. One of these is that it assumes a one-to-one connection between the client and the server, setting up transmission quality at network level. It is also hard to change the characteristics of the transmission in mid-stream. One solution may be to use a protocol that does not use streams, but only ensures that the datagrams – the packets – are delivered, and does the flow control on the application level. This is what IMS enables.

What is IMS?

The Internet Multimedia Subsystem (IMS) was originally developed by a standardization group, which creates standards for mobile networks – hence the clunky name. The idea was to make sure that voice traffic worked well over the Internet, taking the existing VoIP products one step further, to ensure that the user was billed for voice calls, and that they would get corresponding quality in return. As a consequence, the system they created was ideally suited to handle IPTV, even though this was not intentional.

IMS is based on three things:

• The Session Initiation Protocol (SIP).

• A database of users (the Home Subscriber Server, HSS) and a token, which identifies the user (the SIM usually takes the form of a SIM card, which sits in mobile phones and enables the operator to verify that the mobile phone is allowed to connect to the network).

• A violation of the principle of separation between the network and control planes, which enables the IMS nodes to reach down into the network and set the quality of service in the routers.

From the nodes, which connect to the HSS, there are connections to many other nodes in the network, which enable IMS to handle charging, user profiles, and so on. How this works is something we will take a look at later, because the dual use as the control system for quality of service and the control system for user profile management is what makes IMS so useful to IPTV.

One of the keys to understanding IMS is that the use of IMS is based on the idea of a "home" and "visited" network. In other words, in IMS there is not one single network cloud stretching all over the world, instead there are a number of different interconnected networks, and there are services, which are only available in one network, because the services are dependent on the network. This model borders on the nowadays completely discredited "walled garden" model, where the operator tried to lock in their network users, and enforce the use of one single set of the operator-provided services. However, there are differences. To start with, there is no enforcement – the services can be in any network. The networks do have to be connected, and that is something of a rub, since it requires the home and visited network to have interconnection agreements. In the mobile telephone world, there are "roaming agreements" between the mobile networks so that a user can go from one country (and network) to another, and get service. In the IMS world, there is no such thing – although it has been talked about. This means it will be difficult to get to any services that are not in the same network; even though they may be technically reachable. The problem is that there can be no guarantees for the quality of service, and other IMS basic functions such as charging.

There are two basic models in the IMS architecture. The first, illustrated in Figure 12-1, is where the user comes from a visited network (or the home network), and the service is in the home network. The point is that the user goes to the P-CSCF, and that goes to the S-CSCF, and together they enable the user to access the service.

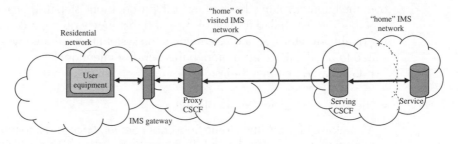

Figure 12-1. Service platform in the IMS network.

There is another way to organize the services: when the services are not in the same network, the access to them goes through the S-CSCF. When the services are located in the same network as the P-CSCF and when they are located in a different network, the P-CSCF passes the control of the session to the S-CSCF, which invokes the service logic – and this is the correct service logic for the user, since the S-CSCF is located in the "home network" of the user. The P-CSCF supports both cases by default, although the deployments may not have happened that way. It is also possible for the user to come from several P-CSCFs at the same time (which may mean they have registered from several terminals, for example, the mobile phone and the television set).

Using an IMS system takes place in two steps: first the registration of the client with the IMS core system; and second the setup of the session used to control the service. The registration means that the user agent finds the P-CSCF closest to it, and it is authenticated and authorized to use the nodes controlled by the IMS system. The P-CSCF connects the user to the S-CSCF, which downloads the filter criteria and other profile information for the user that it may require. Since the setup is now done, and all the information required to manage the user's sessions is available in the S-CSCF, the actual usage – the next step – can be very fast, since processing can be optimized for the SIP message routing based on the filter criteria.

Sometimes, especially when talking to telecoms operators, it is possible to get the impression that a user cannot do anything without authorization in the IMS system – this is not true. The IMS system only registers and controls the user agents that use SIP to set up and handle sessions towards nodes that also have registered in the IMS system. It is quite possible to use other resources on the Internet at the same time; and these resources can be based on SIP, as well as other types of protocols. As long as they are not controlled by the IMS, there is no way that the IMS system can even know about them. IMS uses the underlying Internet-based network (in IMS terminology, the IP-CAN) for the transport and signaling. The terminal can connect from different types of connectivity – this does not matter to IMS.

IMS and the network

The three layers of the IMS system mean that there is a special layer acting on the network. This layer cannot work unless the IMS provider has a relation with the network provider – usually, the IMS and network providers are assumed to be the same. This has a huge impact on IPTV, because the system can use quality of service to create transmissions less sensitive to interruptions and disturbances. Quality of service is not always implemented, however. The alternative is "over-provisioning", in which more resources than required will be used – and then a margin of error is applied. While this is quite possible to do even for a society where all users are watching high definition television over the Internet (there are at least two users in Sweden who have terabit per second connections into their homes, just to prove the point), it would require massive upgrades of so many different parts of the system that it would become very expensive – at least in the short run. A more feasible way of providing good service to all users is to enable better use of the available resources, and upgrade the worst bottlenecks first. This is where IMS becomes useful – among several other ways.

The nodes that work together to handle the session setup, management and user management are called the IMS Core. They combine the media delivery and service provisioning, and they take care of the authentication (checking that you are who you say you are), authorization (checking that you are allowed to do what you want to do), and signaling for the setup of the service. The main parts of the IMS Core are the HSS and the Call Session Control Function (CSCF), of which there are several kinds depending on whether you are attempting to get a service from the same network that you are connected to, or if the services are located in another network. There is also an Authentication Proxy, which interfaces to the system managing user profiles and groups; this is something we will not go deeper into in this chapter.

The HSS stores information about the user, which is used to authenticate and authorize the sessions that the user wants to set up. This can be stored directly in the HSS, or linked through it – the IMS specifications do not define how. What it does define are the session keys, the IMS Private User Identity (IMPI) and the IMS Public User Identities (IMPU) associated to it. The subscriber profile is tied to the user's identity (IMPI) and contains several service profiles. The service profiles consist of the IMPU and a set of filter criteria, which are applied by the S-CSCF before it decides to which AS to send the session request. The filter information is individual to each user (but can, of course, be reused between users). HSS is also the node that generates the key sets used in authentication.

An advantage of IMS being network based is that it is possible for the network to terminate sessions – and also switch off subscriptions. Television sets that use IPTV have to send a signal to terminate the session as well as switch off the display when they are turned off; however, if the electricity is cut, there is no way to manage the session shutdown from the terminal. In this case, based on timers, the CSCF can turn off the session. This not only extends to sessions, which are based on the current ability to receive media; if a terminal is lost (less likely with an IPTV set than a mobile phone), the subscription can be terminated by the IMS operator. This would increase the security for the user, especially in the case of mobile television. Depending on how the work is divided between the service providers, the IMS subscription may not be provided by the same company as the company which provides the IPTV service. If the IMS subscription is shut down, but the IPTV subscription is not, it becomes worthless (on the other hand, the user can freely mix and match providers, at least for IPTV, since the assumption is that to be useful, the IMS provider has to control the network to which the user is connected).

The signaling in the IMS system is done using SIP, but the connections between the different nodes are secured using IPSec, which establishes an encrypted connection at the network level. All signaling is automatically secure, and the same mechanism can be applied to the media transport as well, if there is a desire to connect in a secure way. While security concerns in the network are often overblown (it is much easier to intercept the media in the user's terminal than in the network – and most security violations are probably committed by users who put out content on sharing sites), it is nice to know that messages containing, for example, credit card numbers and similarly sensitive information are secure.

There are three types of CSCF in the IMS system:

- Proxy CSCF

- Interrogating CSCF

- Serving CSCF

In IMS, the assumption is that the CSCF closest to the user handles the routing of the IMS control messages, the rules and triggers that can be applied to messages, the access to the media (which depends on the charging), and the security. The CSCF intercepts all SIP messages that pass through the system – from the originating system, the user agent; to the terminating system, the application server – reads their content, and applies triggers and rules accordingly. Several CSCFs can interact with messages, especially if the user is connecting from a different IMS system than the one where the IPTV service is located.

The user does not have to know anything about the location of the IPTV service, only its address. Since this is a SIP URI, which is independent of the actual location, the CSCF looks up where to send the request for the session, and ensures that the requesting client is allowed to set up the connection. To do this, the CSCF uses the P-Asserted-Identity to set up the user's public identity, based on the P-Preferred-Identity. To identify all messages that relate to IPTV, a mechanism in SIP called feature tags is used; this includes a specific, pre-registered tag in the "contact" header field of the SIP message, making sure that messages, which relate to the same thing, are handled in the same way. In this case, the feature tag is +g.tispan.iptv [RFC 3840].

The SIP URI has a different format from the HTTP URI, but it works in the same way: it points to a logical location of a resource, not a location within a file system (which is what a URL does). This is necessary because there may not be a file behind the URI, instead it probably points to a program in the Application Server (AS), which triggers the session to the particular program the user is looking for.

The first CSCF, which the messages from the user agent encounter, is the Proxy CSCF (P-CSCF). This is used when the user agent registers with the IMS system, since it is the node which intercepts all SIP messages and sets up all the keys used to establish secure connections between nodes in the IMS system. It is responsible for the authentication and authentication – including the charging information.

The second type of CSCF is the Interrogating CSCF (I-CSCF). If the P-CSCF is the first entry point for the user, the I-CSCF is the first entry point for the service provider. It is the CSCF that connects to the HSS and other nodes, using a protocol called Diameter (instead of SIP), which enables a much more fine-grained control. The I-CSCF is also the point that can connect to other HSSs, allowing users from other domains to register in the IMS system.

The third type of CSCF, and crucial to the IMS system, is the Serving CSCF (S-CSCF). The S-CSCF acts as the SIP registrar, maintaining the binding between the user's location (i.e., IP address) and the SIP address. It also uses Diameter to connect to the HSS, and the same protocol to connect to the service and user profile management, and to set up AKA authentication vectors. Authentication of users is done by the S-CSCF based on AKA, a protocol used to manage keys on the Internet [TS 32.203]. If the authentication is successful, the keys are sent to the P-CSCF, from where the user message originated.

All SIP messages traveling to and from a terminal have to pass through the S-CSCF, and it does not just route them onwards, it also acts on their content depending on the triggers and rules it has been given, routing messages to the appropriate application servers. The triggers and rules are set in the user profile, which the S-CSCF gets from the HSS (and the HSS can get from the

XDMS system). Typically, the profile is downloaded during the user registration. One example of routing rules is that "all messages which have a feature tag +g.tispan.iptv in the Contact header should be routed to the IPTV application server".

When the message reaches the application server, in this case the Control Server, it returns a SIP message. Normally, as in all SIP dialogues, this is 200 OK. The message follows the same path through the different CSCFs back to the client who made the invite. There is a function in SIP to set the path to make sure the message takes the same route. Since this does not depend on which IMS provider the user and service provider are located in (as long as they have agreements to interoperate), this gives a lot of flexibility to the system.

The security functions in IMS are not the focus for this book, but we have already mentioned that the control messages and streams are secured using IPSec. There are additional nodes in IMS that enable the setup of the key pairs, the Authentication Vectors, in the system. One of these is the Bootstrapping Server Function (BSF), which authenticates the user equipment using the 3GPP AKA protocol, and provides the keys to the IPTV Control function, which acts as a Network Application Function (NAF) as defined by 3GPP. The same keys can be shared with the User Equipment, and this means the communication between the two can be secured. The BSF acquires the AKA vector and the Generic Bootstrapping Architecture (GBA; a way to authenticate the user when the session is set up) user security settings (GUSS) from the HSS. If requested by the NAF, the BSF also sends application-specific User Security Settings (USS) to the NAF (regulated by BSF policy). How these different nodes work together is shown in Figure 12-2.

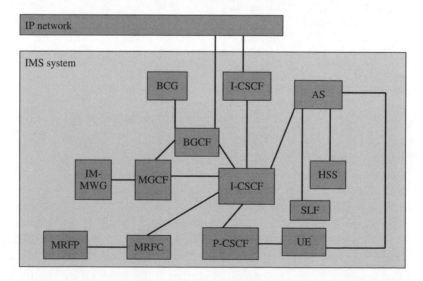

Figure 12-2. Reference architecture of the IP Multimedia Core Network Subsystem.

Registering in IMS

When a user switches on the television set to watch IPTV, this does not only mean that the radio receiver is activated and starts receiving at the preselected frequency, and sends the resulting electric signals to the electron ray gun, which creates three different beams directed to the phosphorized glass screen through the mask that creates the pixels. Nor does it only mean that the electric signals are sent to a decoder, which interprets the bits and creates instructions about which pixels to switch on and off on a LCD screen; it also means getting an IP address, setting up a session to the IPTV service, and start receiving the bits from the multicast channel selected. This does not necessarily imply that there is more delay in starting IPTV (although it may sound like there are a lot of additional operations that have to take place). All digital systems have to decode the received stream and turn it to a signal that can be displayed, and this is typically what takes time in any digital television system, IPTV included. The setup of the session is a very small part of the delay – although anyone who builds an IPTV system has to plan the location and relative position of servers and services so that the potential latencies are minimized.

When the television set is switched on, there has to be a session. If the Internet was like the airwaves, there could be a broadcast signal that anyone could receive, however, IPTV is different, you have to get an address, which tells you where to get the television signal, irrespective of whether the channels are free-to-air or not. There does not necessarily have to be a subscription, but there has to be an authorization to use the system, because there is no public Internet service in the same way as there is public service in radio and television in many countries (where legislation says that the public service channels have to be made available to anyone who wishes to receive them for free – since there is a license fee or tax which the user has paid, although this may not be associated with the consumption of the service itself).

So first, you have to get the IP address; that is part of the connection setup. Then, you have to get the service, and this is where the IMS registration comes in.

How IMS works with SIP

SIP is the cornerstone of IMS. It would not be possible to build an IMS system without a session control protocol like SIP. On the other hand, there are many functions in IMS which are not dependent on SIP as such. Other functions depend on extensions of SIP – created by the IETF, the body that standardized SIP – and which enhance the original standard drastically.

SIP INVITE

The key in any session-based system is a method to set up and tear down the sessions. In SIP, that method is SIP INVITE. This is the command that is sent in the protocol headers to make the receiving side establish a durable state for the communication from that particular IP address and port – a session.

SIP INVITEs can be sent by any SIP protocol stack. Indeed, the very definition of minimum standards compliance is to be able to send and receive SIP INVITE messages, and handle them in a correct way.

SIP SUBSCRIBE and NOTIFY

SIP was originally defined to set up a session for communication, specifically with telephony over the Internet in mind. Like IPTV, telephony is very sensitive to latencies. If a communication is delayed or interrupted, this degrades the user experience – probably so much that they may no longer want to use the system. But while RTSP was created as a session protocol specifically for television-based services, SIP was created as a generic protocol.

A generic protocol can be extended to accommodate new methods, which is exactly what has happened with SIP. The notion of a session can be used not only for communications, but since the session is long-lasting and has a state, changes in that state can also be communicated (instead of the content of the session itself). If you want to get an alert when a session changes, you can subscribe to the alert. That is exactly what the SIP SUBSCRIBE and NOTIFY methods were created for. Although not part of the original SIP specification, they nevertheless extend it and are part of IMS.

Forking and Redirecting Sessions

The CSCFs are proxy servers, and proxy servers can take a request for a SIP session – a SIP INVITE, SUBSCRIBE, or REGISTER – and send it to someone else as well as the original receiver, with modifications if required. When doing so, it is possible to apply a set of rules, in effect deciding to whom a session can be forked. Forking can be very useful if the SIP request is addressed to a PUID, which has multiple contact addresses (e.g., an application server with several IP addresses for load balancing). There is also a way to use the SDP information, which is registered when the session is set up, to select from different alternative forking endpoints, based on the preference information provided by the user. Forking of a session can be done so the new sessions are set up in parallel or sequentially. Redirection can also be done to another system, outside IMS, if the endpoint is an HTTP URI, for example. Of course, it assumes that the content of the session will be meaningful to that endpoint.

Session forwarding in IMS is based on the idea of telephony: the different ways of redirection are based on the intended receiver being busy or not answering. Redirection can also be based on the logic in the user profile, which is retrieved from the HSS. The redirection is done by the P-CSCF and the I-CSCF cooperating.

A session can not only be redirected, it can also be completely transferred. This uses the "Refer" operation in SIP, which has two parts: refer-to, and referred-by. This is used, for example, when the user is logged into a different device (the referral can then be done by using the Globally Routable User Identity (GRUI)). It is important that the referral can be done both to and from an application server, and be based on service logic, which comes from the HSS and is embedded in the user profile. Transfers of sessions in IMS can be done without consulting the originator (blind transfer); with the originator ensuring that the session is transferred (assured transfer); and by consulting the originator (i.e. allowed to approve) before the transfer (consultative transfer); and by a three-way transfer, which typically involves a conference bridge (and may have many more parties than three).

Identity in IMS: the SIP URI, PUID and PSI

In SIP, the resource (i.e. the user agent) is identified by its URI. URIs are familiar to anyone who has used a web browser, since they are the information that goes in the address of a document. However, when you want a website, instead of always having to type the entire address of the starting page (a URL, which gives the address relevant to the file system), you can give the address of the server where the document is, and since you are making an HTTP request, the server automatically serves up a web page.

A SIP URI is a little different from an HTTP URI, except that it is formatted like an e-mail address, because it refers to a user, not a resource such as a program or a file. In IMS, the SIP address is just an address. The identity is tied to the SIM card, and in IMS-based IPTV, this is where subscriptions are tied. There are two types of user identity: the Private User Identity and the Public User Identity.

There are also two other ways of handling addressing in IMS: ENUM (based on telephone numbers); and Public Service Identifiers (PSIs). Users are identified with Public User Identifiers (PUIDs); there can be several of these per SIM card, such as when a family is using the same IMS subscription for all their IPTV services; in the same way, the PSI is tied to the application server, and can be used to identify groups (such as chat services) in the application server.

The PUID is stored in the ISIM, which may be a hard SIM, or a piece of software (although how that is supposed to work is not yet standardized). The same PUID can be registered at many different devices at the same time, but will result in a different registration, and hence a different session, which can be used to separate where data is delivered, even when delivered to the same PUID. It is the service logic in the user profile (and hence the operator) which determines how many simultaneous registrations the user can have. This also applies the other way around: multiple users can be registered at the same device.

The user can have more than one PUID, and they can all be registered at the same time, in an implicit registration. None of them is the master of the others, in this case. This means, for example, that a "family identity" can be registered – and that identity can manage a number of "family member identities". They can have different service profiles, therefore this mechanism can be used for parental control, for example, as shown in Figure 12-3.

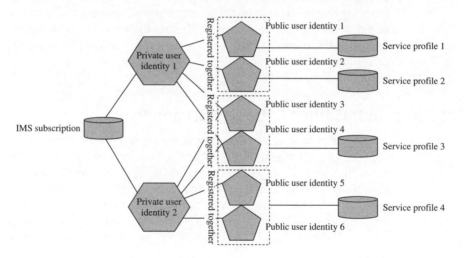

Figure 12-3. Identity hierarchy in IMS.

SDP

When you set up a session, it is useful to specify not just that you want to have a session, but also how you want it. This is the idea behind the Session Description Protocol (SDP), which is used in IMS and other SIP systems to provide a declaration of the capabilities of the sender. There are other methods in other protocols that have similar functions, but which can be much more fine-grained (such as the UAPROF description format from OMA), however, SDP was designed specifically to describe audiovisual capabilities, and to be sent along with the INVITE message. If the capabilities change, an update of the SDP can be sent using SIP UPDATE.

SIP messages do not have a message body, in the sense of HTTP. Instead, a header field is defined to carry the SDP information. True to its legacy as a telephony and video conferencing protocol, the device capabilities focus heavily on codecs, but ignore other things such as screen size.

A session description can be used in two ways: as input to a transcoder; and as the base for selection of variants. For example, if a program is available in two versions, MPEG-2 and MPEG-4, and the codec of the device requesting the program is only MPEG-2, then it is not very hard to select which version to use. It is not difficult to make the transcoding, either, since the transformation between the two is well understood. It is much harder if the program is recorded in 1024×640 pixels, and the screen is 220×480 – not to speak of the differences in going from interlaced to progressive rendering. Television, while having a very well-established model for how to go from one codec to another, misses the details of semantic transcoding. This is a major problem for mobile telephony, if you do not accept that there may be imperfections, and that the only thing you can achieve is a proportional representation (as has happened with mobile television reception in Japan and Korea).

When a user registers in IMS, the device he registers from sends the device characteristics along with the registration. There will be one set of device characteristics for each session the user has, since a user can have more than one session from more than one device. The party that initiates the session (most often the user's device) sends a SIP INVITE message which includes the media characteristics. The parameters are defined in the SDP specification; this is another problem, because the specification has to be updated every time a new codec arrives. There are methods to go around this, such as establishing ontologies (formal descriptions), which can be extended easily by manufacturers, but they have not yet reached complete acceptance.

The receiver of the registration – in our case, when the user registers with an IPTV service – responds with a list of the codecs it is willing to support (i.e., which encodings are available for the media). If it accepts these, the users device sends a 200 OK, but if not, there can be several messages back and forth with different SDPs until the two reach a conclusion about which codecs are supported, and which media are available. Once the session is established, either of the two endpoints can change the session. How this signaling works is shown in Figure 12-4.

This being IMS, the intermediary nodes (the CSCFs) are also involved in the session. They make sure that the resources in the network required to support the flow are available. Here, again, it is a matter of the operator setting the service logic. The user needs to have a good experience, so the bandwidth available needs to be sufficient to deliver the data. If the encoding is MPEG-4, the transmission is much less sensitive than MPEG-2 (and the required bandwidth is lower); if the content is in high definition, it will have completely different requirements than standard definition, and so on. The CSCFs makes this reservation for the network resources, and this is what makes the IMS user experience better than a user experience using HTTP or other unicast protocols.

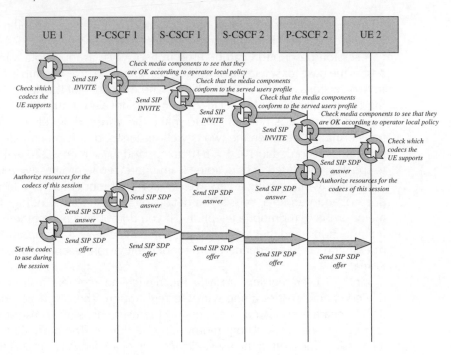

Figure 1.2-4. How SDP is used by IMS.

Another feature that makes IMS-based sessions more powerful is the ability to add another media to the flow. For example, the addition of video to a voice call, or of a voiceover in a different language to a program, are rather simple. In the existing session, an INVITE is sent, which contains an invitation to add the other media. This can be originated by either endpoint. Since the extension of the session passes through the same CSCFs as the previous INVITE, it is easy for the intermediaries to change the resource reservation (which would be required if I added a video to my chat session, for example).

Setting Up and Tearing Down the IPTV Multicast

Using multicast to send and receive content has large advantages in terms of can be achieved, however, it does not mesh that well with the concept of the session that IMS uses. To be able to manage multicast within the sessions, there are some tricks that the service and network provider have to perform – which have been developed by the Open IPTV Forum and ETSI TISPAN, but in slightly different ways (TISPAN looks in greater detail at the network layer).

First, the user in any case has to start the terminal, and set up the session. Then he has to authenticate himself and be authorized. After that, there is a retrieval of the content guide and channel information. This means there is quite a bit of signaling which has to go back and forth before the user can actually set up the session and receive the new channel. This is handled at the network level, so there is no application level processing involved. The idea is that the user gets a "bundle" of services, just like in today's cable-TV services, and that bundle provides the different channels that he has the right to view within the session. The way the signals work is shown in Figure 12-5.

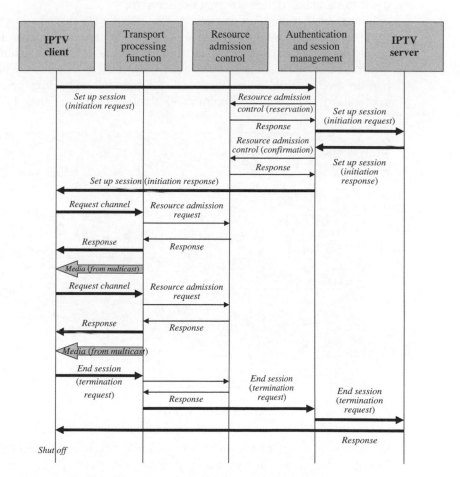

Figure 12-5. Call flow for scheduled content session setup.

This signaling is not only directed at the authentication system in the network. In IMS, it is used to make the reservation in the network, and assigns the appropriate bandwidth. Using sessions for the viewing also means that there has to be a teardown procedure when the user stops watching television – essentially, when the television set is turned off. This releases the network resources (so someone else can use them), and it also provides a trigger to the charging system, so that the user is not charged for watching programs and using network resources he is not accessing – if this is required.

IMS Communications Services

IMS is intended to keep track of users and manage their traffic, not handle the services they use, and their interactions with the services. In order to be able to handle quality of service and other aspects of service management, the IMS standards define a number of services.

In the client, more than one service can use the SIP protocol stack at the same time. This means they have to be separated, otherwise there would be no way to keep track of what should go from the SIP stack to the application. This is also true for IPTV, since it is displayed on end-user equipment, which can handle several communications services at the same time. How the different applications work with the common SIP stack is shown in Figure 12-6.

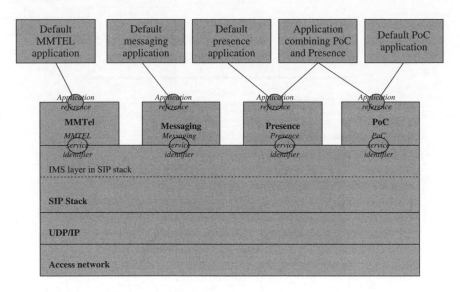

Figure 12-6. IMS communications services.

The way this is done is using the Communications Service Identifier (CSI) in IMS, which works in a similar way to the ports in IP, where a subaddress is attached to the IP address (the :8080 giving the HTTP port address). Each service has its own CSI, but each communications service can be composed of several components in SIP. A SIP application typically consists of a number of SIP functions bound together by service logic; there are APIs to create such applications, but not the CSI. IPTV does not yet have a CSI, since there is no established standard, and this takes time. The CSI is defined in the specification for each individual service, but to ensure that there is no overlap between CSIs, they should be registered. The CSI is sent in the SIP request, and as long as there are applications that can handle the data received, the SIP stack can route it. This works in theory, but not all SIP stacks implement it; if the CSI is not registered by a standards body, it may not route it correctly.

Routing and other handling of SIP signaling based on CSIs can be done by the CSCFs as well, for example, forking a request automatically when a user makes a request to a CSI.

The built-in services that come with IMS are the following:

- Presence

- Multimedia Telephony

- Immediate Messaging

- Push to Talk over Cellular (PoC)

Presence is a built-in feature of IMS, but not of IPTV, although presence is very easy to use together with IPTV. Since the architecture of the Open IPTV Forum is designed to set up a session between the service provider and the user, the user's presence can be registered when the session is set up (i.e., the user has started viewing the television program). Inside this session, there is a "bundle" of multicast channels, so there is no way for the service provider to know which of these channels the user is watching – unless the user's client explicitly informs the service provider, something that can be done using presence. It is different for video on demand, because the user automatically receives only the program he has requested, and the connection is unique to that user.

Presence uses SIP SUBSCRIBE and NOTIFY, but all IMS clients have a built-in messaging feature, which is part of the IMS specification. It allows the exchange of any type of content (e.g., audio, video, sound clips, etc.). It uses the SIP MESSAGE method, which is very limited in terms of the amount of data it can handle. If the data that is to be sent exceeds the limits of a SIP MESSAGE, there are two ways to handle the messaging (make an HTTP request; or send the message over SIP as a FLUTE message. FLUTE is a method that encapsulates a one-way content transfer within a session, and while that may sound useful, there are constraints on this as well. Messages cannot be too large, and because SIP uses UDP for transfer, there have to be consistency checks, which take more time the bigger the message is.

Like all SIP traffic, the SIP MESSAGE passes through the CSCFs in the network, and can be filtered (or forked) according to the logic downloaded from the HSS. It can also be rejected, especially if it is too large. IMS also provides a way to send messages to multiple recipients. This is done by creating a PSI for the group of users. The group can be created using XDMS, but also through other means; in either case, the PSI is bound to an application server, which will receive the message, and distribute it to all the group members. This can be used to trigger a renewal of multicast addresses for a transmission, or to create synchronizations with real-time events (or overlay them on top of existing transmissions, e.g., when an emergency occurs).

The SIP MESSAGE is a very simple messaging mechanism, intended to send short messages without prior notice (within a session). This is exactly how instant messaging (the IMS name is Immediate Messaging) works. When messaging is included in sessions, operators can also charge for the messages, as well as set policies on them, and control the messaging. That means the messages have to be inspected, and the inspection has to be translated into charging records. This is expensive in terms of resources, however.

There can be messaging-only sessions, as well, and messages can go between two or more senders and receivers, in essence creating a conference. In this case, a controller is needed for the conference; this can be an IMS application server, or an MRFC/MRFP, which is a controller for conferences in IMS. Regardless of this, sessions have to be established with the appropriate partners – this is one way to use the PSI.

There is, however, no guarantee for the message reception, since the assumption is that these messages are given only "best-effort" quality. Messages are sent using the messaging method defined in the SDP session description. However, because they are sent using SIP, they can be prioritized, and even though there is no way to guarantee synchronization with actual time in IMS, this means the messages can be sent in something close to real-time, and the control of media sessions can be closer to real time.

Push to Talk over Cellular (PoC), despite the unwieldy name, is a way of transmitting short messages of another kind from one user to a group of users. It was created to build walkie-talkie-like services in mobile phones, but it works with any agent (it is not hard to develop). Like IM, it is not a real-time service – the message can be sent with a small delay, because there is no sustained dialogue between the sender and receiver, it is only a session with voice messages in it.

The key to PoC is the group management, rather than the messaging mechanism itself. As with IM and presence, it assumes that there is a session in which the message is passed; the sender sends the message to the address of a list, rather than to an individual; and the system distributes it to all the receivers on the list (if they are available).

Multimedia telephony (MMTEL), however, is the gold standard of IMS services. It combines both telephony and video conferencing into one package. Unfortunately, many people in the telecommunications industry suffer from the illusion that people want to use this from their television sets, and constrict themselves to bending the system to accommodate this idea.

That users will want video conferencing is probable; that they want telephony is without doubt. But they do not want to use it from their television sets; this has been demonstrated (most often by accident) in several of the trials of interactive systems of cable-TV providers and IPTV service providers. The PC is a different animal and very likely to be a client for IPTV, video conferencing and telephony (it already is to many people). It is also possible to install the necessary client software in a PC (different from television sets). Interactivity with video conferencing is an intriguing idea, but for the next few years it will most likely remain a pipe dream, and we will not discuss it further in this book. There are many other good books about IMS that discuss videoconferencing, and many other books that give a good introduction to MMTEL.

Handling Quality of Service

When the session is set up, media can be delivered. For some types of media, this requires guarantees that there will not be latencies and interruptions, because the data stream is very sensitive to these – even if there is a large buffer, it has to be dimensioned; and having a buffer presents a problem when there is an assumption that all users will receive the service at the same time.

Media streams (because, as we described in previous chapters, it is streams that we are talking about in the IPTV system) can be delivered either directly to one single user (a method called unicast), or to a group of users, who subscribe to a group to which the same stream is delivered (a method called multicast). There are two extremes in this example: a multicast group of one, which gives the same result as unicast; and a multicast group of everyone, which gives the same result as broadcast in a traditional television system. This, however, would be a very bad way of using network capacity.

In IMS, the control over the media server is done through the session, and a control message in IMS has to pass through more nodes than the media session itself does. On the other hand, the streaming session does not have any functions for handling media control, except the most necessary ones. The direct control of the stream (fast-forward, back, skip forward, etc.) goes straight to the media server without passing the control server. This applies both to multicast and unicast.

One consequence of the ability to send streams to many different receivers is that they may not be able to handle the media stream. In a traditional cable television (and IPTV) system this is handled by having a very limited set of capabilities in the set-top box, which handles all the rendering of the media. But if the rendering does not take place in the set-top box, but, for example, in a DLNA television, transcoding will have to take place – probably in the local network, but maybe also in the media server, depending on what transcoding is required.

IMS has three layers, which the nodes of the IMS system map into: the media layer; the session or control layer; and the application layer. These layers roughly map to layers 5, 6 and 7 in the OSI model. The mapping is not straightforward (not in the least because the OSI model does not define the layers that clearly).

Like most standards, the model defines a set of interfaces, not the actual functions (although in keeping with the tradition in the telecom world, the IMS standards build on the interfaces defined in the protocol specifications of the IETF, and add nodes to them). Standards from the ITU and 3GPP typically define the standards in terms of reference points, which not only define the interfaces, but also the nodes between which these interfaces can be applied, in the process effectively defining the system architecture as well as the interfaces.

IMS nodes do not, of course, reach directly into the routers and set the QoS level. They have to go through a number of intermediaries, and the intermediaries have to work though well-defined interfaces. But the end result is the same: based on the user's session, IMS nodes control the quality of service.

IMS separates the media layer from the control layer (this is different from HTTP). This separation – into control layer and media layer – also works for RTSP and RTP, the two other main protocols used in IPTV and television over the web. RTSP is a control protocol, RTP a media transport protocol (which is also used in IMS). This ability to separate the control and transport does not matter today, because the session management is applied to the network, and there are too few users to make a real impact in networks. However, the number of users and the packets they receive can quickly congest the network, which will become especially noticeable if the content received is sensitive to disruptions (as IPTV is), and it becomes even more sensitive if the content has to be received in near real time.

SIP, the control protocol of IMS, is a session protocol. It does not work like HTTP, where the fundamental unit of exchange is the transaction – the query for a resource (the GET) and its response (when the resource is delivered). SIP sets up a session – it says to the node at the other side "establish a connection with me", and when that connection is established, it can hand over to another protocol (e.g., RTP, the streaming protocol most often used for IPTV). IMS is based on SIP, but it is not a part of the SIP family of standards, any more than Web 2.0 is a part of the HTTP standard (it is the other way around, of course).

Streaming using a control protocol works in the same way in Flash, but the protocols used there are not standardized. Standardizing the protocols used has advantages and disadvantages: the main advantage is creating an agreement in the industry, and hence a bigger market for those who have agreed to use it. The disadvantage is the opposite of this: if you agree to use a standard for your products, you are constrained in what you can do – you have to comply to the standard. A proprietary solution can be more efficient (and it often is, because standardized solutions are based on compromises between the participants in the standardization). These tradeoffs are very visible in the IPTV market: there are emerging standards, which are full of compromises; on the other hand, there are a few proprietary solutions, but they do not work with the solutions from other vendors.

This does not matter as long as you buy the entire system from one vendor, then neither user nor service provider can do anything with the client or servers. It starts to matter when the very idea of the system is that components from different vendors should work together. It becomes especially important if the underlying idea is that the network is the only important thing, instead of just one of the components of the system. If the system requires a physical (or at least dedicated) connection between the service provider and the client, assuming the relation to be permanent, the assumption that you are not able to modify any of the components makes sense. But if the assumption is that the network consists of a number of cooperating, but independent, components – then this does not hold any more. And that is what is happening in the new IPTV systems.

When you decouple the different parts of the system from each other, you allow not only for connecting components from different manufacturers to each other – you also make it possible to connect components which are connected in different networks, through different service providers. This is possible using IMS, but a problem is that it requires agreements between the service providers – which are not in place.

When a user is setting up an IMS session, the user's client and the IMS servers exchange identity and security information to create a header, the P-asserted identity header, which is carried in one of the header fields of the SIP protocol. Since this header can only be established by this specific pair of client and server, this also represents a security mechanism, creating some level of trust between the sender and receiver, through the mediation of the network provider.

When a session is set up, the requestor – the computer that requested the session – sends along a description of its capabilities. This is a small text file appended to the header of the request packet (there are other ways of doing this under development, but since they are not yet deployed we do not need to go into them here). The Session Description Protocol (SDP) contains information about the sender, which will be useful to the receiver when creating the session.

In IMS-based IPTV systems – this is a generic feature of IMS, not unique to IPTV – the service is delivered by an application server. IMS application servers are defined in the IMS standard, and use a special interface, the ISC interface (which is based on SIP as is the rest of the IMS system) to connect to the IMS Core. There are two ways which clients in the IMS system can address the AS: using a PSI [3GPP TS 23.228]; or through linking using triggers in the ICS interface (in which case it is the CSCF that decides when to connect to it, and which AS to connect to). IMS application servers can act as a proxy (which works in much the same way as the perhaps more familiar HTTP proxy, passing packets and doing some intermediate storage and limited processing of the signaling); as a User Agent (UA), which is what clients are called in the IMS world; or as a "back to back user agent" (B2BUA) –a user agent with two faces, and some service logic in between.

Any system that delivers a service in IMS can be broken down into a number of more or less native functions, but the IPTV system in particular can be broken down into the following parts, which handle the control of the media delivery.

Service Discovery

This part of the system is used by the user agent to discover the IPTV service, enabling the user agent to find the appropriate application server. Service discovery may be connected to the authorization, which in turn may be connected to the charging part of the system, enabling the discovery of a service to be conditional on whether the user who tries to discover it has paid to use it or not.

Control Function

The service discovery is coupled to the service itself – in this case, the ability to handle the streaming of data. SIP does not have functions to enable signaling between the user and the streaming server in the same way as RTSP, but RTSP can act within a SIP session. The media server can be controlled using RTSP, or there can be a more efficient, specialized protocol to control it. In either case, the control function is the termination point for the SIP traffic – where the session between the IPTV client and the system ends. This means that the control function also keeps track of the P-Asserted-Identity, and how this relates to subscription. That in turns enables user profile management, logging and statistics management related to the IPTV usage – and charging and authorization based on the session.

NPVR Function

RTSP was defined to be a protocol for network personal video recorders; today, the demand for this function has been greatly reduced by the emergence of hard-disc based personal video recorders. However, RTSP can still be a useful function, for example, in cases where the PVR is switched off and the user still wants to record, or where infinite memory is required, or where the content provider requires conditions to be set on the recording, and wants to be sure they are enforced (e.g., when the recording can only be played once). As discussed in Chapters 6 and 7, controlling the replay of what is recorded either requires a trusted agent in the user's system, or that the function is centralized.

The PVR and nPVR are often presented as each other's opposites, but there are many ways in which they can cooperate and the recording improved by the use of the two systems together.

In IMS-based IPTV systems, the nPVR is typically a B2BUA, which sets up the stream towards the IPTV streaming server on behalf of the user; and acts towards the user as a streaming server – either by creating two different sessions, or by inserting itself in the same session.

Since all packets on the Internet are created equally, they all take their place in the queue when transmitted by the router. There is no discrimination: packets containing billions of dollars in payments have to step in line behind lowly pornography videos (actually, the banks have a separate network for their payments).

This is an advantage when building routers, as it lets the router manufacturer focus on the decoding of the packet and the address resolution. It is a problem when it comes to watching television transmitted over the Internet, because your book review discussion comes from the Internet provider to the nearest router mixed up with the celebrity channel of your neighbor, and both of them come mixed up with the chat channels of your teenage daughters, and the games your sons are playing.

If you are watching television, you will not appreciate the screen going blank for a few milliseconds while the packets from the sender are received into the buffer of your receiver and decoded. This may happen if there is a delay, due to your packets being squeezed together with other packets through the router. One solution is to install faster routers and networks; another is to prioritize the traffic so that television packets get through faster. This is one of the most important functions of IMS in IPTV.

To start with, a short explanation of how the networks work. Every network is a combination of the physical characteristics of the medium through which it is transmitted (the air, an optical fiber, or a copper cable), and the router (or bridge), which handles the interface to the network. The physical characteristics limit how fast the transmission can be, since the physical medium limits the speed of the electrical (or optical) signals. This limitation makes it possible to build fiber networks, which are much faster than copper networks (which in turn are faster than radio networks), since the time the bits are transmitted can be shorter.

The second limitation comes from the router, which handles the signaling over the medium. Today, a router (which selects the interface, and hence the medium over which the bits in the packet are to be transmitted based on the network address) is the norm where networks meet. Where the number of users on an interface is large, the router has to have a large capacity to be able to handle them all; as the number of users becomes less, the capacity can be less. The router closest to the user is also the least capable.

In addition, not all network directions are equal. ADSL can transmit data over copper cables at bandwidths which were unthinkable when the traffic in the network was only telephone calls because (among other reasons) it is not symmetric, i.e., the bandwidth of the traffic coming from the sender is much bigger than the bandwidth the user is able to use to send data back to the network.

Often, a generic solution to all network ills is more bandwidth – and if that does not help, more bandwidth. Adding capacity in the network is a cheap way to do away with some problems, especially those which come from bottlenecks in the transmission medium itself. If you open a second hole in a dyke, the water runs out twice as fast. This is a very simple model, and most network designers understand and apply it, sometimes absurdly (such as the researcher who gave his mother a terabit connection).

It works fine for downloading data, because there it is very similar to pumping water: the data goes from the source to the sink, and the important thing is how fast it gets to the sink. But this approach does not work for streaming, in particular real-time streaming, because the problem is different: it is not sufficient to pump water through faster, you have to pump it in the right order, and fast enough so that there is always a little left – but not too much. It is more like emptying a bowl full of numbered golf balls and handing them to a group of players, who have to tee off in the order of their handicap. In this analogy, it will not help if you take out two golf balls instead of one, they will still be played one by one. If you use a vacuum cleaner to suck up the balls and blow them to each player, the speed at which you blow the balls becomes critical. If you replace the golf balls by basket balls, you can get a bigger vacuum cleaner, but the rate at which the players tee them off will still be the same. If you are late, there will be gaps in the synchronized teeing off, and service disruptions. This is what latency is about.

So in this analogy, it is the vacuum cleaner and the players who are the bottleneck, not the size of the hose. In an IPTV network, it is the routers and the receivers – and they are a bottleneck because they have to play out the received data stream at a precise rate for it to look good, without glitches. If there are hang ups, for example, because the vacuum cleaner is too weak to pump the balls at a high enough rate, there will be glitches in the signal. Since the human eye is very sensitive to this kind of thing, this will mean visible disruptions, which users react very badly to.

Latency, not bandwidth, is the biggest problem in IPTV networks. And the key to getting rid of latency is the routers, not the raw network capacity. There are already problems in many IPTV systems, small though they are, in delivering live streams to all users at the same time. Even though live television viewing is shrinking all over the world, having the ability to get live coverage of events is crucial, especially if it is events that cannot be sent by the regular television channels.

Two things are important to guarantee that live video is delivered to the user without interruption: making sure all routers are capable of handling the delivery; and making sure they prioritize the traffic so there are no latencies. The router closest to the user being the least capable in the chain of routers in the network, it is here where the congestion potentially becomes the worst – especially if there are a number of users who want to watch different television programs at the same time. Multicast can solve the issue of the transmission in the network being a source of congestion, but it cannot solve the problem of the games of one player interfering with the television viewing of another.

The solution to this is called quality of service, and it means, put very simply, that certain types of network traffic go ahead in the queue. An IMS system can control the SIP sessions in the network segments leading up to the network in the end-user's home. The traffic management is done by coordinating the routers in the network leading up to the interface towards the end-user's router, including the equipment which handles the ADSL transmission.

You frequently hear the claim that quality of service management is not necessary, since the problem can be solved by adding more network capacity. This is true to some extent, but it is also a matter of tradeoffs between cost and speed of upgrade. It takes time and is costly to change all the network equipment required to make a system ready to handle high definition television, and it may be more expedient to use routers that are capable of handling quality of service (in most cases a software upgrade). Adding IMS control to those routers will probably only require another software upgrade.

In an IMS system, the quality of service control and management is based on the Resource and Admission Control Subsystem (RACS) specified in the ETSI group TISPAN R1. The RACS is responsible for the control of which types of packets are allowed to go ahead (policy control), the reservation of resources for the data transmission, and the admission control – that you are allowed to transmit data over the network.

When you want to use quality of service, you have to make sure that there is sufficient capacity available in the router which is going to handle the traffic. These resources include the capacity on the network, the resources of the router (computational, memory, and otherwise), and are set in the buffer management of the router, where the queuing is handled – effectively shooing the non-IPTV packets to the back of the queue. Going from the SIP protocol in the application layer to the network is not entirely straightforward, since this crosses a number of borders (which are supposed to be isolating the different layers of the network from each other) making a connection between the application and the link layer.

When an IMS service, such as IPTV, wants to reserve resources for quality of service, the P-CSCF (or another node working as an Application Function (AF) such as another CSCF) maps the application layer information about QoS (e.g., described in the SDP header of the SIP packet) to a request for QoS, which is sent through the policy management function, which determines what QoS the requested information requires. The receiver of this is the Access Resource Admission Control Function (A-RACF), which can accept or reject the request for resources in the transport interfaces it has control over. A message is generated back to the AF, which either confirms, rejects or modifies the reservation request. If the resource request is rejected, the P-CSCF simply rejects the SIP INVITE or SIP UPDATE, using, for example, the "Precondition Failure" status code in SIP. How these different nodes work together is shown in Figure 12-7.

Inside the home network, there is typically no QoS management, since this both adds significant extra cost to the router used, and it is easy to provide sufficient bandwidth to the equipment in the home anyway – if everyone is not watching IPTV. The DLNA has produced a set of guidelines for how to handle traffic according to the IEEE 802.1p and WMM [KOLLA] standards, which makes it possible to map the prioritization in the wide-area network

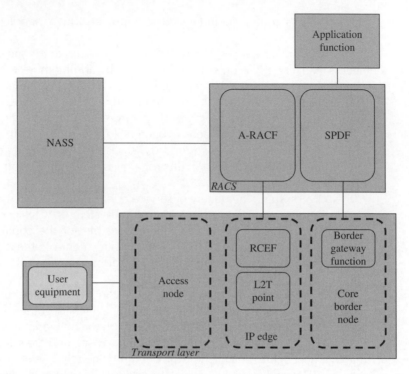

Figure 12-7. RACS and CSCF.

to the prioritization in the home, but this takes resources and hence cost, and is rarely implemented.

All this may seem like a lot to go through to make sure data packets go faster, and you could legitimately ask if it does not in itself slow down the network. The answer is "not really"; the systems have been implemented in

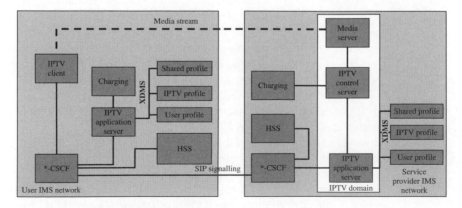

Figure 12-8. How IPTV fits into IMS.

such an efficient way that the routers in the network, which are the parties most concerned, do not see a noticeable delay. This is also true for the other functions of IMS (which, like the QoS reservation, only have to be applied once during a session). How IPTV fits into IMS is shown in Figure 12-8.

Connecting Application Servers: the ISC Interface

The services in an IMS system are executed in application servers. The term may be more impressive than the actual result, since an application server is anything that can provide a service over SIP with the additions that IMS provides. In the architecture for IPTV that we discuss in this book, the IPTV service will be provided by an application server (AS). The AS can be placed in the home network, or in an external network (which may be a data center).

The AS is connected to the S-CSCF through the ISC interface. The AS can be a very simple device (any server), or it can be a whole cluster and subsystem itself. The term "application server" does not have to be a server in the box sense, it refers rather to a server in the client–server model.

A SIP AS can host services and execute them; it can also influence and impact the session on behalf of the services (through the S-CSCF); so the AS does not have to be the endpoint of the session, although the assumption is normally that it is. The ISC interface allows the AS to subscribe to events and other session-related information such as the capabilities included in the SDP. But it is the S-CSCF that decides if the AS should be able to get information, and it does this based on information that the HSS has, which is used to filter the session. The HSS keeps track of the AS as well. This information is related to the user (the PUID), and the filtering mechanism is standardized by the IETF.

When the S-CSCF receives the name and/or address for an AS from the HSS, it contacts the application server. If it gets more than one AS, it contacts them in the order that received from the HSS. It does not stop after going to the first one, but instead includes the answer in the request to the second one. This means that the HSS can chain application servers – this can be used in IPTV, for example, where the first AS is the VoD server and the second is the charging server (the charging information can be conveyed through ISC, although charging servers often have another interface, based on Diameter, to the HSS). This means that when the first AS starts the VoD delivery, it will automatically trigger the charging for the video from another server. An AS can generate SIP requests and dialogues on behalf of a user. These requests are forwarded to the S-CSCF that serves the user, and the S-CSCF will handle these in the same way as any request from the user.

Since IMS comes from the telecoms industry, and the original assumption is that the sessions would be used for telephone calls containing a real-time communication, concepts from the telecom industry feature heavily in the design of IMS – not so much when discussing the design of the system, but when discussing the ways it can be deployed. The specifications make a big deal of how different nodes can route a session, and talks a lot about how different legs of the session can be connected. This may not be very interesting from an Internet point of view, because the ideas behind Internet applications are different, but it is likely that if you are starting to learn IMS, you will most likely start with an application server for media applications such as IPTV, which means you will come across very different aspects of IMS than you would if you were creating a voice application server. The application server can interact with the SIP sessions in various ways, including creating new sessions, as is shown in Figure 12-9.

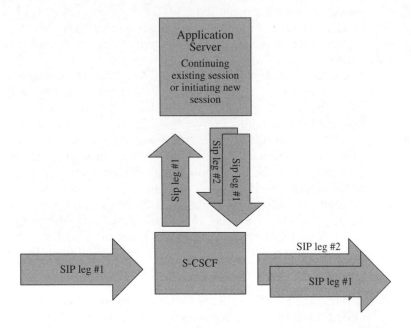

Figure 12-9. The application server and IMS.

When analyzing how IMS works with telephony, knowing how the different sessions work precisely together in real time is very important. But when analyzing how the system would work for IPTV, it is far less relevant –call control is not an issue (video telephony, although widely promoted as the application of the future since the 1950s, has yet to take off). There is still good reason to remember the communications applications of IMS. One reason to build an IPTV system on IMS is the ability to mix the IPTV service with other services, like multimedia telephony (the infamous video telephony being the preferred application, but there are many others that can be realized this way).

The AS uses the ISC interface to communicate with the S-CSCF, and handle the communication with the sessions. The control information can either be carried inside an existing or already initiated SIP transaction (implicitly), or a transaction can be generated specifically for the purpose of communicating the state information (explicitly).

The S-CSCF is in many ways acting as a proxy (another example where the HTTP roots of SIP shines through), by routing requests into or away from the AS, and by keeping state information for the session (something that is not mandatory, however – in the SIP specification, only the endpoints have to keep the state). Since the CSCF is also a B2BUA, which means that it can not only forward and store data like a proxy, it can also terminate a session and continue a different session through the other face of the B2BUA, or just forward the packets – but after modification.

The authentication, authorization and user identity management is done by the HSS, not the S-CSCF. Therefore, to communicate with the HSS the AS has to implement another interface, the Sh interface. The HSS determines what information will be given out to each individual application server. The Sh interface is partly used for the information that the HSS does not handle, such as the service-related signaling, but its main purpose is as a mechanism to handle the user-related information stored in the HSS, such as user service-related data, MSISDN, visited network capabilities, user location (which can be cell global ID/SAI or the address of the serving network element, etc., but not GPS coordinates, since these are not part of the IMS standard, and are handled in the presence information). Location information is sensitive, and may be handled through another interface, the Le interface to the location information server, where the privacy control is better and more appropriate; this means that the HSS must ensure that the location information is not given out over the Sh interface.

The Sh interface can also be used to handle group lists and other standardized data which ASs can use, but ASs must understand the data format. In this way, the Sh interface can be used to manage sharing of data between the ASs. Each AS can also have filter criteria, tied to its identity as an IMS subscriber, which apply to how it handles SUBSCRIBE messages, i.e. the filters can be used to determine when and how notifications should be sent.

Chapter 13: Developing and Deploying IPTV

Television that is not only for couch potatoes, but as interactive as computer games is the theme of this book. It means changing much more than adding a backchannel: displaying the interaction opportunities, capturing them, and sending them to a central service using a trusted and secure method. It means using the users preferences and other information to personalize the content – including the advertising. It means producing a different kind of television show, that will be as different from television of today as "Big Brother" is different from "20 questions". Interacting with television is a way of turning the medium from hot to cold in the sense of Marshall McLuhan; from the user leaning back to the user leaning forward, and taking part in the creation of the TV show, if not the production itself.

Television viewing is full of social conventions, and the way programs are presented has become so ingrained that it is no easier to change than the clutch, brake, and accelerator pedals in a car. Even if there is a better way to do it, the force of habit is hard to break. There are plenty of experiments with different ways of storytelling, for example nonlinear videos (think of it as curved loops of stories turning back at each other), which create a different experience. But the linear flow, where things happen along the sequence of time, will dominate IPTV programming for a long time to come. It is ingrained in the culture, not just in the western world.

But "IPTV" has two parts. If the "TV" part is resistant to change, the "IP" part is not. Even when broadcasts were analog, there were pioneers trying out ways to interact with the audience through chat and web pages – and although the formats were interesting, they were never successes. So what can you do with interactive television?

It's life, Jim, but not as we know it

Television over IP is more efficient than today's data network usage and leverages the installed base of the telecom operator. For the regular viewer, this could hardly be less interesting. The main value of television over IP, as it has been deployed up to now, is completely invisible to the end-user. At the same time, the value proposition is built into the technology: the backchannel is integrated into the distribution network, instead of being separate from the network. That will mean changes to how the viewer interacts with television, and how IPTV is developed and deployed.

IPTV is more efficient as a data transport, but if that was all there was to it, it would hardly motivate the investment. It is the services that can be created using IPTV that will drive the development. Interactivity is an ingrained part of IPTV, and where the most money is to be made in the short term for all the parties involved. One reason is that it can build on models for interactive programs that are already well established – but it can also extend the models considerably, making new kinds of programs possible.

With IPTV, the service provider can also make more money. One example is flat-rate, which always leads to more usage, and even if the revenue per interaction is lower than when you charge the user every time they click, there are different ways of making more money, for example, inserting advertising. That will make users more willing to interact, and make the television experience "stickier". Viewers will be more likely to stay around to watch the continuation of the show since they feel they are part of creating the experience, and this will mean more value in the advertising, because the advertiser has guaranteed viewers.

With IPTV it is possible to track each individual user. Even if that may not be a good idea from a privacy point of view, you can prove who watched what, and by aggregating the data, making them useful to advertisers. Statistics makes it easy to know who watched what when, what demographic they belong to, and who opted out from what. Something, incidentally, many advertisers are willing to pay a premium for.

But if advertisers and viewers stand to benefit, the broadcasters are in for the biggest change. They have to change their mental model from a one-way pipe to a co-creator with their audience. User interaction has already changed television, not only through programs such as **Pop Idol**, **Castaway** and **Big Brother**, but also through the daily interaction with viewers who find that the barriers to communicating with the television studios have become lower since the arrival of the Internet, and that sending emails to a reporter is much simpler than writing letters to the editor.

What is in it for all of us is a lot of change – for the better. A more engaging, yet more convenient user experience; more money for advertisers and broadcasters, and a more interesting evening on the sofa with the family for everyone.

So what does this mean in terms of the technologies described in this book? Here are some examples.

Enhancing Voting

Today, you can vote on the result of shows such as the **Eurovision Song Contest** and **Big Brother**. The viewers have to call or send a text message to a special telephone number. Since this is a special number, the operator charges a premium price for the service. The revenue is shared between the telecom operator, the service provider who manages the service, and the broadcaster. This is one reason why broadcasters are promoting voting so heavily, and it is also the reason it is not likely to disappear – it is very profitable for the broadcaster. One of the challenges of interactive IPTV is how to move this feature into the network (making it a part of the service), instead of doing it out of band (through the third-party connection of the telephone line).

The production is no different from traditional television. Production and encoding are done in the same way as when television was analog. The interaction takes place at a well-defined spot in the show, and the script can easily take it into account. It is no different from having viewers dial in, perhaps during the advertising break. Once, this was an invention, now, it is just run of the mill. Most interactive television applications deployed today work by batching user interactions during a response period for a show, for example, when users are voting (and in those cases the interaction takes place over a completely different system, such as the telephone or mobile (using SMS).

When the user votes with SMS, the mobile phone sends a short message to a server in the network (addressed using the special telephone number). Since this is a "premium SMS" (because the number is special), it is more expensive to send than regular SMSs. In the current interactive television system, the voter pays using his telephone bill; the network operator gets the money and pays a cut to the service provider and the broadcaster. It works the same way when the user calls a number to interact.

However, when the interaction can alter the content of the show, the production has to change. The presentation of the interaction becomes important – and also the technology: the protocols used for the backchannel and the servers which handle the interaction have to be dimensioned to handle the viewers, and they have to do it in real time. If they did not, the program might as well consist of a number of pre-recorded segments automatically combined depending on user feedback.

That is how many experimental formats (and many computer games) work, but here, the interaction might as well be done using email, since there is no live interaction with other people (in the studio and in the audience). The latency in the interaction removes the difference in games and personalized video on demand.

Using IMS and the other systems described in this book, two things happen. First, it becomes possible to reduce the number of middlemen – cutting out the service provider who connects the SMS message system to the broadcaster. Little loved, and the cause of recent scandals in the UK because of rigging the result of contests (to speed up the processing), they have to find a new niche to survive. In IMS, the user has an identity, which is secured by the service provider; and there is a session with the IPTV application server from the user's client, because it is required to watch the content. The server only has to receive the incoming messages, keep track of what interactions they should result in, and send the result on to the studio where it can be displayed on the screen in front of the host, quizmaster, or whatever is appropriate due to the show format. For the developer, this should not be too complex, using the standard interfaces, adding appropriate database requests and other messages. See Figure 13-1.

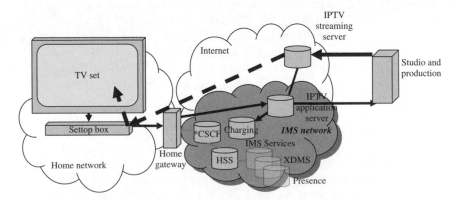

Figure 13-1. Voting without SMS.

Viewers want to interact with television. And they are willing to pay for it. For most people, voting in shows such as the ***Eurovision Song Contest*** is a way of being social. With IPTV, the ability of the user to interact and be social improves drastically. No longer confined to voting, interaction can be anything from requesting statistics, to cheering for the home team, to seeing who else in the fan club is watching at the same time. Since IMS has a messaging system built in, it is very easy for the IPTV application server (or a separate IMS application server) to direct interaction messages to the appropriate service – and to show them on the user's television screen as a result. The user's interactions can be fed back to the television producer, and the server can be the same server that is used to manage the interactive scripts, or it can be another server. IMS and the SIP protocol can easily be used to coordinate the sessions, and hence the way the overlay is produced for the television screen.

Television is social, and so is interaction. Even watching clips on YouTube is not a solitary activity (the comment function is there for the comments to be read by someone else); watching a football game is a way of participating remotely, or at least sharing in the support of the team. It may seem counterintuitive that doing something alone can mean being social, but television (or rather, the shows on television) are social objects, which focus the attention of the group subscribing to them, and makes viewing – even as an individual – a social act. Technology will not change that, but it can be used to enhance it.

Another type of interaction, where the viewer becomes a co-creator, is mashups. Here, the television show is combined with something the user has created, for example, a map or subtitles. The other media stream comes from a different source than the video signal, and the combination creates something completely new. Mashups are not the primary focus of this book, but the distance between user-created interactive television and mashups is minimal.

Implementing IPTV will create new ways of mashing not only web pages with each other, but also television with the web, and other media such as music and radio. While this may be something many wish for, it is a problem for traditional media, where the control of distribution means being able to charge for the media. From the user perspective, interacting with IPTV can be used to rewrite the script of the show for the presenter and the participants – literally.

Automating Scriptwriting

In interactive television productions today, viewers can phone in and vote. The votes are tallied on the screen to the comments from the hosts of the show. On receiving the votes, the participants and presenters take action; these actions are scripted in advance (only rarely do real surprises occur, even on live television). As discussed in Chapter 1, there have been some attempts at creating automatic tools for scriptwriting, such as the LIVE EU research project, where different stories could be brought in through parallel streams, so that the producer can manage and combine them appropriately, but this was for sports events, where the event flow is known from the start (even if you do not know who will win, you know there will be a winner).

Making a television show truly interactive requires three things:

1. A way to capture the user's interactions and send them to the broadcaster.
2. A way for the broadcaster to change the program on the fly.
3. A way for the result to become visible on the television screen.

The hard part is not, as you might believe, to capture the user's interactions and transmit them to the studio. For a producer, and for the creative team devising new television formats, it is much more complicated to understand how to use the "red button", and turn the viewer's key presses into something meaningful. But interaction is not individual, it is interaction as a group, through voting, which makes most sense for both users and broadcasters.

When a user votes on interactive television, this extends the vote of the panel in the studio. The result is the same: one contestant is selected over the others. Depending on the difficulty of the voting, the user can be prompted to vote more or less often. It may not be appropriate to do it too often, since the user may feel stressed by having to act all the time, instead of feeling like a part of the show. The trick here is to make sure the user's voice is being heard, and that the show really changes according to his interactions.

If the format of the television show changes when the users start interacting with the system, it becomes easier to create an application around it, and to create a different kind of show around the application. This is true regardless of whether the interactions are from one individual or from millions. If the interaction comes from a large number of users, it becomes impossible to take their individual actions into account (the opposite of a call-in, whether using video conference, chat, or plain old voice telephony – when one single user is participating in the show, at the potential expense of the others). By tallying the users' votes, users can still feel they are participating in deciding what the next question in a quiz show (or camera angle in a sports event) should be.

Developing interactive television today is like any media development project, and it is not likely to change, since "interactive television" is by definition based on the television experience. The business model is no different from the business models of the industry today; nor is the project management. There may have to be a couple of different players involved, to provide for the difference created by interactivity, but the model does not change.

First, the interactivity has to make the viewers feel more involved in the show. That is not as easy as it sounds, since television traditionally is not interactive. When creating the pitch for the production, there has to be an interactive element for each show, and a way in which the producer can make people know their involvement is wanted. If there is one important aspect about "young people" driving the user experience, it is that they expect to be spoken with, not spoken to. The producer has to listen.

This means the interactivity should not be bolted on as an afterthought. If so, it will become wooden and stiff at best, at worst a limp poke badly masked as a way of selling more advertising (even in that case, however, it has to be included in the budget). If the program does not use the appropriate methods for its particular interactivity, then it has failed from the start. When creating the pitch for an interactive television program, the project manager has to be aware of what interactivity methods he has to choose from, and select the best one. This includes possible extensions of the available material, as well as research notes and other background material (in a medium to large production, this will be important in creating the DVD afterwards). Maybe there isn't, and the show is best without interactivity. Other uses for the production material is as prizes – items from the original **Star Trek** series regularly go for hundreds of thousands of dollars at auctions, and while it may be presumptuous to expect your show to be the next **Star Trek**, owning a piece of the production will help users relate to the next generation of the show. Be careful not to give away props which will make the sequel production easy, however; it will make the sequel more expensive if all props have to be reproduced.

Regardless of the importance of the storytelling and production values, television is all about surface, so it is important how things look and feel. Graphics should be created and included early, but not be locked down until the program is almost finished. Like any creative production, interactive television program productions are likely to change as the work progresses, and recommissioning the artwork can be expensive; it is better to retain a resident artist to make sure the graphics are always appropriate.

Exactly how to do the automatic scriptwriting (or rewriting, as the case may be) is where the creativity of the director and the skill of the scriptwriter comes

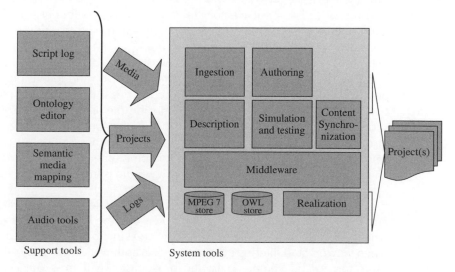

Figure 13-2. Automating scriptwriting for IPTV productions.

in (see Figure 13-2). One consequence is that the interactions have to be part of the script for the show, and that different votes should trigger different results. This has cascading effects. If the voters decide the contestant should get questions about wine, the result has to be wine questions; but the next question could also be decided by the first question (e.g., having asked about wine, the next layer of questions could be about the wine region covered in the question; and the next level of questions about history, and so on).

The script has to be modular and the different modules have appropriate hooks into each other; but the voting has to hook into this as well. For the scriptwriter, this presents an interesting challenge; for the developer, the interactions have to trigger different scripts, which are presented in the studio as the show is broadcasted. This is very similar to any database application – except for the result, of course. For the presenter, the challenge is to follow what is presented at the prompting screen in the studio, without knowing what comes next. Probably presenters will want to learn the different possible directions the script can take, at least in a general way.

Inserting Advertising

While the business model for IPTV has not yet settled down, it looks unlikely to be free. Video on demand is certainly rarely free, and the service provider tries to charge the user either per viewed show, or a monthly subscription. In some countries, there are free-to-air channels, which are financed by license fees or taxes on television sets or by similar means. Often, this means cable systems must carry them; and while the rules are not clear when it comes to IPTV, the IPTV providers will probably also have to carry the free-to-air channels. This is of course a constraint on the business model – on the other hand, the user has to have a network connection, and that you have to get from a network provider.

In the early days of television, there was no way to record programs. This was the true revolution of the VCR, and the reason the television industry opposed its introduction (and that many countries still leverage a tax on video cassettes). With the VCR, users could decide themselves when to watch a show – and to buy and rent shows and movies they could watch later. Passing through a number of technical iterations, the business model of the video store has been trying to move online for some time. With IPTV, it becomes an integrated part of the service. The only hurdle is the broadcasters opening up their libraries.

When viewers start drifting away from the predetermined program times, and begin consuming content at their pleasure, the mass-advertising model which dominates today's television starts creaking; but it really breaks down when viewers start filtering out commercials as they view the show. If enough users forego the commercials, the broadcaster will miss out on advertising income, and the economic basis of broadcast television disappears.

Cable-TV, especially digital cable where users pay a subscription fee for access to channels, provides a reliable alternative – but not if used in combination with a television recorder. While it is possible for the cable-TV provider to check what the viewers have watched and feed this information back to the advertiser, it does not help them if the user is recording the show to view later – and then filtering out the advertisements. Set-top box vendors try to alleviate this by disabling filtering, making the function less popular. IPTV has the potential to break this model too, since the user can filter out whatever they like. And they can freely change service provider, as long as the service provider has an Internet connection.

IPTV also provides a way for the broadcaster and the service provider to get paid, by charging the user for the services he uses within IPTV. Instead of charging the viewers for providing a generic IPTV service, those who do not want interactivity can opt out. IMS makes this possible through the following functions:

- unambiguous identification
- subscriber agreement with service rating agreed
- mechanism to request service
- charging system mechanisms:
  - rating system (providing information about how much each service costs)
  - fulfillment verification mechanism (making sure the user has received the service)

With IMS-based IPTV, the broadcaster has a way of monetizing the user's interactions, which means getting paid more by the user than they would have had they not interacted with the system. It also means giving users a higher degree of freedom, by allowing them to pick their own set of services. And it is also a way for broadcasters to offer different kinds of televisions shows. If the television of today is something to go by, there is a big risk that broadcasters will find it tempting to give away the interactivity instead of charging for it, in order to compete with other broadcasters. Finding out how to make money when the interaction becomes free is going to be a challenge for all parties in the value chain.

There are also other commercial reasons to believe in IPTV. While getting personalized advertisements is something people may welcome, it is difficult for everyone to realize its value. Advertising is geared towards mass media, and even if direct mail could be included in advertising, it would still be a mass distribution medium. Mass distribution means one to many, and in a many-to-many relationship this almost inevitably translates to middlemen inserting themselves to realize some of the value – think of Google, and how it realized the many-to-many values of small advertisements by becoming the aggregator and middleman. In television, there are entrenched commercial structures, and they will make it much harder for anyone who wants to become a new middleman. "Theirs to lose" is the term that springs to mind when it comes to broadcasters and advertising.

But personalization does not have to be done on an individual level. There is an advantage in the advertiser not knowing that it is me (the individual) who is watching, only that this show is popular among white males in Japan aged 40–50 (which is my demographic). On top of that, clever advertisers will want to remember that they are building a relationship with the user, not trying to force him to buy a cubic zirconium necklace.

Targeting not the individual but the generic viewer of a show is precisely what advertising today is about. But selecting the appropriate advertising for the perceived demographic is more or less a black art. In traditional television, statistics is something gathered after the broadcast, by specialized measurement agencies. This is a rather expensive process, and it sometimes yields surprising results – but since advertisers would not place adverts on television without knowing that they had some chance of reaching an audience likely to buy their products, they need an indication from the start that they are not wasting their money.

If the feedback and the selection of the programming can be automated, this means drastic changes for broadcasters and network operators (including Internet service providers), and also new possibilities and challenges for the program creators, and for all the other stakeholders in television – not in the least for the advertising industry, because advertising is likely to benefit the most from the change. It may even lose its position as the most reviled feature of television. This will result in improvements to the bottom line of the broadcasters, because not only does the new interactivity finally enabled by IPTV mean new sources of revenue, it also means new ways for broadcasters to manage that revenue.

For the service provider, it makes sense to ensure users have a subscription and connection to their service, since this means they can get programs in real time. If the program is engaging enough, the user will stay with the show. If subscriber fees are under constant pressure, and producers do not want to make programs which create value from individual interactions, then the hope for service providers is advertisers. Advertising means making money by selling the attention of the audience. But since there is a connection to each individual in IPTV, the demographics can be collected and a real-time view of who is watching right now can be provided. The upside is that television becomes direct marketing. The budget increases from the US$450 billion invested in television advertising every year, to include the US$200 billion invested in direct marketing (which by definition is personal).

In IPTV, the collection of statistics about the user is much easier, since the user (or at least the household) is logged in. The subscription is registered with the service provider, so it is easy to know who watches what, since each user has a session with the IPTV system all the time, and their presence within that session is registered and managed by the service provider. This means that even if the user is not profiled on an individual basis, it is possible to tell the advertisers that the users who watch this show switched channels five times before watching it; or that they made reservations in advance to be able to see the show (both of which is hard to know today, but which will raise the value for the advertiser, since the viewers actively selected the program).

The demographic information about the user is readily available, and easy to integrate in the IMS system. The profile information, which can be managed using XDMS, is not limited to which services a user is subscribing to, it could be anything from shoe size to average annual income. The only constraint is that the information cannot be used other than for the purpose it was collected. Clever aggregation of statistics can lead to much better precision than today, however. It can also be used for data mining and tracing of user behavior; even if this is done at a very high aggregation level, it will still lead to easily managed knowledge.

It is simple to create a group in XDMS that aggregates all user profiles, and calculates the average (and other statistic measures which may be of interest). Since XML processing is not very computation-heavy (the hard work is fetching all the profiles), this can be done quickly, in practice generating a real-time log of who is watching the show. Given the interactivity, which can also be taken into account, it may be interesting to know how active the users are during the show.

For an advertiser, this not only means better feedback on who is watching their advertisements, it also means being able to pull advertisements if the demography and viewer numbers do not fit the requirement; or conversely, it may mean adding advertisements in a show that has attractive demographics. The easiest way to do this is to auction the avails – the advertising slots – in real time. For the advertiser already familiar with auctions for avails through Google, this should not be a surprising, or hard to manage. It is even possible to set up automated auction systems as part of the advertising ingestion system, which take those advertisements which are most appropriate and where the advertiser is willing to pay most.

This is not a very difficult application to write (based on databases). The difficulty is changing the behavior of the advertiser and the advertising agencies, one of the most entrenched and hard-to-change relationship in the industry.

One of those reasons is trust: to capture the interactions of many users, collating it with their demographic profiles, anonymizing the result, and selling it on to an advertiser is something that can be done only when the service provider is trusted by those whose data he sells. If there is no trust, there will be no business. Trust has to be earned, of course, and it has to be respected. It is unfortunate that so many telecom companies have such a bad track record in this regard; those who have earned their users' trust and respect will probably find this aspect of providing an IPTV service rewarding, both for themselves and the end-users. Confidentiality goes hand in hand with it. It is easier to maintain the individual user's anonymity if you can hide him in a crowd; much harder if the user can be traced from his connection.

Another reason is the aggregation itself. While it is possible to get (and even to keep) the profile of each user in the client in his home, the network would be bogged down with requests if everyone who wished to get the user's profile had to ask him directly. Small though it may be, an update of a requested profile is not multiplied by 10 if 10 user profiles are requested by 10 service providers from 10 different clients; it increases tenfold. A few bytes here and there may not sound much, but multiply it with a million users and it will soon become a base load in the network, which has to be prioritized by the routers (increasing their load), and which steals bandwidth from the end-users. More traffic means more lost packets, which may mean latencies at the endpoint, and that translates into a worse user experience.

Aggregation also makes sense based on the services. A recommender engine works best when it has a number of users to make recommendations from; and while personalized television (interactive television shows based on video clips spliced together due to the user's interactions and profiles) may be most interesting on an individual basis, creating a live broadcast which is one to one is an expense most households cannot bear. It is far more efficient, in this case, to produce a broadcast that is customized for the group watching it.

Getting paid for providing videos on demand still requires more than a big server. Above all, it requires a way of tracking the usage. This can also be put to use in tracking how many people watch advertising – and which advertising. Advertising is, simply put, a way to capitalize on the interest created by the editorial team to send a different message to the user. There are border cases, such as infomercials, where the content is paid for by someone, but the "normal" form of television advertising is a way to send a message to many users at the same time.

Television used to be ideal for this, because broadcasts can be immediate and the same message can reach the entire group of receivers. But as television has fragmented – into many different audiences, and into many different timeslots (timeshifting) – advertising has started to suffer. Advertisers have increasingly started to experiment with more interactive formats, such as advertising on the web; but the very nature of advertising is to interrupt and direct interest somewhere else for a short while. The personal nature of the user experience on the web makes it less suitable for advertising which is directed at changing the user's perception of a brand (as opposed to informing them about an opportunity, which the small print ads do – and Google is a living proof of success that this is a viable way of using the web).

Video advertising is effective for certain types of messages, but it is notoriously hard to measure, and while it may be possible to measure the effect of sales the day after an advertisement was shown, the broadcaster has no clue as to whether it was in this program that the advertisement was effective, and even whether it was in his programs at all. This is where measurement agencies come in, trying to gauge the interest of users by sampling the potential audiences.

Personalizing Television

With IPTV, the choices available to the user become easy to display. They can be personalized – including only those options which are of interest to the user who is logged in at that moment. Creating a menu and sending it to the set-top box, capturing the user's actions (pressing buttons on the remote control) and translating the feedback to a SIP MESSAGE, which is sent back to the server inside the same session that the user set up for his original viewing of the television station, are easy. The same goes for interaction with the service. Sending messages is part of IMS; the display on the screen uses an HTML4CE overlay, which essentially means the familiar Ajax methods of an XHTML document with embedded JavaScript. The graphics of the menus and the applications they control have to be created, of course; and there are some niceties which have to be sorted out when it comes to getting the menu to the user. However, the method for capturing user interactions, and including them in user profiles, is a well-known and standardized part of IMS.

There are a few hurdles on the way. The consumer electronics industry has developed tremendously in the past 20 years – but it has been an incremental improvement. A television set is still a screen with which to receive a broadcast signal, even if it is HDTV, plasma and digital. Also consumer electronics now has competition from the computer industry – both the personal computer and the games consoles, which are frenetically turning themselves into set-top boxes. While a home stereo or a television are specialized for what they do, the PC is generic – it can be a stereo, a DVD player and a games console. There is a tradeoff – it becomes good at so many things that it cannot be best at all of them – and the consumer electronics industry has stood up to the competition well so far.

The big threat to the consumer electronics industry is not the PC moving into the home, it is the network connection of their devices. A television set with network access can be used for web browsing (but not very well), and it can be used to interact with television programs. The consumer electronics industry has tried to meet the challenge by creating a set of standards for how a television – or a stereo, or any other device – communicates in the home. The industry has also defined how the devices work over the Internet, and over managed networks.

Apart from Japan, there have so far been very few networked television sets and video recorders – and even in Japan, they follow only the Japanese standards. The consumer electronics industry is remarkably slow in implementing its own standards. This leaves the field open for new players, and there is a ready-made niche for them in the set-top box. But regardless of whether the TV set is controlling itself, or if a PC or set-top box controls it, the interactions of the user are captured and sent to the service provider.

The problem with personalization is not the production – it is not too hard to provide effective programs which can be made individual, even using traditional production methods. It is the distribution that has to change, and how it leverages the available mechanisms for personalization. This is one of the biggest advantages for IPTV, because the personal profiles in IMS can be made much richer than the profiles in other systems. Information from both the telecom provider and other service providers can be integrated into mechanisms which enable the formatting and selection of advertisements for individuals (including combinations of smaller pieces of a program to a larger, personalized one).

While this works very well for video on demand, where the user by definition has logged in and selected the service on an individual basis, it does not work as well for traditional television over the Internet. Traditional television programming means sharing the same program at the time of broadcast with other users. Here, the program has to change when it becomes more interactive, and the part that is most likely to adapt and change is the advertising.

In the future, payment for IPTV is also likely to be more personal. Users will consume different services and have to pay different fees. When asking users to pay for something, especially if the payment is to be received after the service is consumed, it helps knowing with whom you are dealing. It also has to be too hard to fake that identity for the effort to be worthwhile. The identity also has to be easy to communicate in a secure way. If that is true, the same identity can be used for tracking the actions of the user. Getting updates every time someone switches channels is valuable to broadcasters; knowing exactly what demographics (if not individuals) watched an advertisement is valuable to advertisers.

The user identity is the root of the subscription, as well as the personalization. When signing up for a service, the user signs up not only to get the service; he also agrees to pay for receiving it. One crucial aspect of media delivery is knowing whether the user has actually received the service, or not. This is possible to measure in IMS-based IPTV, but more difficult in other solutions. Giving the user credit when the delivery is poor, however, seems to be as difficult as usual for service providers. The same applies when tracking a profile. If someone can be uniquely identified, they can be tracked, and their behavior logged – in addition to the information they provide about themselves.

Users are raising a legitimate concern over personalized television: their privacy is at risk when the service provider knows too much about them. While users are willing to trade their privacy for advantages, this has to be visible in a comprehensive way – and the users have to feel and see that they are getting something tangible in exchange for their personal data.

There are strict laws about which information can be given out to whom in almost all countries around the world, except the US. The strictest laws when it comes to the individual permission are those in Europe. These laws are based on an EU directive, and one of the provisions is that the express permission of the user has to be obtained before any data is used. And data may only be used for the purpose it is collected.

Identity management, in other words, is a crucial component in an IPTV system, at least if it is to offer something better and more interesting than what is available on cable-TV today. And here, IMS actually performs rather well. It provides the mechanisms to create personal television and manage the network at the same time. Once you know what someone has done on the network, and how they characterize themselves, that can be used to tailor the television shows they are receiving – either tailoring in real time, as discussed in Chapter 3; or tailoring in terms of selecting the video clips that will compose the show (or the advertisements inserted into it).

This, however, requires not only knowledge about the demographics and interactions; it also requires a way to match the video clips to the knowledge. This information is metadata, data about data. Metadata is regularly produced as part of any digital video production, but to make it match a user's preferences requires somewhat more effort. If I say I like "beer", there has to be a way for the system to understand that "happoshu" or "ale" belongs to that category, since those are ways in which a video clip (especially an advertisement) may be characterized. There are ways to automate this knowledge matching – either the blunt and not very effective way of a small controlled vocabulary, which while offering serendipity also limits the selection; or by using ontologies, a system which has its roots in artificial intelligence research. Having applied these, the system can draw conclusions and create recommendations, based on the matching of one user's preferences in his profile, with the metadata describing the show.

A second problem is how these metadata are collected. Making the producer of the video clip create them is time consuming and expensive; having archivists doing it no less so. There are standardized formats to characterize television programs, as well as movies. And these are rapidly moving in a way where they can be converted to formats which can be turned into ontologies. However, there is still a huge gap between which metadata can be automatically collected and what can be created by a person.

Electronic Program Guides

When users want to know what is available on television, they look up the program guide. This used to be printed in the newspaper; now it is more likely to be displayed on the TV screen. But it can be combined with other information from the Internet into something completely different. Mashing up video with the web becomes possible with IPTV. Anyone can create their own additions to the user experience; the difficulty being to share them.

The user, however, has another limitation when it comes to interaction with television, and that is the use of the remote control. In television, this is the first and foremost way a user can interact with the service, and the interaction typically consists of adjusting the volume, switching to the channel guide (or EPG), and changing channels – either zapping through them, or using the EPG.

Usually displayed as a table of the current time and the show, EPGs have developed a lot – through the use of XML, and the inclusion of video. They are not yet personal, though, but making a personal EPG once you have the profile information available in an IMS-based IPTV system (the history of the user's interaction, his demographic profile, and the metadata describing the shows) is easy – no more difficult than creating a personalized web page. Whether the system actually uses the web as the base for the EPG (e.g., with cookies triggering the retrieval of the interaction log and personal profile) or whether it uses the session as the trigger is merely a matter of taste and familiarity from the developer.

The original purpose of the metadata accompanying a show was to create the EPG, but today the EPG can be used to refine the presentation – and also to generate more revenue for the service provider (not necessarily by including advertising).

Program guides were originally listings of what was available in the different television channels at different times, but once electronic, there is no need to confine the presentation only to that. Including recommendations in an EPG is simple, since the user's profile and previous actions are known. Recommender engines is a field in itself, both of research and development, and they do not have to be directly included in the IPTV system. The result of the EPG is a set of personalized recommendations. These recommendations can be based on the preferences of the individual user, or a group that the user belongs to. The service provider, however, has another opportunity: to sell a media asset to the user.

If someone likes *24*, they might like **Lost**. And if they like **Lost**, they may want to watch the movie **Castaway** on video on demand; it is free, but also available without advertising for 10 dollars per view. Leveraging such opportunities becomes possible in IPTV, but it not only requires a recommender system which knows about the relations between different media properties from an editorial perspective; the system also has to be programmed with the sales opportunities and prices available. The relations between different media properties can be derived from the log files (someone who watched **Lost** also saw **Castaway**). This is what metadata is for, the only problem being that it cannot determine why. Understanding the plot, and why someone who likes one show might like another, is still beyond the reasoning of recommender systems. That is where the producer has to step in, and tell the system what is relevant, and related.

Using the IPTV Technology

We have already discussed the principles of how interaction can affect programs, and talked a little about how this can become (indeed, has already become) a revenue stream for the broadcaster. However, it is clear that the servers that leverage the IMS system (the application servers) will be crucial.

There are a number of components in the IPTV system, and by now their roles should be clear. Table 13-1 lists what they are, and their role in relation to each others.

What it is	How it works	Why it is needed
Television set/Media renderer	Receives the IPTV signal and renders it on the screen	Makes it possible to see the video content
Set-top box/Interaction device	Captures user input and sends it to a central interaction server	Makes sure the user's interactions get to the IPTV service provider, and can be used to change the content of the show
Home network	Connects the different types of equipment in the home together	Makes it possible for different media stores, renderers and interaction devices to interact with each other, and services on the global network (Internet)
Home gateway	Manages addressing in the home, registration with the service provider, and filtering of content (the last two functions can also be performed by the set-top box)	As a firewall and address management system, and to ensure that the user's actions are authorized
IMS proxy	Captures the request for the video service and makes sure it gets to the right receivers, including the QoS system	Interconnects the network and signaling planes of the system, and makes sure the service requests get to the right nodes. Also connects to the profile management system
QoS system	Instructs routers in the network how their queuing mechanisms should be set up	Without QoS, video can be delayed and result in degraded user experience
IMS identity management	Makes sure the user is who he claims he is, and connects the use of the identity to the relevant subscriptions (and hence charging)	Without identity management, anyone could use anyone else's services; the charging systems would have to work offline and with special tokens to keep track of who should pay for what (as it is now, actually)

Table 13-1. The components of an interactive IPTV system.

What it is	How it works	Why it is needed
IMS presence and profile management	Keeps track of what the user does and has done; makes sure this is registered in the system	Makes it possible to personalize content, and to know what other users are watching (if allowed)
IPTV streaming service control	Manages the video stream, including switching to a different video stream when the user selection demands. Note that this is not the same as channel switching	Makes sure that the program starts when requested, and eventually creates programs from different video sources automatically on the fly
IPTV streaming service delivery	Handles the streaming of the content over the network. Interacts with the QoS management	Makes sure content gets where it is supposed to go
Advertising insertion	At selected points in the media stream, pastes in video sequences which contain commercial messages (although this could be a generic mechanism)	Puts advertising in the right place in the program
Interactivity server	Captures the interactivity requests (from the user's IPTV session), collates them (if required), and sends to the appropriate server(s), such as charging, profile management and streaming service control	Interactions coming from more than one user need to be collated and coordinated, otherwise they will not result in anything

Table 13-1. (Continued).

From the perspective of an IPTV system, applications can be realized in two ways: locally, in the terminal; or remotely, in the IPTV system. This puts different requirements on how the applications communicate with the service provider server. But either way, the application is a function of the protocols used, since protocols both determine what can be done with the application; and applications put requirements on protocols.

Providing video on demand is not difficult nowadays. It used to be the apex of development, and at the end of the 1980s, video servers were all the craze. Moore's law and the expanding bandwidth made it much easier. Just, hook up a number of network connections to a really big server, find a way to make people pay, load a library of movies, and run with it. There are two hurdles, however. The service provider has to pay for the content; and the server has to be really, really big if the model is to be successful.

Charging is easy to do if you have one big server. It is simple to connect a user database to a server system, and many small servers can be used instead of one big one – hundreds, or thousands, even. Connecting them directly to each other, while maintaining control over the directory, is how peer-to-peer file sharing works. The technology has been around since the 1980s, and broadcasters are now becoming interested, especially those that are publicly funded. This makes it possible for users to view programs again, without the broadcaster having to provide the resources. The issue is how to maintain control over the distribution – also there is no way of getting paid, at least not at the moment.

Once the signaling is done (whether between peers, or between client and server), the media should be delivered. It is not just a matter of pushing a button on the TV set, as even when broadcast in digital formats, media has to be encoded. There is no "one size fits all" when it comes to encoding and media transport. How the encoding is done affects how efficiently the distribution can be done, as well as how the video can be mixed with others.

Encoding the video stream in a way that makes it possible to send over the wire is one thing, and this is a well-known technology. Making sure it is sent at the right time, and in the right way, is another. The signaling, confirming that the user is allowed to view the video and that interactivity is working is the core of what is new in IPTV. There are three protocols which are particularly important for the IPTV signaling: SIP, HTTP and RTSP. The Session Initiation Protocol (SIP) is used in IMS – but also handles the crucial presence information, and enables the interactivity communication. The Real Time Streaming Protocol (RTSP) is actually a control protocol for "remote video players" – in other words, video on demand. And the HyperText Transfer Protocol (HTTP) is a way to send documents, such as metadata, around on the network.

These protocols, especially SIP and HTTP, suffer from "feature creep", however. They have been extended, and the methods defined in the protocols were misused (at least compared to the original intentions) to create new services and ways of doing old services. This is especially true for HTTP. This is a bad idea for several reasons (even if HTTP is extensible in a completely different way from other protocols). An especially bad idea is to use HTTP for data transport as well as signaling. This is a very tempting idea from the perspective of developers, since HTTP is probably the most well-known protocol in the industry today, and it was initially intended for data transport, but later was extended to handle signaling.

Using IMS and the functions it offers, such as the signaling, the presence, the profiles and other mechanisms, provides a number of possibilities for the service provider. Current IPTV systems do not provide those possibilities – not without adding functionality. IMS-based IPTV does offer such functions, and as IMS-based IPTV is now being standardized and deployed, it makes sense for developers to prepare.

References

The references in this chapter refer to the entire book. They are ordered by chapter – but mostly they cover everything. Since the market is developing so fast, market reports are not trustworthy. Standards are also not static – they continue to develop and change as the market changes. As far as possible, the documents quoted here are the final versions, however, note that standards bodies often start a second version as soon as they have finished the first.

Standards are not the only important documents, of course, but so far, guidelines from real deployments of IPTV are few and far between. To the extent that they are available and usable outside the particular system of the provider in question, they have been cited here.

A book is a snapshot, but a website can continue to develop; so I have put up a website at www.interactive-iptv.com. Please refer to that for a list of references with hyperlinks, which I will update occasionally.

Chapter 1

The NM2 (New Millennium, New Media) Project: http://www.ist-nm2.org/
The LIVE project: http://www.ist-live.org/
Ericsson MeOnTV: http://www.ericsson.com/

Chapter 2

The standard that describes how IPTV will work with IMS comes from Telecommunications and Internet converged Services and Protocols for Advanced Networking (TISPAN), available from http://portal.etsi.org/Portal_Common/home.asp
IPTV Architecture; Dedicated subsystem for IPTV functions, ETSI TS 182 028 V2.0.0 (2008-01)
IPTV Architecture; IPTV functions supported by the IMS subsystem, ETSI TS 182 027 V2.0.0 (2008-02)
Service Layer Requirements to Integrate NGN Services and IPTV, ETSI TS 181 016 V2.0.0 (2007-11)
Requirements for network transport capabilities to support IPTV services, ETSI TS 181 014 V2.0.0 (2007-11)
Open IPTV Forum architecture specification: Open IPTV Forum – Functional Architecture – V 1.1, Approved Jan 15, 2008. Available at http://www.openiptvforum.org/docs/OpenIPTV-Functional_Architecture-V1_1-2008-01-15_APPROVED.pdf
Frauenhofer Fokus Kompetenzzentren NGNI, available from http://www.fokus.fraunhofer.de/ngni/topics/IPTV.php?lang=de#enabler

Chapter 3

Consumer Electronics Association: CEA-2014 Web-based Protocol and Framework for Remote User Interface on UPnP™ Networks and the Internet (Web4CE), Document number: CEA-2014. Available from http://www.fokus.fraunhofer.de/ngni/topics/IPTV.php?lang=de#enabler

The DLNA Networked Device Interoperability Guidelines, available from http://www.dlna.org/industry/certification/guidelines/

UPnP Forum specifications, available from http://www.upnp.org/standardizeddcps/default.asp

DVB Multimedia Home Platform (MHP), specifications available from http://www.mhp.org/mhp_technology/mhp_1_2/

Chapter 4

BBC Interactive TV standards and guidelines, available from http://www.bbc.co.uk/commissioning/interactive/

Sky Interactive TV standards and guidelines, available from http://www.skyinteractive.com/sky/home/default.htm

Chapter 5

Interactive Advertising Bureau, Broadband Video Commercial Measurement Guidelines, available from http://www.iab.net/iab_products_and_industry_services/1421/1443

ANSI/SCTE 35 2004, Digital Program Insertion Cueing Message for Cable. Available from http://www.scte.org/documents/pdf/ANSISCTE352004.pdf

Chapter 6

BBC Interactive Media Player, available from http://www.bbc.co.uk/iplayer/

EBU P2P Media Portal, available at http://www.ebu.ch/members/EBU_Media_portal_Trial_1.php

Chapter 7

Marlin Developer Community, available at http://www.marlin-community.com/

Creative Commons, http://creativecommons.org/

Chapter 8

The XDMS is documented in a large number of documents from the OMA, available from http://member.openmobilealliance.org/. The most important are:

XML Document Management Architecture

XML Document Management (XDM) Specification

Presence and Group Management are partly created by OMA, partly by IETF. The OMA documents are:

Presence SIMPLE Architecture
SIP/SIMPLE Presence Specification
Presence XDM Specification
Presence SIMPLE Specification
Presence Content XDM Specification
Resource List Server (RLS) XDM Specification

IETF, the Internet Engineering Task Force, defines the basics for presence in a collection of RFCs (Requests for Comment, but actually standards documents). They are available from http://www.ietf.org/rfc.html. The most relevant for this chapter are:

RFC 4662, A Session Initiation Protocol (SIP) Event Notification Extension for Resource Lists
RFC 4661, An Extensible Markup Language (XML)-Based Format for Event Notification Filtering
RFC 4660, Functional Description of Event Notification Filtering
RFC 4480, RPID: Rich Presence Extensions to the Presence Information Data Format (PIDF)
RFC 3903, Session Initiation Protocol (SIP) Extension for Event State Publication
RFC 3863, Presence Information Data Format (PIDF)
RFC 3857, A Watcher Information Event Template-Package for the Session Initiation Protocol (SIP)
RFC 3856, A Presence Event Package for the Session Initiation Protocol (SIP)
RFC 3265, Session Initiation Protocol (SIP)-Specific Event Notification
RFC 2779, Instant Messaging / Presence Protocol Requirements
RFC 2778, A Model for Presence and Instant Messaging
RFC 3859, Common Profile for Presence (CPP)

Chapter 9

XML is the mother of all metadata formats. The standard itself is documented in several W3C Recommendations available from http://www.w3.org/XML/Core/#Publications
OWL, the Web Ontology Language, is documented in a series of W3C Recommendations at http://www.w3.org/2004/OWL/#specs
The TV-Anytime standard is well documented in a large number of documents, available from ETSI, http://portal.etsi.org/Portal_Common/home.asp
IETF RFC 4078, The TV-Anytime Content Reference Identifier (CRID), available from http://www.ietf.org/rfc.html
IPTC news codes is available from http://www.iptc.org/NewsCodes/
NewsML is available from http://www.newsml.org/
SportsML is available from http://www.sportsml.org/. The latest version includes a curling plugin.
Dublin Core Metadata Initiative, http://dublincore.org/
SMPTE-RA Metadata Dictionary is available at http://www.smpte-ra.org/mdd/index.html
MXF, the Material Exchange Format, is available at http://www.ebu.ch/en/technical/trev/trev_291-devlin.pdf

Document number	Name	Contents
ETSI TS 102 822-1 V1.3.1 (2006-01)	Part 1: Benchmark Features	Essentially a set of requirements for TV-Anytime, which the standard is measured against to determine if it has fulfilled its goals
ETSI TS 102 822-2 V1.3.1 (2006-01)	Part 2: System Description	The overall description of how the TV-Anytime system is intended to work
ETSI TS 102 822-3-2 V1.3.1	Part 3: Metadata	Metadata elements, schemas, vocabulary, and structure
ETSI TS 102 822-4 V1.1.2	Part 4: Content Referencing	How the CRID (the URI used in TV-Anytime) works.
ETSI TS 102 822-5-2 V1.2.1	Part 5: Rights Management and Protection (RMP)	How metadata is applied to create content protection
ETSI TS 102 822-6-1 V1.3.1	Part 6: Delivery of Metadata over a Bidirectional Network	Defines a format for SOAP/HTTP bidirectional data transport
ETSI TS 102 822-7 V1.1.1	Part 7: Bidirectional Metadata Delivery Protection	Security and transmission protection for bidirectional data transport
ETSI TS 102 822-8 V1.1.1	Part 8: Phase 2 – Interchange Data Format	Defines retrieval of TV-Anytime data from "alternative sources", essentially website download
ETSI TS 102 822-9 V1.1.1	Part 9: Phase 2 – Remote Programming	Defines how to program a recorder, for instance a PVR, using TV-Anytime

UMID, the SMPTE 330M Unique Material Identifier, available from http://www.smpte.org/standards/smpte_ra/metadata_registries

MPEG-7 is standardized by the Motion Pictures Expert Group, and the standards are approved by ISO. The official MPEG documentation is available from ISO or http://www.chiariglione.org/mpeg/

Chapter 10

Hypertext Transfer Protocol – HTTP/1.1, is documented in RFC 2616, available from http://www.ietf.org/rfc.html.

HTTP State management mechanism (aka cookies), RFC 2965, available from http://www.ietf.org/rfc.html.

SIP: Session Initiation Protocol, RFC 3261, available from http://www.ietf.org/rfc.html.
RTSP: the Real Time Streaming Protocol (RTSP), RFC 2326, available from http://www.ietf.org/rfc.html.
SOAP, standardized by the W3C at http://www.w3.org/2000/xp/Group/
SDP: Session Description Protocol, standardized in RFC 2327

Chapter 11

MPEG-2 and MPEG-4 are standardized by the Motion Pictures Expert Group, and the standards are approved by ISO. The official MPEG documentation is available from ISO or http://www.chiariglione.org/mpeg/
RTP: A Transport Protocol for Real-Time Applications, RFC 3550, available from http://www.ietf.org/rfc.html.

Chapter 12

Internet Group Management Protocol, Version 3 (IGMP), RFC 3376, available from http://www.ietf.org/rfc.html.
IMS is standardized by the 3GPP. There are several specifications relating to IMS, the architecture is documented in TS 23.228, IP Multimedia Subsystem (IMS); Stage 2, available at http://www.3gpp.org/ftp/Specs/html-info/23228.htm
The best IMS reference is however: *The 3G IP Multimedia Subsystem (IMS): Merging the Internet and the Cellular Worlds*, Second Edition by Gonzalo Camarillo and Miguel-Angel García-Martín, John Wiley& Sons, Ltd, 2006, ISBN 978-0470018187.

Index

SMPTE, *see* Society of Motion Pictures and Television Engineers
Society of Motion Pictures and Television Engineers (SMPTE) Metadata Dictionary   224
Stateless vs stateful protocols   258, 266
Streaming video   18, 19, 20
Subscription management   186–7
Superdistribution   176

Taxonomies   221
Telescoping advertisements   131
Testing of interactive applications   116–19
TR-069   82
Transport control in ETSI TISPAN   49
Triple Play   50
TV-Anytime   225–36

UDP, *see* User Datagram Protocol
Universally Unique Identifier (UUID)   60
Universal Plug and Play (UPnP)   59–66
UPnP device control   61
UPnP device description   62
UPnP device discovery   60–2
UPnP Internet Gateway Device (IGD)   70
UPnP QoS   71–2
UPnP, *see* Universal Plug and Play
User Datagram Protocol (UDP)   59, 298

User profiles   10, 45–6, 51, 77, 201–4
and interactivity   104, 341–3
User provided content   123–6, 155–6
copyright on   156–72
UUID, *see* Universally Unique Identifier

VCR, *see* Video Cassette Recorder
Video Cassette Recorder (VCR)   143
Video on Demand (VoD)   41, 51, 143–9
advertising in   148–9
charging for   151–4
constraints on   146–7
user experience of   147
VoD, *see* Video on Demand
Voting in IPTV   15, 331–3

Watermarking   177
Web 2.0   9, 119
Web browser in television set   79–87

XDMS, *see* XML Document Management Server
XHTML   80–4
XML, *see* EXtensible Markup Language
XML Document Management Server (XDMS)   39, 46, 189–204
XML lists   193

YouTube   143, 172, 245–6